T0074845

FRAMEWORKS, TENSEGRITIES, AND SYMMETRY

This introduction to the theory of rigid structures explains how to analyze the performance of built and natural structures under loads, paying special attention to the role of geometry. The book unifies the engineering and mathematical literatures by exploring different notions of rigidity – local, global, and universal – and how they are interrelated. Important results are stated formally, but also clarified with a wide range of revealing examples. An important generalization is to tensegrities, where fixed distances are replaced with "cables" not allowed to increase in length and "struts" not allowed to decrease in length. A special feature is the analysis of symmetric tensegrities, where the symmetry of the structure is used to simplify matters and it allows the theory of group representations to be applied. Written for researchers and graduate students in structural engineering and mathematics, this work is also of interest to computer scientists and physicists.

ROBERT CONNELLY is Professor of Mathematics at Cornell University and a pioneer in the study of tensegrities. His research focuses on discrete geometry, computational geometry, and the rigidity of discrete structures and its relations to flexible surfaces, asteroid shapes, opening rulers, granular materials, and tensegrities. In 2012 he was elected a fellow of the American Mathematical Society.

SIMON D. GUEST is Professor of Structural Mechanics in the Structures Group of the Department of Engineering at the University of Cambridge. His research straddles the border between traditional structural mechanics and the study of mechanisms, and includes work on "morphing" and "deployable" structures.

FRAMEWORKS, TENSEGRITIES, AND SYMMETRY

ROBERT CONNELLY

Cornell University, New York

SIMON D. GUEST

University of Cambridge

CAMBRIDGE
UNIVERSITY PRESS

University Printing House, Cambridge CB2 8BS, United Kingdom

One Liberty Plaza, 20th Floor, New York, NY 10006, USA

477 Williamstown Road, Port Melbourne, VIC 3207, Australia

314–321, 3rd Floor, Plot 3, Splendor Forum, Jasola District Centre, New Delhi – 110025, India

103 Penang Road, #05–06/07, Visioncrest Commercial, Singapore 238467

Cambridge University Press is part of the University of Cambridge.

It furthers the University's mission by disseminating knowledge in the pursuit of education, learning, and research at the highest international levels of excellence.

www.cambridge.org
Information on this title: www.cambridge.org/9780521879101
DOI: 10.1017/9780511843297

© Robert Connelly and Simon D. Guest 2022

First published 2022

Printed in the United Kingdom by TJ Books Limited, Padstow Cornwall

A catalogue record for this publication is available from the British Library

Library of Congress Cataloging-in-Publication Data
Names: Connelly, Robert (Mathematician), author. |
Guest, S. D. (Simon D.), author.
Title: Frameworks, tensegrities, and symmetry / Robert Connelly,
Simon D. Guest.
Description: Cambridge ; New York, NY : Cambridge University Press, 2022. |
Includes bibliographical references and index.
Identifiers: LCCN 2021029104 (print) | LCCN 2021029105 (ebook) |
ISBN 9780521879101 (hardback) | ISBN 9780511843297 (epub)
Subjects: LCSH: Structural analysis (Engineering) | Rigidity (Geometry) |
Engineering mathematics. | BISAC: MATHEMATICS / Discrete Mathematics
Classification: LCC TA645 .C655 2021 (print) | LCC TA645 (ebook) |
DDC 624.1/7–dc23
LC record available at https://lccn.loc.gov/2021029104
LC ebook record available at https://lccn.loc.gov/2021029105

ISBN 978-0-521-87910-1 Hardback

To our wives, Gail and Karen

Contents

Preface

Ultimately, this book will show the reader how to generate the special geometry that allows the structure shown in Figure 0.1 to stand up. This isn't a common-or-garden structure, and understanding how it works turns out to be remarkably instructive.

This book is an attempt to build a bridge between two cultures, one the very practical and concrete, the analysis and synthesis of built structures, and the other a rigorous, but very visual, geometric understanding of the rigidity of configurations of points with distance constraints. The former is associated with an engineering point of view that dates back to Galileo, and the way it designs and analyzes structures to make sure they're safe. The latter is associated with Euclidean geometry and the rigidity of graphs, or what we call frameworks here. From about the time of the early 1970s there has been mathematical interest in the geometric rigidity of frameworks, especially with the work of Janos Baracs, Henry Crapo, and Walter Whiteley in Montreal, Canada. Janos Baracs was an engineer, trained in Hungary, who saw the connection and oneness of geometry and the problems of the rigidity of frameworks and other related structures. Our purpose here is to continue in that tradition from our own point of view as well as present some of the major advances in the theory of rigid structures.

One of us, Connelly, also became interested in the geometry of triangulated surfaces in the early 1970s, but from the point of view of trying to show that any triangulated surface in three-space (a kind of framework) is rigid. It was not the case – there is a counterexample.

Independently of these developments, in the 1940s an artist, Kenneth Snelson, created structures that seemed to magically suspend large metal bars, just with some thin cables. These were called "tensegrities" by R. Buckminster Fuller, who built up an intense following promoting the virtues of tensegrities and other related structures. The problem is, why do they stand up? What are the principles involved? In the 1970s Calladine showed how the standard engineering first-order analysis would not explain why most of Snelson's structures stood up. Of course a blunt analysis could always be applied – if bars and cables are treated like springs an analysis of the inherent potential energy will explain the stability of tensegrities. But this would be blind to the geometry of the structure, and the energy functions that would be employed were clumsy from a mathematical perspective. The simple mathematical "yes or no" question of whether the tensegrity is rigid could actually usually be answered with much simpler "energy" functions, using sums of squared lengths.

(a)

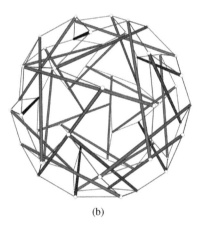

(b)

Figure 0.1 (a) A tensegrity built by David Burnett (with John Wythe) to celebrate the career of his brother Roger Burnett, who worked to elucidate the structure of the adenovirus. To help David with the design, the required geometry of the tensegrity was calculated by Ramar Pandia Raj and the authors, using the tools developed in this book in Part II (Pandia Raj and Guest, 2006). In fact the structure deviates from the ideal icosahedral form because of the effect of gravity, and this effect was later explored by Pandia Raj using the tools described in Part I (Ramar and Guest, 2011). (b) An image of the same tensegrity produced using the software described in Chapter 11.

This brings up the question of what exactly is the problem? From a mathematical point of view, the basic question is whether the configuration of the tensegrity (or bar framework) is fixed – is it rigid? Yes or no? Cables that cannot get longer, and struts that cannot get shorter, give the geometric constraints for the tensegrity. However, from the engineering/physical point of view, that is the very least you would expect to know. Perhaps a more important question is, how does the structure deform under some set of reasonable loads, when there may be some softness in the distance constraints? It seems harsh not to allow a cable not to increase in length at all, for example. Not only that, using sums of squared lengths times a constant to represent physical energies of reasonable cables and struts seems intensely unphysical, especially for struts. It turns out that these two difficulties are in fact intimately married.

This reminds us of an old joke. A mathematician and an engineer are in a hotel in separate rooms. A fire breaks out in the engineer's room, and fortunately there is a bucket of water handy. She puts out the fire with the water and goes to bed relieved. The mathematician sees all this and is quite impressed. As it happens, later, another fire breaks out in the mathematician's room, and sure enough, there is another bucket of water available as well. But he just goes to bed, secure in the knowledge that there is a solution to the "fire" problem.

A key point is that a tensegrity structure can be "self-stressed". Even without external load, each member can carry a force – tension in a cable, compression in a strut. We can associate a scalar value with this force (or the force/length), no matter what the cross-section. An engineer will need to design the member so that it can carry the force without failing, but that is not our task here. We are just interested in how the force might impact the overall rigidity – perhaps naively we assume that the members won't fail.

The approach described in this book has some unexpected and far-reaching benefits beyond buildings and bridges. For example, suppose that you have a framework with its configuration of labelled points and connecting bars, cables, or struts in three-space, say. Is there another configuration, in three-space, not just a congruent copy, that satisfies the same distance constraints? If the answer is no, then we say that the tensegrity is *globally rigid*. This is of great interest for point-location problems, or the uniqueness of protein configurations, for example. It turns out that the Snelson-type tensegrities are globally rigid. The unusual energy functions described above are just what is needed to show the global rigidity property. Indeed, there is more. The stress–energy function that is associated with the stress densities in the members determine a quadratic form that can determine the configuration, not just in the dimension that it sits, but in all higher dimensions. One consequence of this is that if the lengths of the framework are given, then the configuration can be determined by some standard quadratic programming algorithms. In case the configuration is generic (explained below), global rigidity is determined by the rank of the quadratic form. In case the quadratic form is positive semi-definite of maximal rank, then the configuration is determined in all dimensions, which we call *universal rigidity*. All this is determined modulo affine motions, which are usually easy to handle.

It also turns out that the yes-or-no question, "Is it rigid?" can be answered in certain cases to great effect when the configuration of points is generic. Technically this means that there is no non-zero polynomial relation over the integers among the coordinates of the configuration. In effect this means that if you choose the points of the configuration randomly, with probability one, your configuration will be generic. The problem is that often the most interesting structures are not generic, and it is hard to tell whether it is indeed generic. The major result is that if a configuration in the plane is generic, then there is a very efficient algorithm (now known as the pebble game) that can determine whether the bar framework is rigid or not, and it works for graphs with a large number of nodes. This has led to a large industry of determining the generic rigidity of the corresponding frameworks that is completely combinatorial, no matrices, no energies. We give a short introduction to how this combinatorial rigidity works.

Lastly, coming back to Snelson-like tensegrity structures, in Part II, we use some representation theory of finite subgroups of the orthogonal group of three-space to catalog certain classes of symmetric tensegrities, where there is one transitivity class of nodes, two transitivity classes of cables, and one transitivity class of struts.

We thank the US National Science Foundation grant numbers 0209595, 0510625, DMS-0809068, and 1564493 and the UK Engineering and Physical Sciences Research Council grant EP/D030617/1 for support during the creation of this book.

1

Introduction

We see structures all around us, bridges, buildings, furniture, sandcastles, glass, bone, cloth, string, tents, rock piles, for example. Some hold up, some don't. Why? Part of the answer has to do with the materials that make the structure, and part of the answer has to do with the geometry of the structure. A door hinge allows a door to open when pushed, while a cardboard box keeps its shape, unless the cardboard buckles or tears.

We will explain a particular model we call a pin-jointed framework. Joints are regarded as points in space or the plane, and they are connected by inextendible, incompressible bars. The bars are allowed to rotate freely about the joints, and we are seldom concerned whether the bars intersect or not. But we are concerned with rigidity and the behaviour of the framework under loads. There is quite a bit that can be said, not only in calculating forces and displacements, but also in whether the framework is rigid or not.

Another special feature is that after explaining some of the basics in Part I, in Part II we describe the effects of symmetry on the rigidity of the framework. Symmetry can be used to simplify the rigidity calculations described in Part I, but it can also create frameworks that do change their shape.

We have also taken a dual point of view, both from a mathematical/geometric perspective and from an engineering/physical perspective. The geometric problem is: Given the nodes (i.e. points) and distance constraints (i.e. the bars), do these constraints allow the configuration to change its shape. This is a very concrete problem and there are a number of tools that can be used to make this decision. But there are certain demands of rigor and clarity that manifest themselves in the style of exposition. Concepts are defined (usually in italics) clearly and unambiguously, and important statements are put in the form of a "Theorem," helper facts are labelled "Lemmas," and fairly direct consequences of some theorems are called "Corollaries." Justification for these statements are the proof, which is usually set aside and has a little square at the end to let the reader know that the proof has finished, like this. □

When this is done well, it helps the reader see the thread of the argument and see what is important. When it is done poorly it can be tedious, distracting, and confusing. Indeed Gordon (1978) declares

"What we find difficult about mathematics is the formal, symbolic presentation of the subject by pedagogues with a taste for dogma, sadism and incomprehensible squiggles".

We aspire to minimize such tastes, while maintaining our self respect.

The engineering problem is: Given a structure, what is its behaviour under loads? How does the deformation determine the strain (i.e. change in length) of each member? What are the stresses (i.e. forces in the bars) involved? The answer in turn depends on the physical characteristics of the bars, such as Young's modulus that determines their stress–strain ratio. Instead of hard distance constraints, softer more physically realistic elastic behaviour is assumed. Mathematical precision is not necessarily such a concern, but the ultimate justification is the results of an experiment. Indeed, even an experiment in a lab may not be sufficient. Witness the numerous examples of apparently carefully planned structures that have "failed," often with loss of life as described in Levy and Salvadori (2002) and Petroski (2008). Nevertheless, our simple discrete model of a bar and joint framework has been shown experimentally to work for many examples of structures with long relatively thin beams with smaller welded joints. Our purpose here is not to provide complete detail or even provide a totally accurate model of some structure, but instead describe how the analysis of frameworks proceeds and apply this knowledge to a very wide range of circumstances, such as cabled structures (called tensegrities), where some bars are replaced by cables that can only carry tension and not compression, or to packings of spherical balls modelling granular materials, or to large models of glass material, for example. These two approaches are not really separate. The mathematical model effectively assumes some elasticity in the bars in order to prove the framework's rigidity.

But there is more to understand than simply the rigidity or stability of a framework. Another basic question to understand is frameworks that do change their shape even given the hard bar constraints. We call these frameworks *flexible*, but this does not mean that the materials one might use to build them are pliable or soft. It means that there is an exact continuous family of solutions to the equations given by the bars, other than rigid motions of the whole framework. These are often called finite mechanisms in the engineering world.

Chapter 2: Frameworks and rigidity. We introduce the basic framework model, and describe the various different ways in which we can describe the rigidity of the framework, anticipating results in future chapters.

Chapter 3: First-order structures. A natural approach to rigidity is to replace the equations that describe the bar constraints with an appropriate system of linear equations, where the coordinates of the nodes are the variables. There are a wide range of computational and conceptual tools that can be applied. When these techniques apply and show the rigidity of the the framework, it is called infinitesimally rigid or equivalently statically rigid, or kinematically rigid. This is the first requirement for the rigidity or stability in structural engineering. On the hand many structures that are built are not infinitesimally rigid, but nevertheless are still rigid, but not bridges and buildings. They don't fall down, but they tend to be a bit "shaky" in that one can usually feel the play in the framework.

Chapter 4: Tensegrities. In addition to the bar-and-joint model, it is useful to put one-sided constraints between some pairs of joints. If the constraint prevents the end joints from getting further apart, it is called a cable. One can think of this as a string or chain (or cable) connected between its joints. It offers no resistance to the joints coming together, but constrains them from moving apart. It is also useful to have constraints between some joints where the ends are allowed to move apart, but not come together. At first sight this may seem to be strange, but if one imagines hard spherical billiard balls, the centres of any two touching balls form a natural strut. Indeed, this model is natural for some kinds of granular materials and sphere packing problems. See Donev et al. (2004) for how this can be applied. These structures with cables and struts are called tensegrities. Inspired by their "tensional integrity," the name was coined by R. Buckminster Fuller, who saw some of the sculptures/structures that the artist Kenneth Snelson showed him in the 1940s. Snelson (see Snelson, 1948) went on to create many such large tensegrity sculptures all over the world, made of disjoint bars (or struts) suspended with cables in tension – an example is shown in Figure 1.1. The book Edmondson (1987) is an interesting attempt to give a coherent account of how Buckminster Fuller thought about tensegrities. For books with engineering or practical information about tenesgrities there are Pugh (1976); Skelton and de Oliveira (2010); Motro (2003); Gomez-Jauregui (2010); and Juan and Tur (2008).

(a)   (b)

Figure 1.1 Two views of the Snelson sculpture "Needle Tower II," built in 1969 at the Kröller-Müller Museum in the Netherlands. Photographs from: (a) Henk Monster `https://commons.wikimedia.org/wiki/File:Hoge_Veluwe_Kroller_ Muller_sculpture_garden,_the_1971_18_m_tall_Tensegrity_ structure_from_Kenneth_Snelson_-_panoramio.eps`; (b) Onderwijsgek `https://commons.wikimedia.org/wiki/File:Kenneth_Snelson_ Needle_Tower.JPG`.

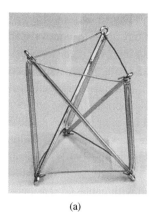
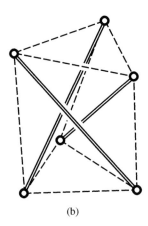

(a)   (b)

Figure 1.2 Perhaps the simplest three-dimensional tensegrity. (a) is a model made by one of the authors from brass tubes, springs and wires. (b) is a simplified representation where members in tension are shown by dashed lines, and members in compression are shown by double lines – this anticipates the convention described in Chapter 4.

Chapter 5: The stress matrix. Many of the tensegrity structures created by Snelson and others are not infinitesimally rigid. They are under-braced in that they have too few cables and struts needed for infinitesimal rigidity. A natural question is: Why don't they fall down? One way to answer this question is to use a "higher-order" model that goes beyond the linear approximations in Chapter 3. As a first step a particular quadratic energy function associated with a stress in the tensegrity can be used to solve the geometric question of whether it is rigid. These are defined by a matrix we call a stress matrix. The energy function is quite unrealistic as a physical energy function, considering Hooke's law for springs that is the usual model for most structures. But interestingly this stress–energy function can be used to show global properties with regard to the one-sided tensegrity constraints. For example, the simplified version of the "Needle" tensegrity shown in Figure 1.2 is not only rigid in space, but it is globally rigid in the sense that any other configuration with the cables not longer and the struts not shorter can be translated and rotated to any other or can be reflected and then translated and rotated to the other. Cables are indicated with dashed line segments and bars (which could alternatively be struts) with doubled solid line segments.

Chapter 6: Second-order mechanism and prestress stability. But what is a more realistic physical analysis of an under-braced structure. How does the tension induce the tensional integrity of tensegrities? One way to understand this is to bring in more realistic energy functions than described in Chapter 5. When this is done, interestingly, the stress–energy given by the stress matrix comes in as one component. Because of a change in coordinate systems used in standard engineering analysis, this decomposition is not usually seen there, but the analysis is finally equivalent. This analysis can be used to show the rigidity of many structures, where it is not clear even from the first-order analysis, whether the tensegrity is rigid or not.

Chapter 7: Generic rigidity. Suppose that one has a framework with a large number of nodes, and one wants to know whether it is rigid or not. Stated this way, the question is clearly too hard. Special positions, where the framework is not infinitesimally rigid, but are rigid, nevertheless, can be quite complicated. One can solve the linear equations in principle, but even that can be too time consuming. One way out is simply to assume that the configuration of nodes is not in any sort of special position. The special positions for infinitesimal rigidity are given by polynomial equations with node coordinates as variables and integer coordinates, so why not just assume that these coordinates do not satisfy any such non-zero polynomial equation. We call those node configurations generic. This can be thought of as randomly choosing the coordinates with a natural continuous distribution. Then the chance of hitting a non-generic configuration is zero. If a generic configuration is chosen, in many cases it is possible to determine infinitesimal rigidity for bar frameworks. A very important case is for generic bar frameworks in the plane. It turns out there is a very efficient purely combinatorial algorithm to decide the generic rigidity of bar frameworks, even for finding the rigid parts, the flexible parts, the over-braced parts, and so on. A popular version of this algorithm is called the pebble game (see Jacobs and Hendrickson, 1997) and this has been used to understand a percolation problem inspired by glass networks, where a large over-braced framework gradually has its bars cut until large parts become flexible as in Jacobs and Thorpe (1996). It is also interesting that the generic hypothesis can be used to determine generic global rigidity Connelly (2005); Gortler et al. (2010); and Jackson and Jordán (2005).

Chapter 8: Finite mechanisms. The other side of structural mechanics is (finite) mechanisms. There is a large variety with a wide range of purposes, applications, and unexpected properties. They can be used to draw curves, open robot arms, and draw straight lines, and there are examples of closed polyhedral surfaces that actually flex, but only with their volume constant. These are exact finite mechanisms, that strictly satisfy the distance constraints. Many of the tools that are used to show frameworks are rigid can be turned around to show that they are not rigid.

Part II is concerned with frameworks that are symmetric. In other words there are rigid motions, such as rotations or reflections that take the nodes and bars of a framework to itself.

Chapter 9: Groups and representation theory. We start with a brief introduction to the theory of groups – which is what we need to deal with symmetric tensegrities, and with representation theory, which explains how groups operate as a set of transformations or matrices. Representation theory turns out to be exactly what we need in the remaining two chapters.

Chapter 10: First-order symmetry analysis. Here we look at how representation theory can be applied to the material from Chapter 3, and how this can be used to predict how to make symmetric structures that are rigid, and how some of the properties will be special.

Chapter 11: Generating stable symmetric tensegrities. Finally we show how representation theory allows us to generate whole families of rigid symmetric tensegrities.

1.1 Prerequisites

We assume that the reader is familiar with a first course in linear algebra including vectors, matrices, linear transformations, etc. such as is in Strang (2009) and is not put off by a formula or two. Of standard books that introduce the reader to the analysis of structures from an engineering perspective, we suggest Parkes (1965) or Livesley (1964). We also have included a few exercises at the end of the chapters for a little work-out for the reader.

For the second part of the book, we have included a brief introduction to group theory and some of the relevant parts of representation theory. The books by Bishop (1973) and James and Liebeck (2001) are good sources for that theory.

1.2 Notation

We denote vectors and matrices in bold. Vectors are thought of as columns, and $()^{\mathrm{T}}$ denotes the transpose operation for both vectors and matrices.

We often have occasion to denote a vector of vectors, which is a string of vectors on top of each other, which we call a configuration. So we use the notation

$$[\mathbf{p}_1; \mathbf{p}_2; \dots; \mathbf{p}_n] = \begin{bmatrix} \mathbf{p}_1 \\ \mathbf{p}_2 \\ \vdots \\ \mathbf{p}_n \end{bmatrix},$$

which is notationally more convenient than writing columns everywhere.

Tensegrities, discussed in Chapter 4, are composed of cables, struts, and bars, where non-extendable cables are denoted by dashed lines, non-compressible struts by double lines, and length invariant bars by solid single lines, as shown in Figure 4.1.

The graph associated with a framework or tensegrity is usually denoted by G, and a framework or tensegrity, which consists of a graph G, as well as a configuration of points $\mathbf{p} = [\mathbf{p}_1; \dots; \mathbf{p}_n]$, is denoted by $(G, \mathbf{p})$.

In Part II a group is usually denoted by $\mathcal{G}$, and a representation of a group as ρ.

Part I

The General Case

2

Frameworks and Rigidity

2.1 Introduction

The basic question of rigidity, where we ask when a series of length constraints (bars) constrain a number of points (joints) to give a "stiff" structure, is likely to elucidate a different response if asked of an engineer or a geometer. There are good reasons why these different groups might think of rigidity in different ways, and the purpose of this chapter is to introduce a number of different definitions of rigidity, which will then be explored in the rest of the book. As an introduction, Figure 2.1 gives a series of two-dimensional examples, one of which is rigid by any definition we give, one of which is not rigid by any definition we give, and the rest of which are rigid by some definitions and not by others.

Before starting on rigidity, however, we will first give a more formal definition of our basic system, the framework, which was first introduced in Chapter 1.

2.2 Definition of a Framework

We start with a finite *configuration* of n labelled points in Euclidean space $\mathbb{E}^d$. We pay special attention to the cases $d = 2$, the plane, and $d = 3$, three-space, although we reserve the right to make forays into higher dimensions. The position vector of a point i is given by $\mathbf{p}_i$, a point in $\mathbb{E}^d$, and we call this a *node*. We denote a configuration by a vector $\mathbf{p}$ in $\mathbb{E}^{nd}$, where

$$\mathbf{p} = [\mathbf{p}_1; \dots ; \mathbf{p}_n]. \tag{2.1}$$

(Note here that we are using the notation described in Section 1.2 where a semicolon implies that the entries are written on top of one another, as a column vector.) We allow some pairs of nodes to coincide (i.e. occupy the same point in space), although physically this may not be what we want.

We next decide which pairs of nodes to connect with a graph G, whose vertices correspond to the nodes, and which has no multiple edges or loops. The edges or G are in general called *members*, and if a member provides an equality constraint on its length that is, it can

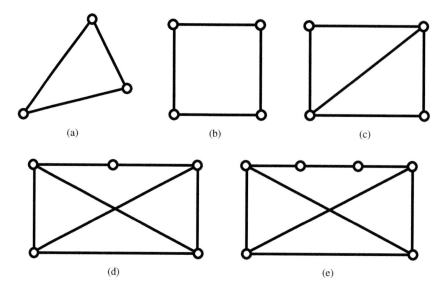

(a) (b) (c)

(d) (e)

Figure 2.1 Five examples of frameworks. By all the definitions we present, the triangle (a) is rigid, while the square (b), called a four-bar linkage by engineers is not rigid (i.e. it is flexible). The other three examples are all rigid (locally rigid) in two dimensions, but neither (d) nor (e) are infinitesimally or statically rigid. Examples (c) and (d) are generically rigid, but (e) is not. While (c) is not globally rigid in two dimensions, it is not rigid at all in three dimensions.

get neither shorter nor longer, we call this a *bar*. (Later, in Chapter 4, members will have the option to be cables, which are not permitted to increase in length, and struts, which are not permitted to decrease in length, as well as bars.) We call the configuration $\mathbf{p} = [\mathbf{p}_1; \ldots; \mathbf{p}_n]$ together with its corresponding graph G a *framework* and denote it by $(G, \mathbf{p})$. If all members are bars we call the framework a *bar framework*.

To draw a picture of a framework, we use small circles for nodes and solid line segments for bars as in Figure 2.1. Note that members can cross each other (i.e. the line segments representing them can intersect) without having a node at the intersection point, as in Figure 2.1(d) and (e).

2.3 Definition of a Flex

We consider a bar framework where each bar cannot change its length even by a very little bit. Of course, any realistic structure will have some member flexibility, and one could argue that such hard constraints are unrealistic. But it will turn out that such constraints are natural and helpful for understanding a wide range of practical structures.

Suppose that each node of a framework is on a differentiable smooth path in a d-dimensional space. The path is parameterized by t, which could be considered to be time. (Note, however, that we will not be considering dynamics, i.e. we are assuming that every part of the system is massless.) The position of node i is given by the vector

$\mathbf{p}_i(t) \in \mathbb{E}^d$, and we define the *initial position* to be $\mathbf{p}_i = \mathbf{p}_i(t)|_{t=0} = \mathbf{p}_i(0)$. Recall that positions of all nodes make up a configuration

$$\mathbf{p}(t) = [\mathbf{p}_1(t); \ldots; \mathbf{p}_n(t)] \in \mathbb{E}^{nd},$$

and the *initial configuration* is $\mathbf{p} = \mathbf{p}(t)|_{t=0} = \mathbf{p}(0)$.

We call the whole path $\mathbf{p}(t)$ a *flex* or *motion* of $\mathbf{p}$ if the length $|\mathbf{p}_i(t) - \mathbf{p}_j(t)| = |\mathbf{p}_i - \mathbf{p}_j|$ is constant for all bars (i.e. members) of the framework. We say a flex $\mathbf{p}(t)$ is a *rigid-body motion* or is *trivial* if for some range of t, say for each $0 \le t \le 1$, there is a rigid-body motion, which is given by a positive-definite orthogonal d-by-d matrix $\mathbf{Q}(t)$ (where $\mathbf{Q}(t = 0)$ is the identity) and a vector $\mathbf{w}(t) \in \mathbb{E}^d$ (where $\mathbf{w}(0)$ is the zero vector), so that the position of each node is given by

$$\mathbf{Q}(t)\mathbf{p}_i + \mathbf{w}(t) = \mathbf{p}_i(t). \tag{2.2}$$

Note that $\mathbf{Q}(t)$ describes a *rigid-body rotation* where the origin is unmoved, and $\mathbf{w}(t)$ describes a *rigid-body displacement* where the framework moves without rotation.

2.4 Definitions of Rigidity

2.4.1 Rigidity, or Local Rigidity

Below, we will introduce three different, but equivalent, definitions of rigidity, that differ basically as to the differentiability assumptions of the flex $\mathbf{p}(t)$. We say that a flex $\mathbf{p}(t) = (\mathbf{p}_1(t); \ldots; \mathbf{p}_n(t))$ is *continuous* if each of the d coordinates of each of the vertices $\mathbf{p}_i(t)$ is continuous in t. Similarly we define $\mathbf{p}(t)$ to be *analytic* if each of its coordinates is an analytic function of t.

Recall that a function, such as $\mathbf{p}(t)$, is *analytic* if it is equal to its power series for t sufficiently small. This is a very useful condition to have available, since such an analytic function $\mathbf{p}(t)$ has derivatives of all orders, and if all those derivatives are 0, for example, then $\mathbf{p}(t)$ itself is 0. On the other hand, it is also useful to have weaker conditions to verify for rigidity.

We say that a framework with n nodes is *rigid* in $\mathbb{E}^d$ if any one of the following properties holds:

Definition 1: Each continuous flex $\mathbf{p}(t)$ of the framework in $\mathbb{E}^d$ is trivial.

Definition 2: Each analytic flex $\mathbf{p}(t)$ of the framework in $\mathbb{E}^d$ is trivial.

Definition 3: There is an $\epsilon > 0$ such that for every configuration of n labelled vertices $\mathbf{q} = [\mathbf{q}_1; \ldots; \mathbf{q}_n]$ in $\mathbb{E}^d$ satisfying the bar constraints for the configuration $\mathbf{p}$ and $|\mathbf{p} - \mathbf{q}| < \epsilon$, then $\mathbf{q}$ is congruent to $\mathbf{p}$.

Theorem 2.4.1 (Equivalence). *All three definitions of rigidity are equivalent.*

A proof of this result relies on some basic results from algebraic geometry, for example Bochnak et al. (1998). Thus, we can use whichever of these definitions is convenient. If a framework in $\mathbb{E}^d$ is not rigid, then we say it is a *finite mechanism*. If a framework is a finite mechanism, by Definition 2, it has an analytic flex that is not a rigid-body motion.

(Commonly in the mathematical literature a framework is called *flexible* if it is a finite mechanism – note that this use of the term flexible does not mean that the members themselves are flexible; rather, flexible is a geometric property of the framework.)

We feel that this very basic concept of rigidity is fundamental to all the others. A physical framework can feel quite "loose" or "floppy" when physically constructed, and still be rigid by the definition here. But few would say that a framework that is a finite mechanism is "rigid."

None of these definitions necessarily explicitly help in determining whether a framework is rigid or not. The analytic definition does allow the possibility of using power series expansions of each of the coordinates, and this can help in rigidity determination. This is in contrast to the case when it has an internal infinitesimal mechanism as defined above, where often only the first term of a power series expansion is determined.

Notice that the ambient space $\mathbb{E}^d$ is part of the definition of whether the framework is rigid. For example, the bar framework in Figure 2.1(c) is easily seen to be rigid in the plane, but it is a finite mechanism when it is considered as being in $\mathbb{E}^3$.

2.4.2 Infinitesimal Rigidity

In fact, most engineers would be unhappy with the description of the frameworks in Figure 2.1(d) and (e) as being rigid. Maxwell (1864b) captured this disquiet with his comment:

A frame of s points in space requires *in general* $3s - 6$ connecting lines to render it stiff. In those cases in which stiffness can be produced with a smaller number of lines, certain conditions must be fulfilled, rendering the case one of a maximum or minimum value of one or more of its lines. The stiffness of such frames is of an inferior order, as a small disturbing force may produce a displacement in comparison with itself.

In order to understand infinitesimal rigidity, one should first understand what an infinitesimal motion is. An *infinitesimal motion* or equivalently an *infinitesimal flex* of a framework is the derivative of a *differentiable motion* (i.e. each coordinate of each node is differentiable), starting at the identity, of the nodes such that the corresponding derivative of the length of each member is zero. So suppose $(G, \mathbf{p})$ is a framework in $\mathbb{E}^d$. Then $\mathbf{p}' = (\mathbf{p}_1, \ldots, \mathbf{p}_n)$ for $\mathbf{p}'_i$ in $\mathbb{E}^d$ is an infinitesimal flex of $(G, \mathbf{p})$ if, for each member $\{i, j\}$ of G,

$$(\mathbf{p}_i - \mathbf{p}_j) \cdot (\mathbf{p}'_i - \mathbf{p}'_j) = 0. \qquad (2.3)$$

Later the notation $\mathbf{p}' = (\mathbf{p}'_1, \ldots, \mathbf{p}'_n)$ for the infinitesimal motion will be replaced by $\mathbf{d} = (\mathbf{d}_1, \ldots, \mathbf{d}_n)$ because $\mathbf{d}$ will be regarded at displacements of the nodes. However, the conditions in Equation (2.3) will be the same, just using a different symbol for the same concept.

If the framework $(G, \mathbf{p})$ has some pinned vertices pinned to the ground, then we say that it is *infinitesimally rigid* if it has only the zero infinitesimal motion.

Quite often we have a framework that has no vertices pinned, and we wish to define what we mean by infinitesimal rigidity in that case. There are always some obvious rigid motions of any unpinned framework, and their derivatives will always be infinitesimal motions for $(G, \mathbf{p})$ for any graph G. In $\mathbb{E}^3$ the derivative of translations are the constant infinitesimal

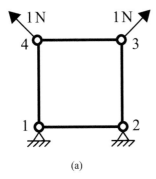 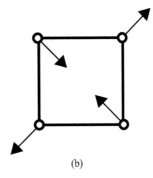

(a)   (b)

Figure 2.2 Framework (a) is pinned at vertices 1 and 2, and the 1 N force at nodes 3 and 4 are indicated (one newton is about the weight of a small apple). In this case that force can be resolved by internal forces $(1/\sqrt{2})\,N$ on all the members except member $\{1,2\}$, where the force carried between rigid supports is undetermined. Nevertheless, Framework (a) is not statically rigid. Framework (b) is similar to Framework (a) except no nodes are pinned. An equilibrium system of forces is indicated that cannot be resolved by any stress in Framework (b).

motion $(\mathbf{b}, \ldots, \mathbf{b})$, and the derivative of rotations can described as $(\mathbf{r} \times \mathbf{p}_1, \ldots, \mathbf{r} \times \mathbf{p}_n)$, where $\mathbf{r}$ and $\mathbf{b}$ are fixed vectors in $\mathbb{E}^3$. The infinitesimal motions $\mathbf{p}' = (\mathbf{p}'_1, \ldots, \mathbf{p}'_n)$, are called *trivial infinitesimal motions* where, for each $i = 1, \ldots, n$, $\mathbf{p}'_i = \mathbf{r} \times \mathbf{p}_i + \mathbf{b}$. In $\mathbb{E}^2$ we can use the same definition of trivial infinitesimal motions as in $\mathbb{E}^3$, except $\mathbf{r}$ must be perpendicular to the plane $\mathbb{E}^2 \subset \mathbb{E}^3$, and $\mathbf{b}$ is in $\mathbb{E}^2$. More generally in any $\mathbb{E}^d$, a trivial infinitesimal motion is given by $\mathbf{p}'_i = \mathbf{V}\mathbf{p}_i + \mathbf{b}$, where $\mathbf{V} = -\mathbf{V}^\mathrm{T}$ is a *skew symmetric* $d \times d$ matrix and $\mathbf{b}$ is a constant vector in $\mathbb{E}^d$. (Skew symmetric matrices $\mathbf{V}$ are the derivative at the identity of orthogonal matrices.) Thus we say that an unpinned framework $(G, \mathbf{p})$ in $\mathbb{E}^3$ is *infinitesimally rigid* if the only infinitesimal motions are trivial.

For example, the frameworks (a) and (c) in Figure 2.1 are infinitesimally rigid in $\mathbb{E}^2$, while the rest are not. The arrows in Figure 2.2(b) interpreted as an infinitesimal motion show a non-trivial infinitesimal motion of the square in the plane.

One of the most basic and commonly used criteria for rigidity is related to the concept of kinematic determinacy. We say that a bar framework is *kinematically determinant* if it is infinitesimally rigid in $\mathbb{E}^d$. In other words every infinitesimal motion is a rigid-body motion. If a framework is not a finite mechanism, but nevertheless is not infinitesimally rigid, the framework itself is often called an *infinitesimal mechanism*, but we shall mostly reserve this word for the motion itself, rather than the framework.

2.4.3 Stress and static rigidity

The internal forces in a framework $(G, \mathbf{p})$ in $\mathbb{E}^d$ are important in understanding its rigidity and flexibility. In its most basic form we consider a *stress* for a framework as simply a scalar associated to each member $\{i, j\}$ of G. (We aren't thinking of stress here as the tensor associated with force/unit area in a mechanics of solids sense – indeed, throughout this book we're not even considering what the cross-sectional area might be!) We are interested in the force at a node, for example, and the *internal force in each member* $\{i, j\}$ is denoted as t_{ij}, and has units newtons, for example. For our purposes of whether a framework is rigid or not, particularly for the definitions here, the stress associated to a member can just as well be regarded as the force per length of the member, and when it is regarded that way, it is usually denoted as ω_{ij} and called a *force density* and has units newtons per centimetre, for instance.

Suppose that a framework $(G, \mathbf{p})$ has some of its nodes pinned to the ground and there is an *external load* or *force* $\mathbf{f}_i$, a vector in $\mathbb{E}^d$ applied to each node $\mathbf{p}_i$, for $i = 1, \ldots, n$. Regarding all the forces as one large vector $\mathbf{f} = (\mathbf{f}_1; \ldots; \mathbf{f}_n)$, we say that a stress with forces $\mathbf{t} = (\ldots, t_{ij}, \ldots)$ *resolves* or *equilibrates* $\mathbf{f}$ if, for each node i that is not pinned,

$$\sum_j t_{ij} \frac{\mathbf{p}_j - \mathbf{p}_i}{|\mathbf{p}_j - \mathbf{p}_i|} + \mathbf{f}_i = \mathbf{0}. \tag{2.4}$$

We assume that there are no zero-length bars. In terms of force density Equation (2.4) becomes

$$\sum_j \omega_{ij}(\mathbf{p}_j - \mathbf{p}_i) + \mathbf{f}_i = \mathbf{0}. \tag{2.5}$$

The forces on the ground do not need to be resolved since it is assumed that any force at those vertices can be resolved. Figure 2.2 shows an example of a framework with an external force and its resolution. If every external force can be resolved by some stress with internal forces $\mathbf{t} = (\ldots, t_{ij}, \ldots)$ (or equivalently stress densities $\omega = (\ldots, \omega_{ij}, \ldots)$) we say this pinned framework $(G, \mathbf{p})$ is *statically rigid*. Note that we consider the vectors $\mathbf{t}$ and ω as row vectors instead of column vectors because of the way that they interact with the configuration vector $\mathbf{p}$, which is a column vector.

Quite often we have a framework that has no vertices pinned, and we wish to define what we mean by static rigidity in that case. But now we have to restrict what the external forces are that we allow. Define an *equilibrium system* of forces $\mathbf{f} = (\mathbf{f}_1; \ldots; \mathbf{f}_n)$ in $\mathbb{E}^3$ (and similarly in $\mathbb{E}^2$) to be vector forces, where each $\mathbf{f}_i$ is applied at the node $\mathbf{p}_i$ and the following equations hold:

$$\sum_i \mathbf{f}_i = \mathbf{0} \quad \text{and} \quad \sum_i \mathbf{f}_i \times \mathbf{p}_i = \mathbf{0}, \tag{2.6}$$

where $\times$ is the usual cross product in $\mathbb{E}^3$. (In higher dimensions the cross product is replaced by the wedge product $\wedge$ as described in Spivak (1979), for example.) Note that the condition for a system of forces $\mathbf{f}$ to be an equilibrium system is that it only depends on the configuration $\mathbf{p}$ and conditions (2.6) and not on any members of any framework G. The conditions of (2.6) are essentially saying that if each force vector $\mathbf{f}_i$ is interpreted as a velocity vector at $\mathbf{p}_i$, with a unit mass, then the system is moving with $\mathbf{0}$ linear and angular momentum. However, we will not be concerned with any issues of dynamics here. This approach is the way static rigidity is explained in Crapo and Whiteley (1982), except we do not concentrate on the "projective coordinates" treated there. Alternatively, we can think of an equilibrium system of forces $\mathbf{f} = (\mathbf{f}_1; \ldots; \mathbf{f}_n)$ as a system of forces that are orthogonal to the space of trivial infinitesimal motions, or equivalently in physical terms, that there is no work done by a trivial infinitesimal motion. Explicitly, this says that for every trivial infinitesimal motion $\mathbf{p}' = (\mathbf{p}'_1; \ldots; \mathbf{p}'_n)$,

$$\sum_{i=1}^{n} \mathbf{p}'_i \cdot \mathbf{f}_i = 0.$$

So for a framework $(G, \mathbf{p})$ with no pinned nodes, we say that it is *statically rigid* in $\mathbb{E}^d$ if every equilibrium system of forces can be resolved with a system of stresses (i.e. forces or stress densities) in $(G, \mathbf{p})$.

2.4.4 Generic Rigidity

Often it is desired to know whether a given graph G is rigid for a "typical configuration" $\mathbf{p}$. This leads to the following. We say that a configuration $\mathbf{p}$ is *generic* if the only polynomial with integer coefficients satisfied by the coordinates of all the nodes $\mathbf{p}_1, \ldots, \mathbf{p}_n$ is the zero polynomial. This is just a strong way of ensuring that there is nothing "special" about the configuration $\mathbf{p}$. So we say that a graph G is *generically rigid in* $\mathbb{E}^d$ if for all generic configurations $\mathbf{p}$, $(G, \mathbf{p})$ is rigid in $\mathbb{E}^d$. It turns out that G is generically rigid in $\mathbb{E}^d$ if and only if there is any generic configuration $\mathbf{p}$, where $(G, \mathbf{p})$ is rigid in $\mathbb{E}^d$ if and only if there is any configuration $\mathbf{p}$, where $(G, \mathbf{p})$ is infinitesimally rigid in $\mathbb{E}^d$ if and only if there is any configuration $\mathbf{p}$, where $(G, \mathbf{p})$ is statically rigid in $\mathbb{E}^d$.

Note that generic rigidity does not depend on the configuration, but only on the graph G. So generic rigidity is a combinatorial property of the graph, and one of the goals of a lot of work (e.g. Graver et al., 1993) is to use combinatorial methods to characterize generic rigidity, with most success with generic rigidity in the plane.

2.4.5 Global Rigidity

Although the concept of (local) rigidity as defined in Section 2.4.1 is quite basic and fundamental, there is another concept that is even simpler and quite natural. We say that a framework (or tensegrity) $(G, \mathbf{p})$ in $\mathbb{E}^d$ with n labelled nodes $\mathbf{p} = [\mathbf{p}_1; \ldots; \mathbf{p}_n]$ is *globally rigid in* $\mathbb{E}^d$ if for any configuration $\mathbf{q} = [\mathbf{q}_1; \ldots; \mathbf{q}_n]$ in $\mathbb{E}^d$ with corresponding n labelled nodes such that $|\mathbf{p}_i - \mathbf{p}_j| = |\mathbf{q}_i - \mathbf{q}_j|$, for all $i = 1, \ldots, n$, then the configuration $\mathbf{q}$ is congruent to the configuration $\mathbf{p}$. In other words, bars of $(G, \mathbf{q})$ have the same length as a the corresponding bars of $(G, \mathbf{q})$. Then the configuration $\mathbf{q}$ is congruent to the configuration $\mathbf{p}$.

Note that the condition of global rigidity is considerably stronger than just rigidity. Global rigidity in $\mathbb{E}^d$ implies rigidity in $\mathbb{E}^d$, but there are many examples of frameworks $(G, \mathbf{q})$ that are rigid but not globally rigid in $\mathbb{E}^d$. For example, the bar framework of Figure 2.3 is globally rigid in $\mathbb{E}^2$ but not in $\mathbb{E}^3$. The proofs that these examples have the properties attributed to them are fairly easy, but when we have the tools from later sections, it will be quite a bit easier still.

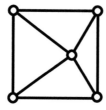

Figure 2.3 A bar framework that is globally rigid in the plane, but not in three-space.

2.4.6 Universal Rigidity

Another related concept that also comes up in this context, is the following. If $(G, \mathbf{p})$ is a framework in $\mathbb{E}^d$ and it is globally rigid in $\mathbb{E}^k \supset \mathbb{E}^d$ for all $k \geq d$, then we say $(G, \mathbf{p})$ is *universally globally rigid* or simply *universally rigid*. For example, a bar triangle, or more generally when G has bars between all pairs of nodes, $(G, \mathbf{p})$ is universally globally rigid for all configurations $\mathbf{p}$. We will develop some tools that can be used to show global rigidity and universal global rigidity in later chapters.

2.5 Exercises

1. Suppose that $(G, \mathbf{p})$ is a bar framework in $\mathbb{E}^2$ as in Figure 2.3, where the four vertices of the square are fixed. Imagine placing the fifth vertex in the plane elsewhere in the plane, not on one of the lines through the edges of the square. For what choices of the fifth vertex is the resulting framework globally rigid?

2. Show that the two definitions of an equilibrium system of forces in Section 2.4.3 are the same.

3. Show that the frameworks of Figure 2.1 (b), (d), and (e) are not infinitesimally rigid by giving a non-trivial infinitesimal flex and an equilibrium system of forces that cannot be resolved by an internal stress.

4. Show that the bar framework as in Figure 2.3 is not globally rigid in the plane, but not universally rigid, when the centre point is chosen inside the square, but not on the diagonals.

5. Find a bar framework in the plane that is universally rigid, but not infinitesimally rigid in the plane.

6. Show that any bar framework in a plane with at least four points is not infinitesimally rigid in $\mathbb{E}^3$.

3

First-Order Analysis of Frameworks

3.1 Introduction

This chapter will describe a simplified analysis of frameworks that we will term a "first-order" analysis. The terminology "first-order" comes from the kinematics that we will consider, where we will make a first-order approximation of the relations between positional changes of vertices and the lengths of members. When we consider statics, the relations between applied loads at nodes, and the forces carried by the members, will turn out to be essentially equivalent to the first-order kinematics. Finally we will combine the ideas of first-order kinematics and statics to define the "first-order stiffness" of a framework, a relationship between the forces applied to the framework and the movement of the nodes. In the limit when members' stiffness becomes infinite, we will be describing the "first-order rigidity" of the framework.

3.2 Kinematics

We start by defining what we mean by infinitesimal displacements or extensions: we want to be more precise than the typical engineer's statement that such things are "small."

Suppose that each node of a framework is on a differentiable smooth path in a d-dimensional space. The path is parameterized by a dimensionless parameter t. (We do not consider dynamics in this text, but if we wished to do so, we could carry over much of what follows by considering t to be a dimensional measure of time.) The position of node i is given by the vector $\mathbf{p}_i(t) \in \mathbb{E}^d$, and we define the *initial position* of a node to be $\mathbf{p}_i = \mathbf{p}_i(t)|_{t=0} = \mathbf{p}_i(0)$. Recall that the position of all n nodes is a configuration,

$$\mathbf{p}(t) = [\mathbf{p}_1(t); \ldots ; \mathbf{p}_n(t)] \in \mathbb{E}^{nd},$$

and we define the *initial configuration* as $\mathbf{p} = \mathbf{p}(t)|_{t=0} = \mathbf{p}(0)$.

We define an *infinitesimal displacement*, or a *flex*, $\mathbf{d}_i \in \mathbb{E}^d$ (often written as $\mathbf{p}'_i$ in the mathematical literature and in Chapter 2) to be

$$\mathbf{d}_i = \left.\frac{d\mathbf{p}_i}{dt}\right|_{t=0}. \tag{3.1}$$

Thus an infinitesimal displacement is a d-dimensional vector associated with every node, which can be written as a single nd-dimensional column vector

$$\mathbf{d} = [\mathbf{d}_1; \ldots; \mathbf{d}_n] \in \mathbb{E}^{nd}.$$

All possible infinitesimal displacements define a tangent space to the space of possible configurations of the framework.

The positions of nodes i and j define the length of the member $\{i, j\}$,

$$l_{ij}(t) = |\mathbf{p}_j(t) - \mathbf{p}_i(t)|,$$

with $l_{ij} = l_{ij}(t)|_{t=0}$, or equivalently (using $(\ )^{\mathrm{T}}$ to denote the transpose),

$$l_{ij}^2(t) = (\mathbf{p}_j(t) - \mathbf{p}_i(t))^{\mathrm{T}}(\mathbf{p}_j(t) - \mathbf{p}_i(t)). \tag{3.2}$$

We use a dual notation for quantities associated with members, referring to them either with a double subscript ij, where, for instance, l_{ij} is the length of the member from node i to node j, or equivalently using a single subscript k, where $l_k \equiv l_{ij}$. There is a unique k associated with every unordered $\{i, j\}$ where there is a member from node i to node j. The member labels k are chosen to run from 1 to b, where b is the number of members.

We define the *rigidity map*, f, to be a function that takes nodal positions to the squares of lengths of members divided by 2, $f : \mathbb{E}^{nd} \to \mathbb{E}^b$, where

$$f(\mathbf{p}_1; \ldots; \mathbf{p}_n) = \left(\ldots; \frac{|\mathbf{p}_i - \mathbf{p}_j|^2}{2}; \ldots \right), \tag{3.3}$$

so each component of f is defined by (3.2) at $t = 0$,

$$f_{ij} = \frac{l_{ij}^2}{2} = \frac{1}{2}(\mathbf{p}_j(t) - \mathbf{p}_i(t))^{\mathrm{T}}(\mathbf{p}_j(0) - \mathbf{p}_i(0)).$$

Notice that when each $\mathbf{p}_i$ in the configuration space $\mathbb{E}^{nd}$ is written in terms of its d coordinates, the function f is a quadratic polynomial in those coordinates. So the rigidity map is continuously differentiable.

We define the *infinitesimal extension* of a member to be the initial rate of change of length with t,

$$e_{ij} = \left. \frac{dl_{ij}}{dt} \right|_{t=0}. \tag{3.4}$$

Thus an infinitesimal extension is a length associated with every member, which can be written as a single b-dimensional vector,

$$\mathbf{e} = [e_1; \ldots; e_b] \in \mathbb{E}^b.$$

We also write

$$\mathbf{l} = [l_1; \ldots; l_b] \in \mathbb{E}^b.$$

All possible infinitesimal extensions define a tangent space to the space of possible member lengths for the framework.

The first-order kinematics describe the relationship between any infinitesimal displacement and the associated infinitesimal extension – we describe the infinitesimal displacement and the infinitesimal extension to be *compatible*. The compatibility relationship is defined by equations (3.1), (3.2), and (3.4). Differentiating (3.2),

$$\frac{df_{ij}}{dt} = \frac{1}{2}\frac{dl_{ij}^2}{dt} = l_{ij}\frac{dl_{ij}}{dt} = \frac{1}{2}(\mathbf{p}_j - \mathbf{p}_i)^{\mathrm{T}}(\mathbf{d}_j - \mathbf{d}_i) + \frac{1}{2}(\mathbf{d}_j - \mathbf{d}_i)^{\mathrm{T}}(\mathbf{p}_j - \mathbf{p}_i).$$

The two terms are equal, $(\mathbf{d}_j - \mathbf{d}_i)^{\mathrm{T}}(\mathbf{p}_j - \mathbf{p}_i) = (\mathbf{p}_j - \mathbf{p}_i)^{\mathrm{T}}(\mathbf{d}_j - \mathbf{d}_i)$, and hence

$$l_{ij}\frac{dl_{ij}}{dt} = (\mathbf{p}_j - \mathbf{p}_i)^{\mathrm{T}}(\mathbf{d}_j - \mathbf{d}_i). \tag{3.5}$$

Assuming that the length of a member is non-zero, we describe a unit vector along member $\{i, j\}$ as

$$\mathbf{n}_{ij}(t) = \frac{(\mathbf{p}_j(t) - \mathbf{p}_i(t))}{l_{ij}(t)} \quad ; \quad \mathbf{n}_{ij} = \frac{(\mathbf{p}_j - \mathbf{p}_i)}{l_{ij}}, \tag{3.6}$$

and the first-order compatibility relationship (3.4) for member $\{i, j\}$ can therefore be written as

$$e_{ij} = \mathbf{n}_{ij}^{\mathrm{T}}(\mathbf{d}_j - \mathbf{d}_i). \tag{3.7}$$

Thus to the first order, relative end displacements of a member $(\mathbf{d}_j - \mathbf{d}_i)$ will cause extension if they are not perpendicular to the member; relative displacements that are perpendicular to the member correspond to an infinitesimal rotation of that member. Figure 3.1 shows some examples of infinitesimal displacements at the ends of a bar.

The compatibility relationships, equation (3.7), can be brought together and written in a matrix form as

$$\mathbf{e} = \mathbf{Cd}, \tag{3.8}$$

where the matrix $\mathbf{C}$ is defined as the *compatibility matrix*. The matrix $\mathbf{C}$ will have b rows (for b members) and dn columns (for n vertices in d dimensions).

An alternative way of writing the compatibility relationship is to define a bar *extension coefficient* directly as the differential of the rigidity map,

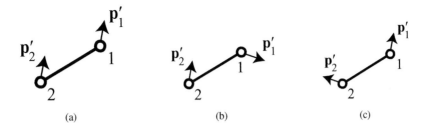

(a)   (b)   (c)

Figure 3.1 A single bar $\{1, 2\}$ with infinitesimal end displacements. (a) A pure infinitesimal displacement of the entire bar, $\mathbf{p}'_1 = \mathbf{p}'_2$, $e_{12} = 0$. (b) An infinitesimal rotation about the bar centre, together with an infinitesimal displacement along the bar length, $\mathbf{p}'_1 - \mathbf{p}'_2$, is perpendicular to a vector lying along the bar, $\mathbf{p}_1 - \mathbf{p}_2$, and so $e_{12} = 0$. (c) A stretch, together with an infinitesimal displacement perpendicular to the bar, $\mathbf{p}'_1 - \mathbf{p}'_2$, is not perpendicular to a vector lying along the bar, $\mathbf{p}_1 - \mathbf{p}_2$, and $e_{12} > 0$.

$$\tilde{e}_{ij} = \left.\frac{df_{ij}}{dt}\right|_{t=0} = \left.\frac{1}{2}\frac{d(l_{ij}^2)}{dt}\right|_{t=0} = l_{ij}\left.\frac{dl_{ij}}{dt}\right|_{t=0} = l_{ij}e_{ij},$$

and equation (3.7) can then be written as

$$\tilde{e}_{ij} = (\mathbf{p}_j - \mathbf{p}_i)^\mathrm{T}(\mathbf{d}_j - \mathbf{d}_i). \tag{3.9}$$

Defining an extension coefficient vector,

$$\tilde{\mathbf{e}} = [\tilde{e}_1; \ldots; \tilde{e}_b],$$

allows the compatibility relationships (equation [3.9]) to be written in a matrix form as

$$\tilde{\mathbf{e}} = \mathbf{Rd}, \tag{3.10}$$

where $\mathbf{R}$ is called the *rigidity matrix*. Notice that $\mathbf{R}$ has the simple form

$$\mathbf{R} = \begin{bmatrix} \cdots & \cdots & & \cdots & & \cdots & & \cdots & \cdots \\ \cdots & 0 & (\mathbf{p}_i - \mathbf{p}_j)^\mathrm{T} & 0 & (\mathbf{p}_j - \mathbf{p}_i)^\mathrm{T} & 0 & \cdots \\ \cdots & \cdots & & \cdots & & \cdots & & \cdots & \cdots \end{bmatrix}.$$

The columns of $\mathbf{R}$ are grouped in n sets of d columns, and each set corresponds to one node of the framework. Each row corresponds to a member of the framework, and all the entries in the row corresponding to the member $\{i, j\}$ are 0 except for the two sets of entries that correspond to the node i and the node j. For node i, the entries in the row corresponding to $\{i, j\}$ are $(\mathbf{p}_i - \mathbf{p}_j)^\mathrm{T}$, and similarly for node j, the entries are $(\mathbf{p}_j - \mathbf{p}_i)^\mathrm{T}$.

The compatibility matrix is in common use in engineering structural mechanics, while the rigidity matrix is common in mathematical rigidity theory. However, they are clearly almost the same thing. The relationship between $\mathbf{e}$ and $\tilde{\mathbf{e}}$ is

$$\tilde{\mathbf{e}} = \mathrm{diag}(\mathbf{l})\mathbf{e},$$

where $\mathrm{diag}(\mathbf{l})$ is the diagonal matrix whose diagonal entries are $\mathbf{l}$. Between $\mathbf{C}$ and $\mathbf{R}$ we have the matrix relation

$$\mathbf{R} = \mathrm{diag}(\mathbf{l})\mathbf{C}.$$

Thus, for any physically meaningful framework (where the member lengths are all finite and non-zero), the rows of $\mathbf{R}$ and the rows of $\mathbf{C}$ are simply rescaled versions of each other. In particular, the matrices will have the same rank, and the nullspaces of $\mathbf{R}$ and $\mathbf{C}$ are identical.

3.2.1 Three Examples of Frameworks

Figure 3.2 shows three frameworks that will be used as examples, and the corresponding compatibility and rigidity matrices are given in Table 3.1.

3.2.2 Mechanisms

A key question in structural mechanics is whether the compatibility or rigidity matrix define a unique infinitesimal displacement $\mathbf{d}$ for any infinitesimal extension $\mathbf{e}$, and in particular, whether there are any non-zero solutions to $\mathbf{Cd} = \mathbf{e} = \mathbf{0}$, or equivalently $\mathbf{Rd} = \tilde{\mathbf{e}} = \mathbf{0}$.

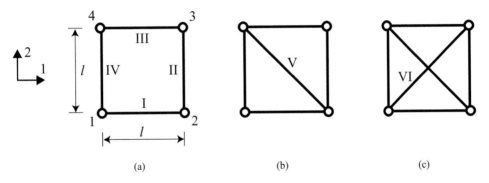

Figure 3.2 Three framework examples in a 2-dimensional space. The four nodes lie on a square grid: each framework has four outer members of length l, and the inner members are of length $\sqrt{2}l$. The nodes and members are numbered: members are numbered with roman numerals. The 1-direction is defined as horizontal, the 2-direction as vertical. In framework (c), the two crossing members are not connected: we shall always denote a connecting node with a circle.

We define an *infinitesimal mechanism* (or, in the mathematics literature, an *infinitesimal flex*) to be any solution of $\mathbf{Cd} = \mathbf{0}$. Clearly all possible infinitesimal mechanisms form a linear subspace of the space of all possible infinitesimal displacements, and this subspace is the *nullspace* (often called the *kernel*) of $\mathbf{C}$, which is equal to the nullspace of $\mathbf{R}$.

Rigid-Body Mechanisms

We often consider a framework without foundation joints, so that there is nothing to prevent a rigid-body motion of the entire framework. A rigid-body transformation will transform a configuration $\mathbf{p}$ to a configuration $\mathbf{q}$, leaving the distance between every node unchanged (whether they are connected by a member or not), and the transformation can be defined by a $d \times d$ orthogonal matrix $\mathbf{Q}$, $(\mathbf{QQ}^{\mathrm{T}} = \mathbf{I})$, where $\mathbf{I}$ is the identity matrix. This is because a transformation $\mathbf{Q}$ is a (proper) rotation or a reflection composed with a rotation (an improper rotation). In these sections, we will be concerned with that component of those orthogonal matrices that do not involve any reflections. In particular, for the time being, this means we will insist that the determinant of Q is $+1$. (The other component has a determinant of -1.) Then the rigid-body motion can be described as the transformation described by $\mathbf{Q}$, together with a vector $\mathbf{u} \in \mathbb{E}^d$, defining displacement of the origin, so that every node $\mathbf{q}_i$ is defined by

$$\mathbf{q}_i = \mathbf{Qp}_i + \mathbf{u}.$$

An infinitesimal rigid-body displacement will be denoted as $\mathbf{d}_R$, where the subscript R (for rigid) is not a numerical index, but a symbol. $\mathbf{d}_R$ will lie in the tangent space to the space of rigid-body displacements, at the current configuration, in the sense that it can be regarded as the derivative of a smooth one-parameter family of rigid-body displacements. Any rigid-body infinitesimal displacement can be defined in terms of a skew-symmetric $d \times d$ matrix $\mathbf{V}$ (i.e. $\mathbf{V}^{\mathrm{T}} = -\mathbf{V}$), defining an infinitesimal rotation around the origin, and a

Table 3.1. The compatibility and rigidity matrices for the three examples of frameworks shown in Figure 3.2. Note that C is dimensionless, while R avoids the need for surds.

$$\mathbf{C} = \begin{bmatrix} -1 & 0 & 1 & 0 & 0 & 0 & 0 & 0 \\ 0 & 0 & 0 & -1 & 0 & 1 & 0 & 0 \\ 0 & 0 & 0 & 0 & 1 & 0 & -1 & 0 \\ 0 & -1 & 0 & 0 & 0 & 0 & 0 & 1 \end{bmatrix}$$

$$\mathbf{R} = l \begin{bmatrix} -1 & 0 & 1 & 0 & 0 & 0 & 0 & 0 \\ 0 & 0 & 0 & -1 & 0 & 1 & 0 & 0 \\ 0 & 0 & 0 & 0 & 1 & 0 & -1 & 0 \\ 0 & -1 & 0 & 0 & 0 & 0 & 0 & 1 \end{bmatrix}$$

(a)

$$\mathbf{C} = \begin{bmatrix} -1 & 0 & 1 & 0 & 0 & 0 & 0 & 0 \\ 0 & 0 & 0 & -1 & 0 & 1 & 0 & 0 \\ 0 & 0 & 0 & 0 & 1 & 0 & -1 & 0 \\ 0 & -1 & 0 & 0 & 0 & 0 & 0 & 1 \\ 0 & 0 & 1/\sqrt{2} & -1/\sqrt{2} & 0 & 0 & -1/\sqrt{2} & 1/\sqrt{2} \end{bmatrix}$$

$$\mathbf{R} = l \begin{bmatrix} -1 & 0 & 1 & 0 & 0 & 0 & 0 & 0 \\ 0 & 0 & 0 & -1 & 0 & 1 & 0 & 0 \\ 0 & 0 & 0 & 0 & 1 & 0 & -1 & 0 \\ 0 & -1 & 0 & 0 & 0 & 0 & 0 & 1 \\ 0 & 0 & 1 & -1 & 0 & 0 & -1 & 1 \end{bmatrix}$$

(b)

$$\mathbf{C} = \begin{bmatrix} -1 & 0 & 1 & 0 & 0 & 0 & 0 & 0 \\ 0 & 0 & 0 & -1 & 0 & 1 & 0 & 0 \\ 0 & 0 & 0 & 0 & 1 & 0 & -1 & 0 \\ 0 & -1 & 0 & 0 & 0 & 0 & 0 & 1 \\ 0 & 0 & 1/\sqrt{2} & -1/\sqrt{2} & 0 & 0 & -1/\sqrt{2} & 1/\sqrt{2} \\ -1/\sqrt{2} & -1/\sqrt{2} & 0 & 0 & 1/\sqrt{2} & 1/\sqrt{2} & 0 & 0 \end{bmatrix}$$

$$\mathbf{R} = l \begin{bmatrix} -1 & 0 & 1 & 0 & 0 & 0 & 0 & 0 \\ 0 & 0 & 0 & -1 & 0 & 1 & 0 & 0 \\ 0 & 0 & 0 & 0 & 1 & 0 & -1 & 0 \\ 0 & -1 & 0 & 0 & 0 & 0 & 0 & 1 \\ 0 & 0 & 1 & -1 & 0 & 0 & -1 & 1 \\ -1 & -1 & 0 & 0 & 1 & 1 & 0 & 0 \end{bmatrix}$$

(c)

vector $\mathbf{w} \in \mathbb{E}^d$, defining infinitesimal displacement of the origin so that the infinitesimal displacement of every node is defined by

$$\mathbf{d}_{R,i} = \mathbf{V}\mathbf{p}_i + \mathbf{w}.$$

In dimension three, and implicitly in dimension two, this is equivalent to

$$\mathbf{d}_{R,i} = \mathbf{v} \times \mathbf{p}_i + \mathbf{w},$$

where $\times$ is the cross product, and $\mathbf{v}$ is a vector in three-space (in the nullspace of $\mathbf{V}$ in terms of the skew-symmetric matrix version). Restricting to a plane in three-space gives a similar formula for the case where the configuration is in a plane.

Any infinitesimal displacement defined in this way must satisfy $\mathbf{C}\mathbf{d}_R = \mathbf{0}$. Thus any infinitesimal rigid-body motion is an infinitesimal mechanism. In the mathematical literature, rigid-body infinitesimal mechanisms are called *trivial* infinitesimal motions, although this is not meant to denigrate their importance.

For most structures in dimension d, the infinitesimal rigid-body displacements form a $d(d + 1)/2$-dimensional linear subspace of the space of infinitesimal mechanisms. The exception is when the nodes of the structures span no more than a $(d-2)$-dimensional affine linear subspace (a linear subspace is a line, plane, etc. that contains the origin; an affine linear subspace is a line, plane, etc. that does not have to contain the origin). Thus, in the plane, for most structures the rigid-body displacements are a 3-dimensional subspace (spanned by e.g. two translations and a rotation), but if all nodes lie on top of one another (a 0-dimensional subspace) the rigid-body displacements are 2-dimensional – the translations of that point. In 3-space, for most structures, the rigid-body displacements are a 6-dimensional subspace, but if all the nodes lie along a single line (a 1-dimensional affine subspace), rotation around that line is the same as a $\mathbf{0}$ displacement.

We will say that any configuration that does not lie on a $(d-1)$-dimensional affine linear subspace is *full dimensional*. For frameworks that are not full dimensional, the dimension of the space of rigid-body motions has to be adjusted to $d(d+1)/2-(d-k)(d-k-1)/2$, where k is the dimension of the affine span of the configuration, because $(d-k)(d-k-1)/2$ is the dimension of the space of rigid-body motions that leave the affine k-dimensional subspace invariant. (When $k = d - 1$, no such adjustment is necessary.)

The three framework examples in Figure 3.2 all have the same rigid-body motions, which form a 3-dimensional subspace. This space is defined as any linear combination of three basis vectors:

$$\mathbf{d}_{R,1} = \frac{1}{2}\begin{bmatrix} 1 \\ 0 \\ 1 \\ 0 \\ 1 \\ 0 \\ 1 \\ 0 \end{bmatrix} \quad ; \quad \mathbf{d}_{R,2} = \frac{1}{2}\begin{bmatrix} 0 \\ 1 \\ 0 \\ 1 \\ 0 \\ 1 \\ 0 \\ 1 \end{bmatrix} \quad ; \quad \mathbf{d}_{R,3} = \frac{1}{2\sqrt{2}}\begin{bmatrix} 1 \\ -1 \\ 1 \\ 1 \\ -1 \\ 1 \\ -1 \\ -1 \end{bmatrix}.$$

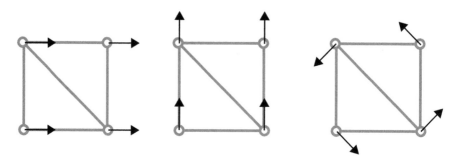

Figure 3.3 Three infinitesimal rigid-body motions for the framework of Figure 3.2(b).

The three basis vectors are illustrated in Figure 3.3. A rigid-body displacement in the 1-direction is denoted by $\mathbf{d}_{R,1}$; a rigid-body displacement in the 2-direction is denoted by $\mathbf{d}_{R,2}$; and an infinitesimal rigid-body rotation is denoted by $\mathbf{d}_{R,3}$. Recall that displacement vectors $\mathbf{d}$, such as the one above, consist of a sequence of (in this case 2-dimensional) vectors $\mathbf{d}_i$, one at each node, with $\mathbf{d} = [\mathbf{d}_1; \ldots ; \mathbf{d}_n]$, and Figure 3.3 shows these vectors as arrows attached to the relevant node i.

Internal Mechanisms

We define an *internal infinitesimal mechanism* to be any infinitesimal mechanism that is orthogonal to all the infinitesimal rigid-body motions: the direct sum of internal infinitesimal mechanisms and the infinitesimal rigid-body motions is the space of infinitesimal mechanisms.

The dimension of the nullspace (the nullity N) of a matrix with nd columns and a rank r is given by

$$N = nd - r.$$

Subtracting the dimension of the infinitesimal rigid-body motions gives the dimension of the space of internal infinitesimal mechanisms, m, sometimes loosely called the number of internal infinitesimal mechanisms. For a full-dimensional configuration, the dimension of the subspace of infinitesimal rigid-body motions is $d(d + 1)/2$, giving

$$m = nd - r - \frac{d(d + 1)}{2}.$$

For the framework examples, the compatibility matrices all have $n \times d = 4 \times 2 = 8$ columns. The compatibility matrices for Figure 3.2(b) and Figure 3.2(c) both have a rank r of 5, and the dimension of the nullspace is $nd - r = 3$, which is fully accounted for by the 3-dimensional space of infinitesimal rigid-body displacement. Thus, both frameworks of Figure 3.2(b) and Figure 3.2(c) have no internal infinitesimal mechanisms, and $m = 0$. The compatibility matrix for framework Figure 3.2(a), however, has a rank of 4 (clearly the rank cannot be greater than the number of rows). Thus the nullspace of $\mathbf{C}$ is 4-dimensional, $m = 1$, and there is a 1-dimensional space orthogonal to the infinitesimal

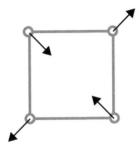

Figure 3.4 The internal infinitesimal mechanism for the framework of Figure 3.2(a).

rigid-body displacements. This is the space of internal infinitesimal mechanisms for this framework, and it is determined by the single basis vector

$$\frac{1}{2\sqrt{2}}\begin{bmatrix} -1 \\ -1 \\ -1 \\ 1 \\ 1 \\ 1 \\ 1 \\ -1 \end{bmatrix}.$$

This basis vector is illustrated in Figure 3.4. Bear in mind that this single vector in configuration space is represented by j distinct vectors in the plane, each pictured as an arrow bound to its corresponding node.

In mechanical engineering terms, Example 3.2(a) is a four-bar linkage.

Kinematic Determinacy

We define a framework where $m = 0$ to be *kinematically determinate*, and a framework where $m > 0$ to be *kinematically indeterminate*. We also define a framework where $m = 0$ to be *infinitesimally rigid*. For a framework that is infinitesimally rigid, any infinitesimal displacement that is not a rigid-body motion will cause extension of at least one member.

We have concentrated on the uniqueness of solutions to $\mathbf{Cd} = \mathbf{e}$ by examining the nullspace of $\mathbf{C}$, but have not yet considered the existence of solutions, i.e. whether it is possible to achieve a particular set of extensions $\mathbf{e}$ with any displacement $\mathbf{d}$. For a solution to be possible, $\mathbf{e}$ must lie in the column-space of $\mathbf{C}$; there will then be a solution to $\mathbf{Cd} = \mathbf{e}$, and $\mathbf{e}$ is defined as a *compatible extension*.

If $\mathbf{e}$ does not lie entirely in the column-space of $\mathbf{C}$, then it must partly lie in the orthogonal complement of the column-space. The orthogonal complement of the column-space of $\mathbf{C}$ is the left-nullspace of $\mathbf{C}$, equal to the nullspace of $\mathbf{C}^{\mathrm{T}}$ (and sometimes called the cokernel). Thus, for any b-dimensional vector $\mathbf{x}$ in the left-nullspace of $\mathbf{C}$ (i.e. where $\mathbf{C}^{\mathrm{T}}\mathbf{x} = \mathbf{0}$), if $\mathbf{x}^{\mathrm{T}}\mathbf{e} \neq 0$, then $\mathbf{e}$ does not entirely lie in the column-space of $\mathbf{C}$; there will be no solution to $\mathbf{Cd} = \mathbf{e}$, and $\mathbf{e}$ is defined as an *incompatible extension*.

For a framework with b members, $\mathbf{C}^{\mathrm{T}}$ has b columns, and the dimension of the left-nullspace is $b - r$. If $b - r = 0$, all possible infinitesimal extensions $\mathbf{e}$ are compatible. If $b - r > 0$, however, there are extensions $\mathbf{e}$ for which there is no solution to $\mathbf{Cd} = \mathbf{e}$.

For frameworks that are kinematically determinate, with $m = 0$, the frameworks are sometimes called *just rigid* or *isostatic* when $b - r = 0$, and *kinematically overdetermined* or equivalently *redundant* when $b - r > 0$.

3.2.3 Henneberg Constructions

One source of examples of kinematically determinate, equivalently infinitesimally rigid frameworks is to take one that is known and build onto it in such a way that the resulting framework is also kinematically determinate. We describe a construction for dimensions 2 and 3 that is from Henneberg (1911) but was surely well-known before that.

Henneberg Type I Construction in the Plane

Suppose that we have an infinitesimally rigid framework in the plane, where $\mathbf{p}_1$ and $\mathbf{p}_2$ are distinct nodes. Create an additional node, say $\mathbf{p}_3$, not on the line through $\mathbf{p}_1$ and $\mathbf{p}_2$, and place a bar from $\mathbf{p}_3$ to $\mathbf{p}_1$ and $\mathbf{p}_2$. This new bar framework is easily seen to be infinitesimally rigid as well. This can create several more complicated frameworks, as in Figure 3.5. Frameworks that are constructed using only Henneberg Type I operations, starting from pinned nodes or some simple structure such as a triangle or tetrahedron in three-space, are often called *simple trusses*. They make the calculation of forces (as in Section 3.7.1) straightforward by hand.

Henneberg Type II Construction in the Plane

Suppose that we have an infinitesimally rigid framework in the plane, where $\mathbf{p}_1$ and $\mathbf{p}_2$ are distinct nodes on a bar. Create an additional node, say $\mathbf{p}_3$, on the line through $\mathbf{p}_1$ and $\mathbf{p}_2$ but distinct from $\mathbf{p}_1$ and $\mathbf{p}_2$. Remove the bar between $\mathbf{p}_1$ and $\mathbf{p}_2$, and connect $\mathbf{p}_3$ to $\mathbf{p}_1$ and $\mathbf{p}_2$ as well as to another node, say $\mathbf{p}_4$, of the framework not on the line through $\mathbf{p}_1$ and $\mathbf{p}_2$. Any infinitesimal motion not including the vector $\mathbf{p}_3$ is a rigid-body motion, and so we may assume it is 0, say. Then the infinitesimal motion at $\mathbf{p}_3$ must be perpendicular to the line through $\mathbf{p}_1$ and $\mathbf{p}_2$. But when $\mathbf{p}_3$ is attached to $\mathbf{p}_4$, the infinitesimal motion at $\mathbf{p}_3$ must be 0. So the resulting framework will be infinitesimally rigid. This is shown in Figure 3.6. Note that the final framework in Figure 3.6 cannot be obtained by Henneberg Type I operations since

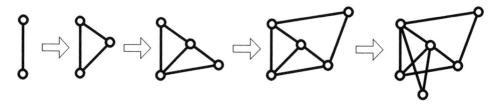

Figure 3.5 An example showing four successive Henneberg Type I operations for a plane framework.

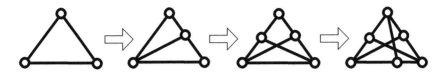

Figure 3.6 An example showing three successive Henneberg Type II operations.

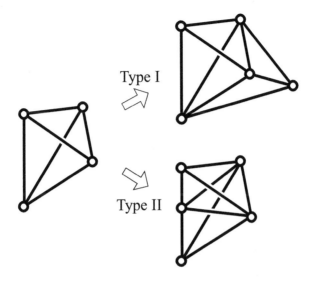

Figure 3.7 An example of 3-dimensional Henneberg operations.

each node is connected to three others. After any sequence of Henneberg Type I operations, there is always a node connected to only two other nodes.

3.2.4 Henneberg Constructions in Three-Space

There are similar constructions in three-space, as indicated in Figure 3.7. Note that the 3-dimensional Type II operation must connect the new vertex to two others in such a way that the four vertices of the starting framework involved do not lie in a plane.

3.3 Statics

The statics of a framework are motivated by the physical concepts of potential energy and work, where work is the change of potential energy of a system as it moves. We are interested in considering frameworks that are in *static equilibrium*, where there is no change in the total potential energy of the system if the structure moves – physically, if a framework is not in static equilibrium, it would start to move in search of a lower energy state. The work done by external loads (the energy put into the system) must match the work done by internal loads (the increase in energy stored in the framework).

We define an *external force* for the framework as a d-dimensional vector associated with every node. Row vectors with d coordinates and column vectors with d coordinates are said to be in *dual vector spaces*. So vectors $\mathbf{f}_i^{\mathrm{T}}$ and $\mathbf{d}_i$ are in dual spaces. If it turns out that $\mathbf{f}_i^{\mathrm{T}} \mathbf{d}_i$ can be regarded as work, then they are said to be *work-conjugate*. In other words, force vectors (in the dual space) are work conjugate with displacements of nodes.

For an infinitesimal displacement $\mathbf{d}_i$ of node i, the *infinitesimal external work done*, $w_{X,i}$, by the external force at that node, $\mathbf{f}_i$ is given by

$$w_{X,i} = \mathbf{f}_i^{\mathrm{T}} \mathbf{d}_i .$$

(In structural engineering, the external forces are the loads that are applied to the system, due to, for example, self-weight or environmental factors.)

All of the forces applied to the framework can be written as a single nd-dimensional vector,

$$\mathbf{f} = [\mathbf{f}_1; \dots ; \mathbf{f}_n],$$

and the total infinitesimal external work done, w_X, is thus given by

$$w_X = \mathbf{f}^{\mathrm{T}} \mathbf{d}. \tag{3.11}$$

We also define a set of *internal forces* for the framework. An internal force is a scalar associated with every member that is work-conjugate with the infinitesimal extension of that member. For an infinitesimal extension e_k of member k, the *infinitesimal internal work done*, $w_{I,k}$ by the internal force in the member, t_k, is given by

$$w_{I,k} = t_k e_k .$$

(In structural engineering, the internal force t_k is often called the tension in member k, with an understanding that this quantity could be negative, and hence actually a compressive force.)

We can alternatively define a set of *internal forces* that are work-conjugate with the infinitesimal extensions coefficients of the members that were defined from the differential of the rigidity map. (In the mathematical literature, these internal forces are called stresses, but that is inconsistent with the use of the word in physics and engineering, and we will not use the word "stress" in that sense here.) For an infinitesimal extension coefficient $\tilde{e}_k$ of member k, the infinitesimal internal work done, $w_{I,k}$ by the internal force in the member, ω_k, is given by

$$w_{I,k} = \omega_k \tilde{e}_k$$

and, as $\tilde{e}_k = l_k e_k$, this implies that $\omega_k = t_k / l_k$. The quantity ω_k is called a *force density*, using a notation introduced by Schek (1974). (In rigidity theory, the force density would sometimes be called a stress, but that notation is so confusing for engineers or physical scientists that we will endeavour to not use that notation here.)

We will use a dual notation for internal forces, with a double subscript t_{ij} referring to the internal force in the member from node i to node j, or equivalently using a single subscript, $t_k \equiv t_{ij}$, where there is a unique k associated with every unordered $\{i, j\}$, where $\{i, j\}$ is a member. All the internal forces in the framework can be written as a column vector $\mathbf{t}$, which we define as a *internal force* of the framework,

$$\mathbf{t} = [t_1; \ldots; t_b],$$

and the internal force densities can be written as

$$\boldsymbol{\omega} = [\omega_1; \ldots; \omega_b],$$

which, as noted in Section 2.4.3, we will term the *stress* in the framework ("stress" here capturing the idea of the framework being "stressed", not the force/area in a particular member) so that

$$\mathbf{t} = \text{diag}(\mathbf{l})\,\boldsymbol{\omega},$$

and the total infinitesimal internal work done, w_I, is given by

$$w_I = \mathbf{t}^\mathrm{T}\mathbf{e}, \tag{3.12}$$

or equivalently,

$$w_I = \boldsymbol{\omega}^\mathrm{T}\tilde{\mathbf{e}}.$$

We will consider two formulations for equilibrium, one a local, "strong" form at individual nodes, the other a "weak", global statement, before showing that they are equivalent.

3.3.1 Nodal Equilibrium: a "Strong" Statement of Equilibrium

Consider making a "free body" from each individual node. To do so, we "cut" through each member that attaches the node to other nodes. In doing so, we destroy the internal force in the member; in order to consider equilibrium, we then explicitly restore the force acting on the free body, now as an external force acting with a magnitude equal to the internal force in the member, parallel with that member, and acting away from the node under consideration.

We are thus considering a free body that is made from a single node i, that is acted upon by a force $\mathbf{f}_i$ and forces $t_{ij}\mathbf{n}_{ij}$ for every member ij attached to node i. All of these forces act through the node i. To satisfy force equilibrium requires that the forces acting on the body sum (in a vector sense) to zero:

$$\mathbf{f}_i + \sum_j t_{ij}\mathbf{n}_{ij} = 0, \tag{3.13}$$

where the summation is over all nodes such that $\{i, j\}$ is a member. The equilibrium of an example node is shown in Figure 3.8.

An alternative form of the equilibrium relationship in (3.13), using the definition of $\mathbf{n}_{ij}$ from (3.6), is

$$\mathbf{f}_i + \sum_j \omega_{ij}(\mathbf{p}_j - \mathbf{p}_i) = 0, \tag{3.14}$$

where again the summation is over all nodes such that $\{i, j\}$ is a member.

Geometry   Forces

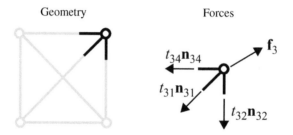

Figure 3.8 A free body diagram of node 3 of example framework in Figure 3.2(c). An external force $\mathbf{f}_3$ is applied, as are the forces $t_{31}\mathbf{n}_{31}$, $t_{32}\mathbf{n}_{32}$, and $t_{34}\mathbf{n}_{34}$ which reinstate the internal forces destroyed when the associated members were cut.

Note that, using the definition of the compatibility matrix $\mathbf{C}$ (3.8) and rigidity matrix $\mathbf{R}$ (3.10) given in (3.7) and (3.9), the nodal equilibrium equations (3.13) and (3.14) can be written together in matrix form as

$$\mathbf{f} - \mathbf{t}^{\mathrm{T}}\mathbf{C} = \mathbf{0}$$

or

$$\mathbf{f} - \boldsymbol{\omega}^{\mathrm{T}}\mathbf{R} = \mathbf{0}.$$

Here we are anticipating the duality results described explicitly in Section 3.4.

3.3.2 Virtual Work: a "Weak" Statement of Equilibrium

An alternative starting point for the equilibrium relationship is to require that, for any infinitesimal displacement of the nodes, the work done by the external loads, w_X (3.11), is equal to the change in energy stored in the members of the framework, or the work done by the internal forces, w_I (3.12).

For equilibrium, we equate w_X and w_I, to give the *equation of virtual work for a framework*. If a set of external forces $\mathbf{f}$ are in equilibrium with internal forces $\mathbf{t}$, then for any displacements $\mathbf{d}$ and compatible extensions $\mathbf{e}$, i.e. extensions defined by (3.8),

$$\mathbf{f}^{\mathrm{T}}\mathbf{d} = \mathbf{t}^{\mathrm{T}}\mathbf{e}. \tag{3.15}$$

Equivalently, written in terms of internal forces $\tilde{\mathbf{t}}$ and compatible extension coefficients $\tilde{e}$ defined by (3.10),

$$\mathbf{f}^{\mathrm{T}}\mathbf{d} = \boldsymbol{\omega}^{\mathrm{T}}\tilde{\mathbf{e}}.$$

3.3.3 Equivalence of Equilibrium Relationships

Proposition 3.3.1. *Consider a framework with a given set of internal forces* $\mathbf{t}$ *and external forces* $\mathbf{f}$. *The equation of virtual work (3.15) is true for any compatible* *infinitesimal displacements* **d** *and internal extensions* **e** *if and only if nodal equilibrium* (3.13) *holds for every node.*

Proof. We will first show that the proposition holds for an infinitesimal displacement of any single node, and then show that this implies that the proposition is true generally.

Consider initially the displacement of a single node $\mathbf{d}_i$, and the compatible extensions in any connecting member, e_{ij}, given from (3.7) as

$$e_{ij} = -\mathbf{n}_{ij}^{\mathrm{T}}\mathbf{d}_i.$$

The internal work done is given by summing over the members $\{i, j\}$ connected to node i, as these are the only members that extend,

$$w_I = \sum_j t_{ij} e_{ij} = -\sum_j t_{ij} \mathbf{n}_{ij}^{\mathrm{T}} \mathbf{d}_i.$$

The external work done is given by

$$w_X = \mathbf{f}_i^{\mathrm{T}} \mathbf{d}_i.$$

The equation of virtual work, $w_I = w_X$, thus gives

$$\mathbf{f}_i^{\mathrm{T}} \mathbf{d}_i = -\sum_j t_{ij} \mathbf{n}_{ij}^{\mathrm{T}} \mathbf{d}_i,$$

and this is true for any displacement of the node, $\mathbf{d}_i$, if and only if

$$\mathbf{f}_i = -\sum_j t_{ij} \mathbf{n}_{ij}$$

i.e. if and only if nodal equilibrium (3.13) is true for this node.

Thus the proposition is valid for displacements of, and nodal equilibrium of, any single node. However, because all of the relationships are linear, any general displacement can simply be considered as the sum of displacements of individual nodes, and if the proposition is true for individual nodes, it is therefore valid for the whole system. Thus, the proposition is proved. □

A similar proposition and proof could be straightforwardly constructed using force densities instead of internal forces.

3.3.4 Matrix Form of Equilibrium Equations

The equations (3.13) for each node i together define the equilibrium relationship between the nd-dimensional vector **f**, and the b-dimensional vector **t**,

$$\mathbf{f} = \mathbf{A}\mathbf{t}, \tag{3.16}$$

where the matrix **A** is defined as the *equilibrium matrix*. The matrix **A** will have nd rows (for n vertices in d dimensions) and b columns (for b members).

It is similarly possible to write the equilibrium relationship between the nd-dimensional vector $\mathbf{f}$, and the b-dimensional vector ω (3.14) as

$$\mathbf{f} = [\mathbf{A}\,\mathrm{diag}(\mathbf{l})]\,\omega. \tag{3.17}$$

3.3.5 Framework Examples

For the three frameworks used as examples, shown in Figure 3.2, the equilibrium matrices defined by (3.16) and (3.13) are given in Table 3.2.

Table 3.2. The equilibrium matrices for the three examples of frameworks shown in Figure 3.2.

$$\mathbf{A} = \begin{bmatrix} -1 & 0 & 0 & 0 \\ 0 & 0 & 0 & -1 \\ 1 & 0 & 0 & 0 \\ 0 & -1 & 0 & 0 \\ 0 & 0 & 1 & 0 \\ 0 & 1 & 0 & 0 \\ 0 & 0 & -1 & 0 \\ 0 & 0 & 0 & 1 \end{bmatrix}$$

(a)

$$\mathbf{A} = \begin{bmatrix} -1 & 0 & 0 & 0 & 0 \\ 0 & 0 & 0 & -1 & 0 \\ 1 & 0 & 0 & 0 & 1/\sqrt{2} \\ 0 & -1 & 0 & 0 & -1/\sqrt{2} \\ 0 & 0 & 1 & 0 & 0 \\ 0 & 1 & 0 & 0 & 0 \\ 0 & 0 & -1 & 0 & 1/\sqrt{2} \\ 0 & 0 & 0 & 1 & -1/\sqrt{2} \end{bmatrix}$$

(b)

$$\mathbf{A} = \begin{bmatrix} -1 & 0 & 0 & 0 & 0 & -1/\sqrt{2} \\ 0 & 0 & 0 & -1 & 0 & -1/\sqrt{2} \\ 1 & 0 & 0 & 0 & 1/\sqrt{2} & 0 \\ 0 & -1 & 0 & 0 & -1/\sqrt{2} & 0 \\ 0 & 0 & 1 & 0 & 0 & 1/\sqrt{2} \\ 0 & 1 & 0 & 0 & 0 & 1/\sqrt{2} \\ 0 & 0 & -1 & 0 & 1/\sqrt{2} & 0 \\ 0 & 0 & 0 & 1 & -1/\sqrt{2} & 0 \end{bmatrix}$$

(c)

3.3.6 Self-Stress

A central question in structural mechanics is whether the equilibrium relationships define a unique set of internal forces $\mathbf{t}$ for any external loads $\mathbf{f}$; and in particular, whether there are any non-zero solutions to $\mathbf{At} = \mathbf{f} = \mathbf{0}$.

We define a *state of self-stress* to be any solution of $\mathbf{At} = \mathbf{0}$. Note that in terms of force densities, we say that ω is a *self-stress* if $[\mathbf{A}\mathrm{diag}(\mathbf{l})]\omega^{\mathrm{T}} = \mathbf{0}$. Here one must be careful to distinguish between the internal stress $\mathbf{t}$ and the internal force densities ω, since, unlike infinitesimal displacements, they may not be the same for the same framework.

Clearly all possible states of self-stress form a linear subspace of the space of all possible internal forces, and this subspace is the nullspace of $\mathbf{A}$ (or the nullspace of $[\mathbf{A}\mathrm{diag}(\mathbf{l})]$ for force densities). We denote this dimension of the space of states of self-stress as s, sometimes loosely called the number of states of self-stress. If $\mathbf{A}$ has rank r, then

$$s = b - r.$$

For the examples of frameworks, the equilibrium matrices for Figures 3.2(a) and (b) both have rank equal to the number of columns and $s = 0$. The equilibrium matrix for Figure 3.2(c), however, $b = 6$ columns, but a rank $r = 5$, and thus framework Figure 3.2(c) has a single state of self-stress, and $\mathbf{A}$ and $\mathbf{A}\mathrm{diag}(\mathbf{l})$ respectively have a 1-dimensional nullspace defined by their respective single basis vectors:

$$\mathbf{t} = \begin{bmatrix} 1 \\ 1 \\ 1 \\ 1 \\ -\sqrt{2} \\ -\sqrt{2} \end{bmatrix}, \quad \omega = \begin{bmatrix} 1 \\ 1 \\ 1 \\ 1 \\ -1 \\ -1 \end{bmatrix}.$$

This basis vector, up to scaling, is illustrated in Figure 3.9.

It is useful to consider transformations of a framework that leave aspects of the framework unchanged. An *affine transformation* of a configuration is a function that transforms each node of the configuration $\mathbf{p}_i$ to a position $\mathbf{Vp}_i + \mathbf{w}$, where $\mathbf{V}$ is a d-by-d matrix, $\mathbf{w}$ is a d-component column vector, and the matrix $\mathbf{V}$ and the vector $\mathbf{w}$ are the same for all nodes of the configuration.

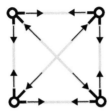

Figure 3.9 The state of self-stress for examples framework in Figure 3.2(c). A free-body diagram is drawn for each node, showing the internal forces acting on the node.

Proposition 3.3.2. *If a state of self-stress exists for a framework, then following any affine transformation of the nodes, the transformed framework will also have a state of self-stress, with unchanged force densities ω.*

Proof. Suppose a configuration $\mathbf{p} = [\mathbf{p}_1; \ldots; \mathbf{p}_n]$ has a self-stress ω, i.e. for each node i, the following equilibrium equation holds,

$$\sum_j \omega_{ij}(\mathbf{p}_j - \mathbf{p}_i) = 0, \tag{3.18}$$

where the sum is taken over all j such that $\{i, j\}$ is a member of the framework. Under an affine transformation, in the new position, (3.14) gives the external force as

$$\mathbf{f}_i + \sum_j \omega_{ij}((\mathbf{Vp}_j + \mathbf{w}) - (\mathbf{Vp}_i + \mathbf{w})) = \mathbf{0},$$

hence

$$\mathbf{f}_i = -\mathbf{V} \sum_j \omega_{ij}(\mathbf{p}_j - \mathbf{p}_i)$$

which, from (3.18), gives $\mathbf{f}_i = \mathbf{0}$, and hence ω is also a self-stress for the transformed structure.   □

Note that the proof is valid even for projections, where the matrix $\mathbf{V}$ is singular.

This proposition shows that sometimes the internal force densities ω are convenient, while the corresponding internal forces $\mathbf{t}$ behave in a more complicated way, since the length of the members may change and even become 0 under an affine transformation.

3.3.7 An Equilibrium System of External Forces

For a framework that is not connected to a foundation, the applied loads must be an equilibrium system. We define a set of external forces $\mathbf{f}$ to be *equilibrium system* if there is no external work done for any rigid-body infinitesimal mechanism of the framework, $\mathbf{d}_{R,i}$ i.e.

$$\mathbf{d}_{R,i}^{\mathrm{T}}\mathbf{f} = 0.$$

There certainly can be no solution to the equilibrium equations, $\mathbf{At} = \mathbf{f}$, if the external forces are not self-equilibrating, as can clearly be seen from the virtual work form of equilibrium (3.15).

The rigid-body infinitesimal mechanisms form a $d(d+1)/2$-dimensional subspace when the configuration is full dimensional, and hence loads that are not self-equilibrating are defined by a $d(d+1)/2$-dimensional subspace of the space of all possible loads.

3.3.8 Static Determinacy

We define a framework where $s = 0$ to be *statically determinate*, and a framework where $s > 0$ to be *statically indeterminate*.

We have concentrated on the uniqueness of solutions to $\mathbf{At} = \mathbf{f}$ by examining the nullspace of $\mathbf{A}$, but have not yet considered the existence of solutions, i.e. whether it is possible to find a set of internal forces $\mathbf{t}$ in equilibrium with an external forces $\mathbf{f}$. For a solution to be possible, $\mathbf{f}$ must lie in the column-space of $\mathbf{A}$; then there will be a solution to $\mathbf{At} = \mathbf{f}$, and we say that $\mathbf{f}$ can be *resolved*.

If $\mathbf{f}$ does not lie entirely in the column-space of $\mathbf{A}$, it must partly lie in the orthogonal complement of the column-space, which is the left-nullspace of $\mathbf{A}$, equal to the nullspace of $\mathbf{A}^T$ (and sometimes called the cokernel). Thus, if there is any nd-dimensional vector $\mathbf{x}$ where $\mathbf{A}^T\mathbf{x}=0$, for which $\mathbf{x}^T\mathbf{f} \neq 0$, then $\mathbf{f}$ does not lie entirely in the column-space of $\mathbf{A}$: there will be no solution to $\mathbf{At} = \mathbf{f}$, and $\mathbf{f}$ cannot be resolved. For a framework in d dimensions with n nodes, $\mathbf{A}$ has nd columns, and the dimension of the left-nullspace is $nd - r$.

We know that, unless $\mathbf{f}$ is an equilibrium system, there can be no solution to $\mathbf{At} = \mathbf{f}$, and the loads that are not resolvable form a $d(d + 1)/2$-dimensional subspace of the space of all possible loads, assuming that the configuration is full dimensional: this subspace must be in the left-nullspace of $\mathbf{A}$. If this subspace fully defines the left-nullspace of $\mathbf{A}$ we say that the framework is *statically rigid*: there are internal forces in equilibrium with any self-equilibrated externally applied loads. If the left-nullspace is bigger than this, we say that the framework is *statically overdetermined*. Note that in the case of a single member connecting two distinct nodes in three-space, it is also statically rigid even though the configuration is degenerate. This and the case of a single node and no members are the only cases with degenerate configurations that are statically rigid in three-space.

A framework that is statically rigid with $s = 0$ is said to be *isostatic*: there is a unique equilibrium solution for the internal forces, for any self-equilibrating applied load.

3.4 Static/Kinematic Duality

The previous sections have considered statics and kinematics independently, but these are simply different viewpoints of common underlying material. This static/kinematic duality can be straightforwardly shown by starting with the virtual work statement of equilibrium, (3.15),

$$\mathbf{f}^T\mathbf{d} = \mathbf{t}^T\mathbf{e},$$

and substituting the compatibility relationship (3.8), $\mathbf{Cd} = \mathbf{e}$,

$$\mathbf{f}^T\mathbf{d} = \mathbf{t}^T\mathbf{Cd}.$$

As this is true for any $\mathbf{d}$, and comparing this with the equilibrium relationship (3.16), $\mathbf{At} = \mathbf{f}$, we can see that

$$\mathbf{A} = \mathbf{C}^T, \tag{3.19}$$

and similarly that

$$[\mathbf{A}\text{diag}(\mathbf{l})] = \mathbf{R}^T.$$

Table 3.3. Alternative, but equivalent, terms for frameworks with m internal mechanisms, and s states of self-stress. For a framework in d dimensions, with n nodes, b members, and an equilibrium matrix A (and compatibility matrix C = A^T) with rank r, m = nd − d(d + 1) / 2 − r and s = b − r.

		Examples from Figure 3.2
$m = 0$	infinitesimally rigid, statically rigid, kinematically determinate	(b),(c)
$m > 0$	kinematically indeterminate, statically overdetermined	(a)
$s = 0$	statically determinate	(a),(b)
$s > 0$	statically indeterminate, kinematically overdetermined	(c)
$m = 0, s = 0$	statically and kinematically determinate, just rigid, isostatic	(b)
$m > 0, s > 0$	statically and kinematically indeterminate	—

Based on this static/kinematic duality, we can see that many of the concepts of rigidity that we have mentioned are equivalent. The full range of possibilities depend on the number of states of self-stress, $s = b − r$, and the number in infinitesimal internal mechanisms, $m = dn − d(d + 1)/2 − r$ (for a full dimensional configuration). Table 3.3 summarizes the earlier descriptions.

3.4.1 Maxwell Counting Rule

Simple ideas about determinacy can be explored by just counting the numbers of joints and members; here we assume that the configuration is full dimensional.

For a framework to be isostatic, then both $m = s = 0$, where

$$m = nd - \frac{d(d + 1)}{2} - r \tag{3.20}$$

and

$$s = b - r. \tag{3.21}$$

Substituting $r = b$ as the solution to (3.21) when $s = 0$ into (3.20), with $m = 0$, gives

$$nd - \frac{d(d + 1)}{2} - b = 0.$$

This necessary (but not sufficient) condition for an isostatic framework is known as Maxwell's rule, after Maxwell, who first wrote the version for $d = 3$ in 1864 (Maxwell, 1864b). A more general version of the rule is due to Calladine (1978): by eliminating r (which in general depends on the geometry of the framework) between (3.20) and (3.21) we get

$$m - s = nd - \frac{d(d + 1)}{2} - b, \tag{3.22}$$

which holds in general for any full dimensional configuration.

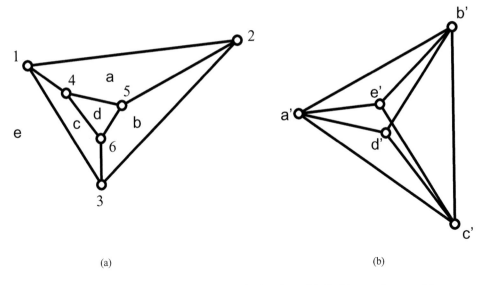

Figure 3.10 A pair of reciprocal figures. Nodes in (a) are labelled by numbers, and faces by letters. Each face i has a corresponding node i' in (b). Similarly, each node in (a) has a corresponding face in (b), but these are not shown as the faces overlap. If figures with the same connectivity but different geometry are considered, it should be noted that there is a reciprocal diagram for (b) for any configuration, but that a reciprocal diagram only exists for (a) under a certain geometric condition, that lines 1–4, 2–5, and 3–6 meet at a point.

3.5 Graphical Statics

There is a long history of the use of drawing to solve problems in statics, particularly before the days of easy access to electronic computation. Indeed, this would have been the standard way for engineers to calculate internal forces in structures, and description of these methods will still be found in standard structures textbooks, such as Coates et al. (1988). More recently, the implementation of graphical statics in a computer package has given a visual tool that has been found to be of great use in structural and architectural design (see, for example, chapters in Adriaenssens et al., 2014).

Early work on graphical statics was described by Rankine (1858), but the real break-through came with the work of Maxwell (Maxwell, 1864a, 1872). We start with the defini-tion of reciprocal figures taken from Maxwell (1864a).

> **Definition 3.5.1.** Two plane figures are reciprocal when they consist of an equal number of lines, so that corresponding lines in the two figures are parallel, and corre-sponding lines which converge to a point in one figure form a closed polygon in the other.

An example of a pair of reciprocal figures is given in Figure 3.10.

Reciprocal figures are of interest here when we consider one of the figures to represent a framework, and the other (its *reciprocal diagram*, or *Maxwell diagram*) to represent the forces carried in the framework. Consider a framework with no external forces applied. At every node i, the equation of nodal equilibrium (3.13) gives

$$\sum_j t_{ij} \mathbf{n}_{ij} = \mathbf{0},$$

and these equations are simultaneously solved by vectors of magnitude t_{ij} and unit direction $\mathbf{n}_{ij}$ forming a closed polygon around each face i' in the reciprocal diagram. Thus, the existence of a reciprocal diagram shows that the framework is able to carry a state of self-stress.

For engineering analysis of structures with external loads applied, the method can be straightforwardly extended by including the lines of applied force on the original figure, and considering the reciprocal diagram for this extended diagram. Then, the labeling of faces in the extended diagram is known as "Bow's notation".

3.5.1 The Maxwell–Cremona Correspondence

An important extension of the ideas of reciprocal diagram related internal forces in a framework with the motion of a triangulated surface. The germ of this idea is in Maxwell (1872), and is extended in Cremona (1890).

We regard a triangulated surface in $\mathbb{E}^3$ as a framework of triangles connected along their edges such that the neighbours of any node form a cycle, or in the case of a surface with boundary, the neighbours of the boundary nodes form a polygonal path. This definition is concerned with the underlying graph G of the framework rather than for a particular configuration, where self-intersections could occur. We will assume that the edge lengths of our surface are non-zero. A triangulated surface is called *oriented* if there is a cyclic direction associated to the three edges of each triangle such that adjacent triangles (ones that share a common edge) have opposite orientations, as shown in Figure 3.11.

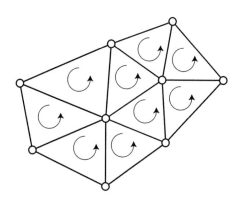

Figure 3.11 An example of an oriented triangulated surface.

We use the orientation to define unit vectors normal to the planes of the triangular faces; as shown in Figure 3.12 the direction of these normal vectors is given by the right-hand rule, where it is such that, looking down along the normal line, the orientation vectors in the triangle rotate counterclockwise, and the normal vector point up toward you. Similarly, at any vertex $\mathbf{p}_i$, let $\mathbf{n}_1$ and $\mathbf{n}_2$ be the normal vectors to successive faces in counterclockwise order according to the orientation around $\mathbf{p}_i$. The angle θ from $\mathbf{n}_1$ to $\mathbf{n}_2$ is the (external) *dihedral angle* along that edge, which we choose to be in the range $-\pi \leq \theta \leq \pi$. When $\pi > \theta > 0$, we say that the surface is *convex at that edge*, and when $-\pi < \theta < 0$, we say that the surface is *concave at that edge*.

Theorem 3.5.2. *Suppose that there is an infinitesimal flex* $\mathbf{d}$ *of an oriented triangulated surface* $(G, \mathbf{p})$ *in* $\mathbb{E}^3$. *Then there is a corresponding self-stress* $\boldsymbol{\omega}$ *for* $(G, \mathbf{p})$, *such that* ω_{ij} *is positive if the corresponding infinitesimal change in the oriented dihedral angle between adjacent triangular faces is increasing, negative if it is decreasing, and zero if it is zero.*

Roughly speaking the internal force $t_{ij} = \omega_{ij} l_{ij}$ in each edge $\{i, j\}$ is given by the relative angular velocity of the "hinge" between faces. More precisely, consider any node $\mathbf{p}_i$ and let $\theta_1, \ldots, \theta_k$ be the dihedral angles in cyclic order around $\mathbf{p}_i$, and $\phi_1, \ldots, \phi_k$ be the corresponding internal angles at $\mathbf{p}_i$ of the triangle from the j-th edge to the $(j + 1)$-st edge. Place a coordinate system with the origin $\mathbf{p}_i$, the x-axis along $\mathbf{n}_k$ and the z-axis along the edge 1. Let $\mathbf{A}(\theta_j)$ be the matrix that rotates by $-\theta_j$ about the z-axis and $\mathbf{B}(\phi_j)$ be the matrix that rotates by $+\phi_j$ about the x-axis. Then the condition that the surface stays "closed up" is equivalent to the matrix equation $\mathbf{B}(\phi_1)\mathbf{A}(\theta_1)\ldots \mathbf{B}(\phi_k)\mathbf{A}(\theta_k) = \mathbf{I}$. By taking the derivative of this expression in the flat state, we find that

$$\sum_{j=1}^{k} \theta'_j \frac{(\mathbf{p}_j - \mathbf{p}_i)}{|(\mathbf{p}_j - \mathbf{p}_i)|} = \sum_j \theta'_j \mathbf{n}_{ij} = \mathbf{0}.$$

Comparing with (3.13) we see that θ'_j is the desired internal force in the edge ij. This is essentially the argument in Gluck (1975), who used it to show that a non-infinitesimally rigid triangulated sphere must have a non-zero equilibrium self-stress in his proof of Dehn's Theorem (3.9.4). In Whiteley (1982) the correspondence is presented entirely in a projective

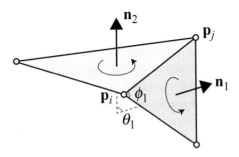

Figure 3.12 Two faces of an oriented triangulated surface.

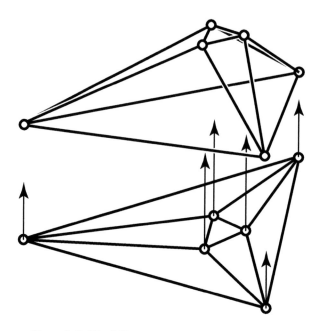

Figure 3.13 The lifting of a planar triangulated surface.

setting. See also Crapo and Whiteley (1982, 1993, 1994a), where, among other things, it is pointed out that if the surface is simply connected, i.e. it is a triangulated sphere or disk, then there is a converse to the Maxwell–Cremona correspondence. Namely, if there is an equilibrium stress ω on the surface, then there is a corresponding infinitesimal mechanism $\mathbf{d}$ that gives that stress. If the surface is not simply connected, such as a torus, as one proceeds around the surface, the motion can return with a different unwrapping motion.

There are some interesting special cases. Consider, for instance, if an originally planar polygon moves as a rigid plate: the change of each of the dihedral angles is 0, and the corresponding internal force coefficients on the interior edges found from the Maxwell–Cremona correspondence will be 0 as well.

An important special case is when the surface lies in a plane. Each dihedral angle is either 0 or π, and any infinitesimal mechanism can be chosen so that all of the displacements are perpendicular to the plane. Consider the new surface obtained by translating each node $\mathbf{p}_i$ in the plane to $\mathbf{p}_i + \alpha \mathbf{d}_i$, for any positive α, to obtain a new *lifted surface* $(G, \mathbf{p} + \alpha \mathbf{d})$. Figure 3.13 shows an example of this, where the lifted surface is a convex octahedron.

Note that when the two adjacent triangles to an edge are on opposite sides of the edge in the plane, the original dihedral angle was 0: then, when the lifted surface is convex at that edge, the change in dihedral angle, and hence the corresponding internal force coefficient, is positive. Alternatively, when the two triangles are on the same side of their common edge, the original dihedral angle was π: then, if the surface is convex at the lifted edge, the change in the dihedral angle, and the corresponding change in the internal force coefficient, is negative.

3.6 First-Order Stiffness

The concepts introduced in this chapter can be used to give an estimate of the *stiffness* of a framework, the rate of change of loading with displacement of nodes. As we shall see later, this will give an exact expression when the framework is initially unstressed. We write

$$\mathbf{f} \approx \mathbf{K}_m \mathbf{d},$$

where $\mathbf{K}_m$ is an $nd \times nd$ matrix, sometimes called the material stiffness matrix, that we shall refer to as the "first-order" stiffness matrix.

To find the first-order stiffness matrix requires the equilibrium relationship (3.16), the compatibility relationship (3.8), and additionally, knowledge of the material properties for the members of the framework. We assume for the material properties that there is a linear relationship between the extensions $\mathbf{e}$ and the internal forces $\mathbf{t}$, which we write as

$$\mathbf{t} = \mathbf{Ge}, \tag{3.23}$$

where $\mathbf{G}$ is a $b \times b$ matrix. Commonly it is assumed that the internal force in member k is a function of only the extension of that member, and hence

$$t_k = g_k e_k.$$

Writing

$$\mathbf{g} = [g_1; \dots ; g_b],$$

then we have

$$\mathbf{G} = \mathrm{diag}(\mathbf{g}).$$

We will normally assume that each member has *positive axial stiffness*, $g_i > 0$: in this case $\mathbf{G}$ will be positive definite.

The first-order stiffness is found by substituting the compatibility matrix (3.8), $\mathbf{Cd} = \mathbf{e}$ into (3.23), premultiplying by the equilibrium matrix $\mathbf{A}$, and then using the equilibrium relationship (3.16), $\mathbf{At} = \mathbf{f}$ to give

$$\mathbf{f} = \mathbf{At} = \mathbf{AGCd}$$

and, using the static-kinematic relationship that $\mathbf{C} = \mathbf{A}^{\mathrm{T}}$, we define the first-order stiffness as

$$\mathbf{K}_m = \mathbf{AGA}^{\mathrm{T}}. \tag{3.24}$$

The matrix $\mathbf{K}_m$ is clearly symmetric.

Proposition 3.6.1. *For any framework where each member has positive axial stiffness, the nullspace and the column space of $\mathbf{K}_m$ are equal to the nullspace and the column space of $\mathbf{C} = \mathbf{A}^T$.*

Proof. Consider a vector $\mathbf{x}$ in the nullspace of $\mathbf{C} = \mathbf{A}^{\mathrm{T}}$. That vector must also lie in the nullspace of $\mathbf{K}_m$, as $\mathbf{K}_m \mathbf{x} = \mathbf{AGA}^{\mathrm{T}} \mathbf{x} = \mathbf{0}$. Also consider a vector $\mathbf{y}$ in the nullspace of $\mathbf{K}_m$.

$\mathbf{K}_m\mathbf{y} = \mathbf{0}$ implies that $\mathbf{y}^T\mathbf{K}_m\mathbf{y} = (\mathbf{A}^T\mathbf{y})^T\mathbf{G}(\mathbf{A}^T\mathbf{y}) = 0$. $\mathbf{G}$ has full rank, as $g_i > 0$, and so $\mathbf{A}^T\mathbf{y} = \mathbf{0}$, and $\mathbf{y}$ must lie in the nullspace of $\mathbf{C} = \mathbf{A}^T$.

Thus a vector lies in the nullspace of $\mathbf{K}_m$ if and only if it lies in the nullspace of $\mathbf{C} = \mathbf{A}^T$. Thus the nullspaces of $\mathbf{K}_m$ and $\mathbf{C} = \mathbf{A}^T$ are equal. Further, as the column space is the orthogonal complement of the nullspace, the column spaces of $\mathbf{K}_m$ and $\mathbf{C} = \mathbf{A}^T$ are also equal. □

One key consequence is that, for any internal or rigid-body mechanism $\mathbf{d}_{mi}$, $\mathbf{K}_m\mathbf{d}_{mi} = \mathbf{0}$.

Proposition 3.6.2. *For any framework, where each member has positive axial stiffness, $\mathbf{K}_m$ is positive semi-definite.*

Proof. Consider an infinitesimal displacement $\mathbf{d}$ and compatible extensions $\mathbf{e}$,

$$\mathbf{d}^T\mathbf{K}_m\mathbf{d} = (\mathbf{A}^T\mathbf{d})^T\mathbf{G}(\mathbf{A}^T\mathbf{d}) = \mathbf{e}^T\mathbf{G}\mathbf{e}.$$

If $\mathbf{d}$ is an infinitesimal mechanism, $\mathbf{e} = \mathbf{0}$ and $\mathbf{e}^T\mathbf{G}\mathbf{e} = 0$. Otherwise, as $\mathbf{G}$ is positive definite, $\mathbf{e}^T\mathbf{G}\mathbf{e} > 0$. Thus, for any $\mathbf{d}$, $\mathbf{d}^T\mathbf{K}_m\mathbf{d} \geq 0$. □

3.7 Example: Structural Analysis of a Pin-Jointed Cantilever

This section will provide a description of how the methods described in this chapter might be used by structural engineers. We will show an analysis of the cantilever structure shown in Figure 3.14. We shall be primarily concerned with the forces carried by the members due to the applied load (perhaps to ensure that they don't exceed the safe limit for the member), but will also be interested in the deflections of the structure.

The structure shown in Figure 3.14 is an abstract representation of a real structure. The real structure would be made in practice from long thin members, probably rigidly welded

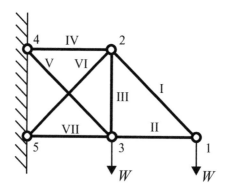

Figure 3.14 A cantilever structure. All the members are of length L or $\sqrt{2}L$. Loads of W are applied at two of the nodes. Nodes 4 and 5 are attached to a foundation, and are unable to move. All members are made from material with a Young's modulus E, and have a cross-sectional area A.

or bolted together. However, experiments show that the pin-jointed model gives a good estimate of the forces carried in the members, because these members are much more flexible in bending than they are for axial deformation.

We will show two methods for the analysis of the cantilever. The first is called the "force method," where the primary calculation finds forces in equilibrium with the applied loads. The second is called the "displacement method," where the primary calculation finds the displacement of the nodes. For hand calculation, the force method would probably be the method of choice for the engineer, as the analysis can be broken down into a sequence of small calculations. However, the advent of the computational analysis has led to the displacement method becoming ubiquitous.

The calculations will be performed using internal forces $\mathbf{t} = [t_1; \dots ; t_{VII}]$, externally applied loads $\mathbf{f} = [\mathbf{f}_1; \mathbf{f}_2; \mathbf{f}_3]$ and nodal displacements $\mathbf{d} = [\mathbf{d}_1; \mathbf{d}_2; \mathbf{d}_3]$. Note that the foundation joints, nodes 4 and 5 are not considered. The external loads at the non-foundation joints are $\mathbf{f}_1 = [0; - W]$, $\mathbf{f}_2 = [0; 0]$, and $\mathbf{f}_3 = [0; - W]$, giving

$$\mathbf{f} = W[0; -1; 0; 0; 0; -1].$$

The equilibrium matrix for the structure is given by

$$\mathbf{A} = \begin{bmatrix} 1/\sqrt{2} & 1 & 0 & 0 & 0 & 0 & 0 \\ -1/\sqrt{2} & 0 & 0 & 0 & 0 & 0 & 0 \\ -1/\sqrt{2} & 0 & 0 & 1 & 0 & 1/\sqrt{2} & 0 \\ 1/\sqrt{2} & 0 & 1 & 0 & 0 & -1/\sqrt{2} & 0 \\ 0 & -1 & 0 & 0 & 1/\sqrt{2} & 0 & 1 \\ 0 & 0 & -1 & 0 & -1/\sqrt{2} & 0 & 0 \end{bmatrix}.$$

To find a solution by either the displacement or the force method requires the axial stiffness of the members to be known. The usual assumption is that for each member i, the axial stiffness is given by $g_i = A_i E_i / l_i$, where A_i is the cross-sectional area of the member and E_i is the Young's modulus of the material from which it is made. For the example structure, all bars have a cross-sectional area A and a Young's modulus E, and so the axial stiffness and its inverse is given by

$$\mathbf{G} = \frac{AE}{L} \begin{bmatrix} 1/\sqrt{2} & & & & & & \\ & 1 & & & & & \\ & & 1 & & & & \\ & & & 1 & & & \\ & & & & 1/\sqrt{2} & & \\ & & & & & 1/\sqrt{2} & \\ & & & & & & 1 \end{bmatrix};$$

$$
\mathbf{G}^{-1} = \frac{L}{AE}
\begin{bmatrix}
\sqrt{2} & & & & & & \\
 & 1 & & & & & \\
 & & 1 & & & & \\
 & & & 1 & & & \\
 & & & & \sqrt{2} & & \\
 & & & & & \sqrt{2} & \\
 & & & & & & 1
\end{bmatrix}.
$$

3.7.1 Force Method

An appropriate version of Maxwell's rule (3.22) for the planar case, where rigid-body motions are suppressed by connection to a foundation is

$$
m - s = 2n - b,
$$

where $b = 7$ is the number of members; $n = 3$ is the number of non-foundation joints; and the triangulated nature of the construction guarantees that there are no mechanisms, $m = 0$. Thus there is a single state of self-stress, $s = 1$.

A single state of self-stress shows that there is a line of solutions in internal force space, and all are in equilibrium with the applied load,

$$
\mathbf{t} = \mathbf{t}_0 + x\mathbf{t}_s, \tag{3.25}
$$

where $\mathbf{t}_0$ is a particular equilibrium solution ($A\mathbf{t}_0 = \mathbf{f}$), $\mathbf{t}_s$ is a basis vector describing the state of self-stress ($A\mathbf{t}_s = \mathbf{0}$), and x is here a scalar having units of force, as W is in units of force, say newtons.

To find a suitable choice of $\mathbf{t}_0$, a particular member is chosen to be redundant: the internal force in that member is taken to be zero, and the internal forces are then calculated in the other members. The equilibrium matrix $\mathbf{A}$ will usually not be formed, rather equilibrium calculations proceed on a node-by-node basis, using (3.13). (In a worst case the complete set of equilibrium equations will have to be solved simultaneously, but practically this is rare. If the structure is a simple truss, generated from type I Henneberg moves as described in Section 3.2.3, it is always possible to work back through the Henneberg construction, satisfying equilibrium node-by-node.) Here, if we choose member V to be redundant ($t_V = 0$), t_I and t_{II} can be found from equilibrium of node 1, t_{III} and t_{VII} can then be found from equilibrium of node 3, and finally t_{IV} and t_{VI} can be found from equilibrium of node 2, giving

$$
\mathbf{t}_0 = W
\begin{bmatrix}
\sqrt{2} \\
-1 \\
1 \\
3 \\
0 \\
-2 \\
\sqrt{2} \\
-1
\end{bmatrix}.
$$

A state of self-stress can be found as an equilibrium solution where the tension in the redundant member is set to 1, and the nodal equilibrium equations are solved in the absence of any external loads. Thus, setting $t_V = 1$, and the external loads to zero, t_I and t_{II} can be found from equilibrium of node 1, t_{III} and t_{VII} can then be found from equilibrium of node 3, and finally t_{IV} and t_{VI} can be found from equilibrium of node 2, giving

$$\mathbf{t}_s = \begin{bmatrix} 0 \\ 0 \\ -1/\sqrt{2} \\ -1/\sqrt{2} \\ 1 \\ 1 \\ -1/\sqrt{2} \end{bmatrix}.$$

An incorrect choice of a redundant member (in this case neither member I or II could successfully be chosen) will result in it being impossible to find $\mathbf{t}_s$, and to there being no unique solution for $\mathbf{t}_0$.

Any value of x will give, from (3.25), internal forces $\mathbf{t}$ in equilibrium with the applied load. To find a unique solution that models the behaviour of an initially unstressed structure requires us to consider the compatibility of the resultant extensions. The extension $\mathbf{e}$ is given by substituting (3.25) into (3.23) to give

$$\mathbf{e} = \mathbf{G}^{-1}\mathbf{t} = \mathbf{G}^{-1}\mathbf{t}_0 + x\mathbf{G}^{-1}\mathbf{t}_s,$$

where $\mathbf{G}^{-1}$ exists if every member has non-zero axial stiffness. To find values of x that give a compatible $\mathbf{e}$, we can make use of the results in Section 3.2.2. The resultant $\mathbf{e}$ must lie in the column space of $\mathbf{C} = \mathbf{A}^T$, i.e. it must be orthogonal to any vector in the left nullspace of $\mathbf{C}$, which is equal to the nullspace of $\mathbf{A}$, and is fully defined by the basis vector $\mathbf{t}_s$. Thus we can write

$$\mathbf{t}_s \mathbf{e} = \mathbf{t}_s \mathbf{G}^{-1}\mathbf{t}_0 + x\mathbf{t}_s \mathbf{G}^{-1}\mathbf{t}_s = 0 \tag{3.26}$$

and hence, as $\mathbf{G}^{-1}$ is positive definite, there is a unique solution x to

$$(\mathbf{t}_s\, \mathbf{G}^{-1}\mathbf{t}_s)x = -\mathbf{t}_s\mathbf{G}^{-1}\mathbf{t}_0$$

and thus the tensions in each of the members are found. For the example structure,

$$\mathbf{t}_s\mathbf{G}^{-1}\mathbf{t}_s = \frac{3+4\sqrt{2}}{2}\frac{L}{AE}; \quad \mathbf{t}_s\mathbf{G}^{-1}\mathbf{t}_0 = -\frac{3+4\sqrt{2}}{\sqrt{2}}\frac{WL}{AE}$$

and hence

$$x = \sqrt{2}W$$

and

$$
\mathbf{t} = W
\begin{bmatrix}
\sqrt{2} \\
-1 \\
1 \\
3 \\
0 \\
-2\sqrt{2} \\
-1
\end{bmatrix}
+ \sqrt{2}W
\begin{bmatrix}
0 \\
0 \\
-1/\sqrt{2} \\
-1/\sqrt{2} \\
1 \\
1 \\
-1/\sqrt{2}
\end{bmatrix}
= W
\begin{bmatrix}
\sqrt{2} \\
-1 \\
0 \\
2 \\
\sqrt{2} \\
-\sqrt{2} \\
-2
\end{bmatrix}.
$$

(Equation (3.26) can also be found by appealing directly to the equation of virtual work (3.15), written here as

$$(\mathbf{f}^*)^{\mathrm{T}}\mathbf{d} = (\mathbf{t}^*)^{\mathrm{T}}\mathbf{e},$$

and choosing a virtual force system $\mathbf{t}^* = \mathbf{t}_s$ in equilibrium with $\mathbf{f}^* = \mathbf{0}$.)

If displacements are required, they can straightforwardly be found from the solution of (3.8), $\mathbf{Cd} = \mathbf{e}$, which is guaranteed to have a solution for the particular choice of x. If (3.8) does not have a unique solution, then the structure is kinematically indeterminate, which for conventional structures would be considered as a bad thing. For hand calculation of the displacement of practical structures, the displacements can usually be found on a node-by-node basis (guaranteed for a simple truss), starting from the known zero displacements of the foundations, rather than simultaneously solving $\mathbf{Cd} = \mathbf{e}$.

3.7.2 Displacement Method

The displacement method relies on the formation of the first-order stiffness matrix (3.24):

$$
\mathbf{K}_m = \mathbf{AGA}^{\mathrm{T}} = \frac{AE}{L}
\begin{bmatrix}
1 + \frac{1}{2\sqrt{2}} & -\frac{1}{2\sqrt{2}} & -\frac{1}{2\sqrt{2}} & \frac{1}{2\sqrt{2}} & -1 & 0 \\
-\frac{1}{2\sqrt{2}} & \frac{1}{2\sqrt{2}} & \frac{1}{2\sqrt{2}} & -\frac{1}{2\sqrt{2}} & 0 & 0 \\
-\frac{1}{2\sqrt{2}} & \frac{1}{2\sqrt{2}} & 1 + \frac{1}{\sqrt{2}} & -\frac{1}{\sqrt{2}} & 0 & 0 \\
\frac{1}{2\sqrt{2}} & -\frac{1}{2\sqrt{2}} & -\frac{1}{\sqrt{2}} & 1 + \frac{1}{\sqrt{2}} & 0 & -1 \\
-1 & 0 & 0 & 0 & 2 + \frac{1}{2\sqrt{2}} & -\frac{1}{2\sqrt{2}} \\
0 & 0 & 0 & -1 & -\frac{1}{\sqrt{2}} & 1 + \frac{1}{2\sqrt{2}}
\end{bmatrix}
$$

If the structure is statically determinate, rigid-body modes are fully suppressed because the structure is attached to a foundation, and every member has positive axial stiffness (as for the example structure), then $\mathbf{K}_m$ will be positive definite, and there will be a solution to $\mathbf{K}_m\mathbf{d} = \mathbf{f}$ for any loading $\mathbf{f}$. Having found displacements of nodes, it is straightforward to find the extensions of the members using (3.8):

$$\mathbf{e} = \mathbf{Cd}$$

and hence the tensions in the members can be found using: (3.23)

$$\mathbf{t} = \mathbf{Ge}$$

Thus the displacement method requires fewer steps, and less fundamental understanding of the structural response of the system. However, it does require the formation and solution of the first-order stiffness equation, $\mathbf{K}_m\mathbf{d} = \mathbf{f}$. This is in general not amenable to hand calculation, but is straightforward using a computer package.

3.8 The Basic Rigidity Theorem

Having considered some engineering aspects of frameworks, we now return to a key geometrical concept.

Theorem 3.8.1 (Basic Rigidity). *If a framework is infinitesimally rigid in $\mathbb{E}^d$, then it is rigid in $\mathbb{E}^d$.*

At first sight, this result may see obvious, but there are pitfalls. For instance, it is possible that the path of a finite mechanism may, at a particular point, involve no first-order changes in position – some of these pitfalls are well illustrated by the "cusp mechanism," that will be introduced in Section 8.3.

There are at least two approaches to proving the Basic Rigidity Theorem. The method that we will employ here depends on the inverse function theorem from advanced calculus and will occupy the next few subsections. This also has the advantage of pointing to some of the basic theoretical tools that can be used to get insight into the nature of this subject. The other method uses Definition 2. The idea is to use the derivative of the motion, associated with any possible finite mechanism, to show that that derivative is an infinitesimal mechanism, and indeed, an analogous theorem for tensegrities will be proved using this approach in Chapter 4. As will be seen, however, there is a difficulty with this method because of peculiarities of the way the motion may be forced to start.

In the previous notation, a framework is infinitesimally rigid if and only if the number of motions $m = 0$. Indeed, using Calladine's formula (3.22), for a full dimensional configuration, this is equivalent to the condition $s = b - nd + d(d + 1)/2$, which emphasizes the role of s, the number of states of self-stress in the framework. With this in mind and the discussion in Section 3.4, we have the following.

Corollary 3.8.2. *A framework is infinitesimally rigid if and only if it is statically rigid.*

One should also be aware that the converse of the Basic Rigidity Theorem is not true. Figure 3.15 shows an example which is an infinitesimal mechanism in the plane, and yet is rigid.

Recall that function $f : \mathbb{E}^N \to \mathbb{E}^M$ is *continuously differentiable* near $\mathbf{p}$ if its Jacobian of partial derivatives, also called its differential, $[\partial f_i/\partial x_j] = df_{\mathbf{p}}$ exists and is continuous for all points sufficiently near $\mathbf{p}$ in $\mathbb{E}^N$. Also for a set X in the domain $\mathbb{E}^N$, the notation $f(X)$ is used to denote the set of points $f(\mathbf{q})$, where $\mathbf{q}$ is in X. The following is a simple version of the inverse function theorem that we will need.

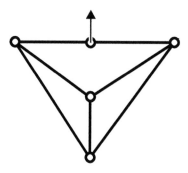

Figure 3.15 The vectors defining the infinitesimal mechanism are shown. This is an example of an infinitesimal mechanism that is nevertheless rigid.

Theorem 3.8.3 (Inverse Function). *Suppose that $f : \mathbb{E}^N \to \mathbb{E}^M$ is continuously differentiable and that for a point $\mathbf{p}$ in $\mathbb{E}^N$, $df_{\mathbf{p}}$ has rank N. Then for points U sufficiently near to $\mathbf{p}$ in $\mathbb{E}^N$ there is an (inverse) function $g : f(U) \to \mathbb{E}^N$ such that for all $\mathbf{q}$ in U, $g(f(\mathbf{q})) = \mathbf{q}$.*

The proof of this result can be found in most books on advanced calculus such as Spivak (1965) or Krantz and Parks (2002).

Proof of Theorem 3.8.1. We assume that the configuration in $\mathbb{E}^d$ is full dimensional, since if it is not, it is easy to handle that case separately. For example, in dimension 2, the only framework with a degenerate configuration that is infinitesimally rigid is a single vertex. In dimension 3 the only frameworks with a degenerate configuration that are infinitesimally rigid are a single vertex and single bar. In each of those cases, it is clear that those frameworks are, indeed, rigid.

The idea is to apply the Inverse Function Theorem 3.8.3 to the rigidity map. But one must deal with the rigid-body mechanisms and motions, which reduce the dimension of domain space. In the space of all configurations in $\mathbb{E}^d$, since the configuration $\mathbf{p}$ is full dimensional, the space of rigid-body motions is $d(d + 1)/2$-dimensional. These rigid-body motions applied to $\mathbf{p}$ have a $d(d + 1)/2$-dimensional tangent space. Choose an $nd - d(d + 1)/2$-dimensional affine linear subspace X orthogonal and complementary to this tangent space at the configuration $\mathbf{p}$, where n is the number of nodes of the configuration. The rigidity map, when restricted to X at $\mathbf{p}$, by the definition of infinitesimal rigidity, has rank $nd - d(d+1)/2$. By the Inverse Function Theorem 3.8.3, there is a map $g : f(U) \to U$ such that $g(f(\mathbf{q})) = \mathbf{q}$ for all $\mathbf{q}$ in U which are points that are sufficiently close to $\mathbf{p}$ in X.

With Definition 2, say, in mind, suppose $\mathbf{p}(t)$ is an (analytic) flex for $t \geq 0$, where $\mathbf{p}(0) = \mathbf{p}$. For t sufficiently small, we can alter $\mathbf{p}(t)$ by a rigid-body motion so that $\mathbf{p}(t)$ lies in X. But since $\mathbf{p}(t)$ is a flex of the framework starting at the configuration $\mathbf{p}$, this means that $f(\mathbf{p}(t)) = f(\mathbf{p})$ a constant. Applying the map g we get that $\mathbf{p}(t) = \mathbf{p}$. In other words, $\mathbf{p}(t)$ is a rigid-body motion, and thus the framework is rigid. $\qquad\square$

3.9 Another Example of Infinitesimal Rigidity

In Section 3.2.3 we saw a way of creating many examples of infinitesimally rigid framework. In fact, we will show in Chapter 7 that, in the plane, all infinitesimally rigid bar frameworks can be created by the Henneberg operations described, together with an appropriate generic perturbation, described later. This will be fundamental to proving Laman's theorem, giving a combinatorial characterization of rigidity in the plane.

This section will discuss a fundamental result about convex polyhedral surfaces in three-space that originates with Cauchy (1813).

3.9.1 Convex Polyhedral Shells

One very useful class of infinitesimally rigid structures has to do with what are called "polyhedral surfaces" made from rigid plates hinged on their edges. These structures can be modelled very nicely as bar frameworks. We start with the definition of a convex polyhedron, together with a little topology concerning the Euler characteristic. We continue with a combinatorial lemma in Cauchy (1813). Next we describe another simple lemma that describes the sign pattern of self-stresses in a surface of a convex polytope. Putting these results together gives the infinitesimal rigidity of convex triangulated polyhedral surfaces.

Convex Polyhedra

A subset X in $\mathbb{E}^d$ is called *convex* if for every pair of points $\mathbf{p}$ and $\mathbf{q}$ in X, the line segment connecting $\mathbf{p}$ and $\mathbf{q}$ is also in X. For example, the cube or tetrahedron is convex, but the boundaries of a cube or tetrahedron are not convex. Note that the intersection of any collection of convex sets is convex (considering the empty set as convex).

Consider a finite collection of points X in $\mathbb{E}^3$ not contained in a plane. The intersection of all the closed half-spaces that contain X is called a *convex polytope P*, and P is called the *convex hull* of X. If L, a 2-dimensional plane, is the boundary of a half-plane containing X, and L contains a point of X, then we say that L is a *support plane* for the polytope P. If a support plane L is such that $L \cap P = F$ is a single point $\mathbf{p}$, then $\mathbf{p}$ is called a *vertex* of P. If F is not a point but contained in a line, it is called an *edge*. If F is not contained in any line, then it is called a *face* of P. It is easy to show that a convex polytope is the convex hull of its vertices. For vertices, edges, and faces, we say a pair of such sets are *adjacent* if one is a subset of the other.

For example, a tetrahedron is the convex hull of four points in $\mathbb{E}^3$, when the four points do not lie in a plane. It has four vertices, six edges, and four triangular faces. Each vertex is adjacent to three edges and three faces. For any convex polytope, each edge is adjacent to exactly two faces. In Figure 3.16 we see a convex octahedron (i.e. it has eight faces), where the support planes are indicated.

The following, Theorem 3.9.1, is Cauchy's result which has some very interesting and original ideas that are quite powerful. We will discuss some of his ideas in the next few subsections, but the completion of his proof will be delayed until Chapter 5, Section 5.14.3,

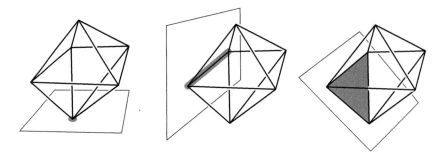

Figure 3.16 A convex polytope, an octahedron, is indicated with support planes positioned so that they intersect the polytope in a typical vertex, edge, and face.

so that some newer results about stress matrices can be brought to bear. See also comments there about troubles with the details of his original proof.

Theorem 3.9.1. *Let P and Q be two convex polytopes in $\mathbb{E}^3$ with boundaries ∂P and ∂Q. Let $f : \partial P \rightarrow \partial Q$ be a continuous one-to-one correspondence, such that restricted to each face of ∂P, f is a rigid congruence (and similarly for each face of ∂Q and f^{-1}). Then f extends to a congruence of all of P.*

The following is a classical result that was apparently known to Descartes, although it is usually attributed to Euler.

Theorem 3.9.2. *Suppose a convex polytope has v vertices, e edges, and f faces. Then*

$$v - e + f = 2.$$

A proof of this in a very general context can be found in any book on algebraic topology such as Hatcher (2002). There are quite elementary proofs such as Lakatos (1976).

Convex Polytopes with All Faces Triangles

When a convex polytope is such that each face is a triangle, it is possible to find a relation between the number of vertices v and the number of edges e. In this case, each face is adjacent to exactly three edges, and each edge is adjacent to exactly two faces. Thus the number of face-edge adjacencies is $2e$ on the one hand and $3f$ on the other hand. Thus $2e = 3f$. Combining this with the Euler formula of Theorem 3.9.2, we get

$$6 = 3v - 3e + 3f = 3v - 3e + 2e = 3v - e.$$

In other words we have the following:

Corollary 3.9.3. *If a convex polytope has v vertices, e edges, and all faces are triangles, then*

$$e = 3v - 6.$$

For example, the regular octahedron has $v = 6$ and $e = 3 \cdot 6 - 6 = 12$. The regular icosahedron has $v = 12$ and $e = 3 \cdot 12 - 6 = 30$.

Bar Frameworks for Convex Polytopes with All Faces Triangles

For any convex polytope P with all of its faces triangles, associate a bar framework, where the vertices of P are the nodes, and a bar is placed between two nodes if their vertices are adjacent to the same edge. The following is a result due to Dehn (1916). We provide a proof that follows some ideas of Cauchy (1813), Alexandrov (2005), and Gluck (1975).

Theorem 3.9.4 (Dehn). *A bar framework associated to a convex polytope with all faces triangles is infinitesimally rigid in $\mathbb{E}^3$.*

The proof of this very basic result will follow from two lemmas that we describe next. But we observe that by Corollary 3.9.3 and the Calladine formula (3.22), we have $m = s$, since $v = n$, and $b = e$, where m is the number of infinitesimal motions and s is the number of internal stresses in the associated bar framework. So in order to show that such a framework is infinitesimally rigid, it is enough to show that $s = 0$.

Geometric Local Lemma

Suppose that a bar framework in $\mathbb{E}^3$ has a self-stress with internal force densities ω. We concentrate on one node, say $\mathbf{p}_0$ of the configuration, and the signs of the ω_{ij}. Consider any plane L through $\mathbf{p}_0$. If the members where $\omega_{ij} > 0$ lie on one side of L, and the members where $\omega_{ij} < 0$ lie on the other side, the equilibrium relation (3.18) cannot hold, since the vectors all sum to one side of L.

Lemma 3.9.5. *If a a bar framework, associated to a convex polytope P with all faces triangles, has a self-stress, then for each node $\mathbf{p}_0$, the internal forces in adjacent bars are either all 0, or there are at least four changes in sign proceeding cyclically around $\mathbf{p}_0$.*

Proof. Suppose that the internal forces are such that there are exactly two changes in sign. Consider the cycle of edges of P not adjacent to $\mathbf{p}_0$, but which are adjacent to the triangular faces that are adjacent to $\mathbf{p}_0$. Each vertex of this cycle corresponds to an edge of P adjacent to $\mathbf{p}_0$. Choose two points in the cycle $\mathbf{q}_1$ and $\mathbf{q}_2$, say, that separate the vertices that correspond to the bars with opposite signs. Then the plane determined by $\mathbf{p}_0$, $\mathbf{q}_1$, and $\mathbf{q}_2$ separates the bars with opposite sign. But from the comments above, this is impossible, and so there cannot be just two changes in sign. Figure 3.17 shows this argument graphically. This ends the proof. □

Combinatorial Global Lemma

For this lemma we will need to appeal to a somewhat more general version of Euler's Theorem. We consider not just P, but also any connected subset of P in order to deal with cases where only some of the bars carry non-zero force.

Theorem 3.9.6. *Let G be a connected subset of the edges of a convex polytope P. Then this set separates the surface of P into say f connected regions, and*

$$v - e + f = 2,$$

where e is the number of edges of G, and v is the number of vertices of P that are in G.

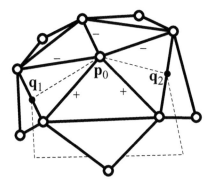

Figure 3.17 The region of a convex polytope P that has all triangular faces around a vertex p_0. The figure shows that for a state of self-stress, there cannot be just two changes in sign for the internal forces. Nodal equilibrium cannot be satisfied perpendicular to the plane (indicated with dashed lines) determined by the points that separate the edges with opposite signs of internal force.

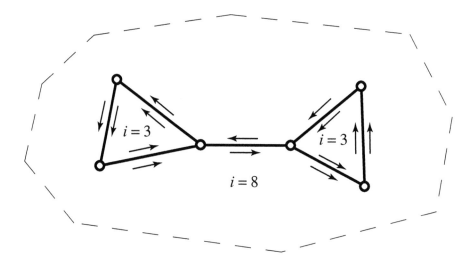

Figure 3.18 A graph is shown embedded on the surface of some polytope. This graph has three regions, two with three edges, $f_3 = 2$, and one with eight edges, $f_8 = 1$ (but notice that one edge is counted twice for that region). We can verify that $3 f_3 + 8 f_8 = 2 \cdot 3 + 8 \cdot 1 = 6 + 8 = 14 = 2e$, where $e = 7$ is the number of vertices.

Note that we must assume that the graph G is connected. If not, then we must replace the constant 2 by 1 plus the number of components.

The "connected regions" might be described as "topological faces." They may be non-convex, and may even wrap around an island of one or more connected regions so that the same edge bounds the region twice, once on either side. Each connected region will have a number of *sides i*, which counts the number of edges traversed as one proceeds around the boundary. This is shown in Figure 3.18.

Lemma 3.9.7. *Let G be a subset of the edges of a convex polytope P, such that each edge is labelled with a + or −. Then some vertex of G has no changes in sign or exactly two changes in sign as one proceeds cyclically around it.*

Proof. Assume that there are at least four changes in sign at every vertex, and we will arrive at a contradiction. If G is not connected, we will restrict to one of the connected components. Let f_i, for $i = 3, 4, \ldots$ be the number of connected regions that have i sides; let $\mathbf{e}$ be the total number of edges in G; and let n be the total number of vertices of G. Then we have one equation from Theorem 3.9.6:

$$n - e + (f_3 + f_4 + f_5 + f_6 + \cdots) = 2$$

and a second because each edge is counted twice in the edge-region adjacencies:

$$2e = 3f_3 + 4f_4 + 5f_5 + 6f_6 + \cdots + kf_k + \cdots .$$

Solving for $4n$ in terms of the f_i gives

$$4n = 8 + 2f_3 + 4f_4 + 6f_5 + 8f_6 + \cdots + (2k - 4)f_k + \cdots$$

Let c be the total number of times there is a change in sign between two edges that are both adjacent to a common vertex in a region determined by G (see Figure 3.19 for an example of counting changes in sign). The assumption that there are at least four changes in sign around every vertex implies that $c \geq 4n$, and so

$$8 + 2f_3 + 4f_4 + 6f_5 + 8f_6 + \cdots + (2k - 4)f_k + \cdots \leq c.$$

Furthermore, counting around the edges of a region with i edges can give at most i sign changes if i is even and at most $i - 1$ sign changes if i is odd. Thus

$$c \leq 2f_3 + 4f_4 + 4f_5 + 6f_6 + \cdots$$

So we get the following inequality:

$$8 + 2f_3 + 4f_4 + 6f_5 + 8f_6 + \cdots + (2k - 4)f_k + \cdots \leq c \leq 2f_3 + 4f_4 + 4f_5 + 6f_6 + \cdots$$

This implies the following inequality, which is clearly impossible:

$$8 + 2f_5 + 2f_6 + \cdots \leq 0.$$

See Figure 3.19 for an example of the placement of the + and − signs. This finishes the lemma. □

Proof of Dehn's Theorem

Proof of Theorem 3.9.4. As we mentioned earlier, since a triangulated convex polytope has $3n - 6$ edges, by Calladine's equation (3.22), the infinitesimal rigidity of the bar framework is equivalent to showing that there are 0 states of self-stress in the framework. But by Lemma 3.9.5, if there is a non-zero self-stress, there must be four changes in sign at each vertex adjacent to a stressed edge. By Lemma 3.9.7 this is combinatorially impossible.

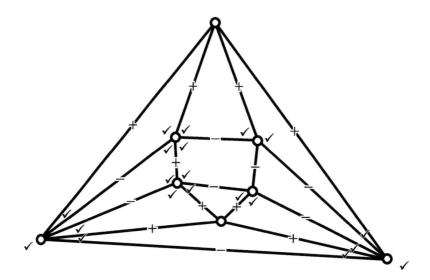

Figure 3.19 A graph is shown on the surface of a polytope: each edge is assigned a + or −.
A tick mark indicates where there is a change in sign. Note that there are two vertices with
no changes in sign, two vertices where there are only two changes in sign, and four vertices
where there are four changes in sign. Lemma 3.9.7 implies that it is impossible to have four
or more changes in sign at every vertex. In the argument for Lemma 3.9.7, $c = 20$ for this
example.

Thus the only possibility is that there are 0 states of self-stress, and the framework is
infinitesimally rigid.   □

Alexandrov's Theorem Generalizations of Dehn's Theorem

An interesting extension of Dehn's Theorem 3.9.4 is when some of the faces of the convex
polytope P are not triangles. The idea is to prove infinitesimal rigidity, and if some of the
vertices of the triangulated surface lie in the relative interior of a face in the boundary of
P there are very natural non-trivial infinitesimal flexes. In Alexandrov (2005) there is a
discussion of what to do when extra edges, which are part of a triangulation of the surface
of P, have nodes only on the vertices of P and on the edges of P.

> **Theorem 3.9.8.** *Let $(G, \mathbf{p})$ be the framework coming from a triangulation of the
> boundary of a convex polytope, but with no vertex in the relative interior of any face.
> Then $(G, \mathbf{p})$ is infinitesimally rigid in $\mathbb{E}^3$.*

There have been various extensions of Dehn's Theorem 3.9.4 and Alexandrov's
Theorem 3.9.8 to other surfaces that are not necessarily convex, but still have some
"convex properties." See Stoker (1968); Izmestiev and Schlenker (2010); and Connelly
and Schlenker (2010). Also, for higher dimensions, a theorem of W. Whiteley states that if

the 2-dimensional faces of convex polytope in $\mathbb{E}^d$ are triangulated with only vertices from the edges and natural vertices, then that framework is infinitesimally rigid in $\mathbb{E}^d$.

3.10 Projective Transformations

One important aspect of the analysis of structures is to determine what happens when they are transformed in some particular way. For example, if you translate or rotate the structure the kinematic and static properties stay the same. But there are other transformations that also preserve such properties. One such is non-singular affine transformations, such as shear, uniform expansion, or uniform contraction. But there are other transformations that are useful and informative. These are called projective transformations, which we describe here.

For this description it is helpful to use an extra dimension. For example, to describe projective transformations of the plane, it is helpful to regard the Euclidean plane as sitting in Euclidean three-space. Indeed, regard the plane as the set of points $[x; y; 1]$. In other words, it is the plane determined by $z = 1$. Suppose that the plane is "projected" from the origin onto another plane not through the origin, say the plane given by the equation $x + y + z = 1$. Explicitly this function is given by

$$[x; y; 1] \rightarrow \left[\frac{x}{x+y+1}; \frac{y}{x+y+1}; \frac{1}{x+y+1} \right].$$

Notice that the function involved is not linear, and that the function is not defined for a line $(x + y + 1 = 0)$ in the plane. If we now project orthogonally back onto the $z = 1$ plane, we get a function from the plane $z = 1$ (minus that line) to itself. In Euclidean coordinates this is just the transformation

$$[x; y] \rightarrow \left[\frac{x}{x+y+1}; \frac{y}{x+y+1} \right].$$

Any such function that is the composition of projections is called a *projective transformation*. Such a transformation is defined on all points in $\mathbb{E}^d$ except possibly some hyperplane. Those points where it is not defined are said to be *transformed to infinity*. The inverse transformation is similar with some points being transformed to (and from) infinity. In other words, for the original transformation, there are some points that come from infinity. The crucial point is that projective transformations take straight lines to straight lines.

There is a very convenient description of such projective transformations in terms of matrices. For any point $[x; y; z; \ldots] = \mathbf{p}_i$ in $\mathbb{E}^d$ consider a non-singular $(d + 1)$-by-$(d + 1)$ matrix $\mathbf{V}$. Then the last coordinate $f(\mathbf{p}_i)$ of $\mathbf{V}[\mathbf{p}_i; 1]$ is an (affine) linear function in the coordinates of $\mathbf{p}_i$. Then the first d coordinates of $\frac{1}{f(\mathbf{p}_i)} \mathbf{V}[\mathbf{p}_i; 1]$ gives a projective transformation of $\mathbb{E}^d$ to itself. For example, for the projective transformation described above, the matrix is

$$\mathbf{V} = \begin{bmatrix} 1 & 0 & 0 \\ 0 & 1 & 0 \\ 1 & 1 & 1 \end{bmatrix}.$$

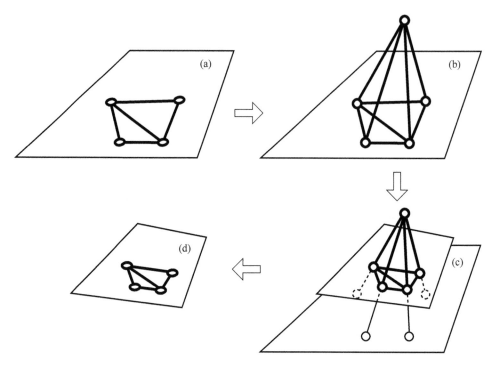

Figure 3.20 Projective invariance and cones on frameworks: (a) a framework in 2D; (b) a cone over the framework in 3D; (c) the intersection of the cone with another 2D plane; (d) a framework in 2D that is a projection of the original in (a).

3.10.1 Cones on Frameworks

Suppose that $\mathbf{p}$ is the configuration for a framework in $\mathbb{E}^d$. Choose a point $\mathbf{p}_0$ outside of $\mathbb{E}^d$ but in $\mathbb{E}^{d+1} \supset \mathbb{E}^d$, and connect it with a bar to all the points in $\mathbf{p}$. Call this the *cone* on the original framework (Figure 3.20(a), (b)).

There is a natural correspondence between the self-stresses of the framework in $\mathbb{E}^d$ and the self-stresses of cone over the framework in $\mathbb{E}^{d+1}$. Namely any self-stress in the framework in $\mathbb{E}^d$ will automatically be a self-stress in the cone over the framework in $\mathbb{E}^{d+1}$, where the stresses on the bars adjacent to the cone point are 0. Conversely, any self-stress on the cone over the framework in $\mathbb{E}^{d+1}$, will automatically be 0 on the bars adjacent to the cone point, because of the equilibrium condition at vertices in $\mathbb{E}^d$ by considering the extra coordinate in $\mathbb{E}^{d+1}$. Thus the number of self-stresses s is the same for both frameworks.

By Calladine's Formula (3.22) the number of motions of a full dimensional framework in $\mathbb{E}^d$ is $m = s + nd - d(d+1)/2 - b$, where b is the number of bars. The cone over the framework in $\mathbb{E}^{d+1}$ has the same value for s from the above argument; n and d both increase by 1, so nd increases by $n + d + 1$; $d(d+1)/2$ increases by $d + 2$; and b increases by n. So the net change in m is $n + d + 1 - (d+1) - n = 0$. If the framework in $\mathbb{E}^d$ is degenerate, then $d(d+1)/2$ is replaced by $d(d+1)/2 - (d-k)(d-k-1)/2$, where k is the dimension of the affine span of the configuration. The correction term $(d-k)(d-k-1)/2$ is the same for both the original framework in $\mathbb{E}^d$ and the cone over it in $\mathbb{E}^{d+1}$, so again the number of motions m is the same for both frameworks.

This shows the following.

> **Proposition 3.10.1.** *The dimension m of the space of internal infinitesimal mechanisms of a framework in $\mathbb{E}^d$ is the same as the dimension of the space of internal infinitesimal mechanisms of the cone over the framework in $\mathbb{E}^{d+1}$. The same is true for s, the number of states of self-stress.*

It is also interesting to see directly the kinematic connection between the infinitesimal mechanisms of the framework, and those of the cone. Consider now the cone as a collection of rigid plates, each plate corresponding to each bar of the planar framework, where the plates have hinges corresponding to each of the nodes of the planar frameworks (the bars of the cone, together with the original bar of the framework make a rigid triangle in a 2D hyperplane corresponding to the plate). Then there will be a one-to-one correspondence between non-trivial in-plane infinitesimal motions of the framework, and non-trivial infinitesimal motions of the plate structure (where the nodes will additionally move out of the original plane). And indeed, this motion would not be altered by the addition of any number of bars within the plates, nor by nodes added along the hinges.

3.10.2 Projective Invariance

An important aspect of configurations, well-known in the nineteenth century, was that if it is changed by a projective transformation its static rigidity properties remained the same. Indeed Crapo and Whiteley (1994b); Penne and Crapo (2007); and Sturmfels and Whiteley (1991) go so far as to write the configuration in what they call "projective" coordinates and use something called a "Cayley Algebra," where the projective invariance of statics is obvious and built in to the definitions. We will not go that far. We use Proposition 3.10.1 to show the projective invariance of infinitesimal rigidity.

> **Theorem 3.10.2.** *Suppose that a configuration $\mathbf{p}$ in $\mathbb{E}^d$ is transformed to a configuration $\mathbf{q}$ by a projective transformation to $\mathbb{E}^d$ such that no point of the configuration $\mathbf{q}$ is at infinity. Then any framework with nodes at $\mathbf{p}$ has the same dimension m of the space of internal infinitesimal mechanisms as the corresponding framework at $\mathbf{q}$. The same is true for s the number of states of self-stress.*

Proof. By Theorem 3.10.1, the framework with nodes at $\mathbf{p}$ has the same dimension m of the space of internal infinitesimal mechanisms as the cone over it in $\mathbb{E}^{d+1}$. We can now intersect the cone with a different d-dimensional hyperplane, and place bars in that hyperplane connecting bars on the cone in the same way as the original. But by the plate construction described above, this additional new structure will not change the infinitesimal motions of the cone, nor will the subsequent removal of the original framework. Thus the new cone has the same dimension m of the space of internal infinitesimal mechanisms, and again by Proposition 3.10.1 the new framework considered in its own hyperplane has the same dimension m of the space of internal infinitesimal mechanisms. By composing

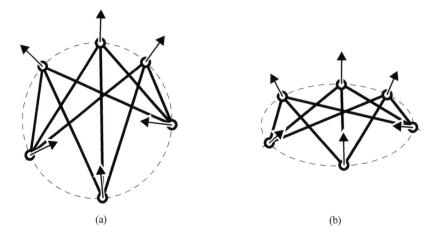

(a)   (b)

Figure 3.21 Here the infinitesimal displacement of the bipartite graph is shown in (a). Then a projective transformation of the configuration is shown in (b). Both frameworks have infinitesimal mechanisms. Note that the displacements in (a) are perpendicular to the circle, and it turns out that in (b) the displacements are perpendicular to the ellipse.

several of these transformations we can achieve any given projective transformation of $\mathbb{E}^d$, thus giving the desired invariance. An example is shown in Figure 3.20.   □

3.10.3 Example of Projective Invariance

The following example due to Whiteley (1984) demonstrates the usefulness of projective invariance to use one framework to calculate and then transforming it to find several others.

Choose six points on a circle in the plane, partition the nodes into two sets, and place bars from each node from one set to each node in the other set. (This describes the bipartite graph $K_{3,3}$.) Then there is an infinitesimal displacement that is an internal mechanism, where the displacement vectors are the same length, normal to the circle, pointing in or out depending on which partition the node is in. This is shown in Figure 3.21. Note that after the projective transformation, the six points lie on a conic (i.e. ellipse, hyperbola, or circle) but there is still an infinitesimal internal mechanism.

3.10.4 Bipartite Frameworks

One very rich source of examples of frameworks where infinitesimal rigidity can be determined exactly, essentially in all cases, is when the underlying graphs are *complete bipartite graphs* $K(m,n)$, for m,n positive integers. There are two sets of nodes, one set of m nodes and another with n nodes in $\mathbb{E}^d$ and all the nodes in one partition are connected to all the nodes in the other partition. The surprising thing about these graphs is that their infinitesimal rigidity for all d can be determined, especially in the case the configuration is generic.

The main source for this is the paper by Bolker and Roth (1980). Figure 3.21 shows an example of $K(3, 3)$ in the plane that is included in the following theorem.

Theorem 3.10.3 (Bolker–Roth). *In E^d suppose the nodes of the both partitions of $K(d + 1, d(d + 1)/2)$ are such that they do not all lie in a $(d - 1)$-dimensional hyperplane. Then the resulting framework is infinitesimally rigid if the $(d+1)(d+2)/2 = d+1+d(d+1)/2$ nodes do not lie on a quadric surface.*

A *quadric surface* in Euclidean space is a surface defined by a second-degree polynomial. For example, in the plane, it is a conic such as an ellipse, circle, hyperbola, parabola, or two lines. Another corollary of this theory is the following:

Theorem 3.10.4. *In E^3, for $n \geq 6$, $K(4,6)$ and $K(m,n)$, for $m, n \geq 5$, are generically rigid.*

Notice that bipartite graphs have no triangles that one can use to pin to start the rigidity analysis. Later we will see that there several other applications, in particular with respect to stresses.

3.11 Exercises

1. Find a non-trivial infinitesimal displacement and a non-zero stress for the Desargues' configuration of Figure 3.22. The lines through the three bars I, II, and III are parallel, and the bars I and II are part of a square, and all the bars have the same length.
2. Show that the framework of Figure 3.22 is a finite mechanism.
3. In Figure 3.23 there are two frameworks supporting a load between two fixed nodes.

   (i) Calculate the internal forces in each framework.
   (ii) Is this a unique equilibrium solution?

4. In Figure 3.14, if member V was removed, the rest of the members could be generated from the fixed nodes 4 and 5 by a sequence of Henneberg type I moves. What are

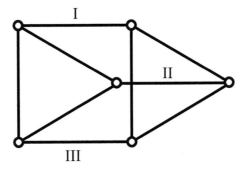

Figure 3.22

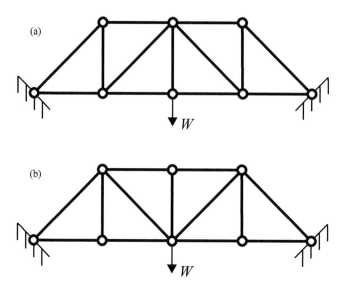

Figure 3.23

the other members that can be removed besides member V so that the rest can be constructed by Henneberg type I moves?

5. An *Assur graph* (Servatius et al., 2010) is a pinned isostatic framework in $\mathbb{E}^2$ that has no strict subgraph containing the pinned nodes that is infinitesimally rigid.

   a. Show that there is no unpinned node connected to two pinned nodes unless there is just that one unpinned node.
   b. Show that any isostatic pinned graph can be decomposed into a nested sequence of Assur graphs $G_1, \subset G_2 \ldots \subset G_n = G$ such that the pinned nodes of G_i are contained in the nodes of G_{i+1}, for $i = 1, \ldots, n - 1$.
   c. Decompose the graphs in Figure 3.24 into Assur components.

6. Suppose that one has an $(n + 1)$-by-$(m + 1)$ rectangular grid of bars and nodes, as in Figure 3.25, where some of the small rectangles are braced as shown. There are n rows and m columns of small rectangles. Create a bipartite graph with n vertices in one partition and m vertices in another partition, where vertex i is joined to vertex j if the small rectangle in row i and column j is braced. Show that the framework with the bracing is infinitesimally rigid and rigid if and only if the bipartite graph is connected. See Bolker and Crapo (1979) for a discussion of this problem.

7. One way of misstating Cauchy's Theorem 3.9.1 is the following:

   Given two convex polytopes P and Q in $\mathbb{E}^3$, if there is a one-to-one onto incidence preserving correspondence between the vertices of P and the vertices of Q such that corresponding facets are congruent, then P and Q are congruent.

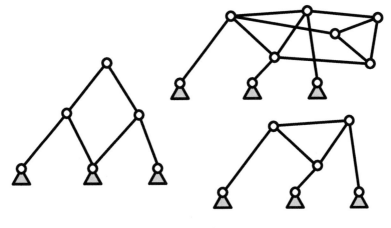

Figure 3.24

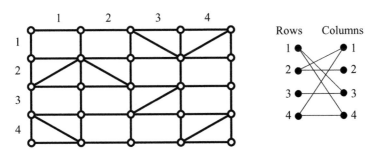

Figure 3.25

Incidence preserving means that if a set of vertices are the vertices of an edge or face of P there is a corresponding edge or face of Q with the corresponding vertices.

Find a counterexample to this statement. See Grünbaum and Shephard (1975) for one example.

4

Tensegrities

4.1 Introduction

Tensegrity structures were named by Buckminster Fuller, after he had seen some examples constructed by Kenneth Snelson: it is a form of structure that has become very popular, both for engineering and artistic structures, and also as a model, for example for some biological systems. However, the term tensegrity is not always clearly defined, particularly in the engineering and artistic literature. Certain elements are always present: compression elements are clearly differentiated from tension elements, which are typically cables, and the whole structure is prestressed to make it stiff; but there may be a debate about, for instance, whether more than one compression elements may be connected at a node. We shall skirt around such debate by instead using a simple definition of a tensegrity that will certainly encompass almost all "classic" tensegrity structures, but may not be as specific as some uses of the term.

There is a basic distinction between the basic geometric model and the physical/ engineering model. The geometric model is not directly concerned with the behaviour of the materials that are used to construct any particular structure. It only involves the configuration of nodes, and the distances between them. The physical model is concerned with forces within the structure and how they behave. Our definitions reflect these two points of view. For example, when two nodes are forced to stay at a fixed distance apart, that is a concrete geometric condition, but physically we would expect that if a member between the nodes carries load, there will be some change in length.

We define a *tensegrity* as a finite configuration of nodes $\mathbf{p} = [\mathbf{p}_1; \ldots; \mathbf{p}_n]$ as with the definition of a framework in Section 2.2, but the edges of G are labelled as a *bar*, *cable*, or *strut*.

So we define a *tensegrity* (sometimes a *tensegrity structure*) to be a framework where the members may be of three different types. A *cable* is a member that can only carry tension, i.e. a positive internal force, using the definition in Section 3.3. A *strut* is a member that can only carry compression, i.e. a negative internal force. And a *bar* can carry either tension or compression. Figure 4.1 shows an example of a tensegrity structure in 2D, showing the graphic notation that we will use for cables, struts, and bars. Cables have a clear physical meaning interpretation as a member that will buckle up under compression; similarly, struts can be considered as, for instance, a physically discontinuous member that is held together in compression but could pull apart in tension.

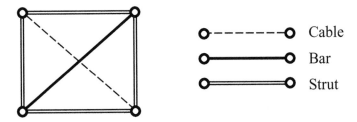

Figure 4.1 An example of a tensegrity. When we show a tensegrity graphically, we denote a cable by a dashed line, a strut by a double line, and a bar by a single line.

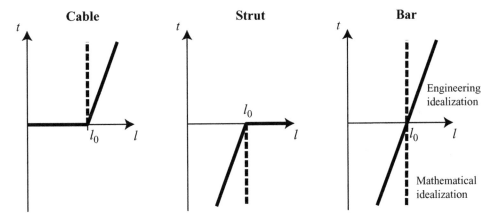

Figure 4.2 The idealised internal force (t) vs. length (l) relationships for a cable, strut, and bar, with initial length l_0.

We will often further idealize cables, struts, and bars using either a physical idealization, or a geometric idealization. Consider members that have a rest-length l_0, and carry some internal force t. In the physical idealization, we shall assume that the members have some stiffness k, and then:

cable: $t = k(l - l_0)$ for $l \geq l_0$, $t = 0$ for $l < l_0$;
strut: $t = -k(l_0 - l)$ for $l \leq l_0$, $t = 0$ for $l > l_0$;
bar: $t = k(l - l_0)$ for all l.

In the geometric idealization, we effectively assume that the members are infinitely stiff, and so cables may become shorter, and can carry any positive internal force; struts may become longer, and can carry any negative internal force; and bars cannot change length, but can carry any internal force, positive or negative. Figure 4.2 shows graphically the physical and geometric idealizations.

This chapter will only discuss the first-order theory of tensegrity structures: in fact, many of the interesting aspects of these structures are only revealed by higher-order analysis, but that will have to wait until Chapter 6.

4.2 Rigidity Questions

One important property of a tensegrity is its rigidity, and extending the results from Chapter 3, we can apply a first-order analysis. We first more formally describe our model for a tensegrity. The underlying combinatorics is given by a graph G, with n vertices, without loops or multiple edges, where each edge of G is labelled as a cable, bar, or strut. Furthermore, there is a configuration of nodes $\mathbf{p} = [\mathbf{p}_1; \ldots ; \mathbf{p}_n]$, where each $\mathbf{p}_i$ is in d-dimensional Euclidean space for $i = 1, 2, \ldots, n$. The node $\mathbf{p}_i$ corresponds to the i-th vertex of G. We combine the discrete combinatorial information in the graph G together with the geometric information of the configuration $\mathbf{p}$ using the notation $(G, \mathbf{p})$.

If $(G, \mathbf{q})$ is another tensegrity, but with the configuration $\mathbf{q} = [\mathbf{q}_1; \ldots ; \mathbf{q}_n]$, we say that it satisfies the *tensegrity constraints* if the following conditions are satisfied.

cable: Cables are allowed to become shorter; for all $\{i, j\}$ cables, $|\mathbf{p}_i - \mathbf{p}_j| \geq |\mathbf{q}_i - \mathbf{q}_j|$.
bar: Bars stay the same length; for all $\{i, j\}$ bars, $|\mathbf{p}_i - \mathbf{p}_j| = |\mathbf{q}_i - \mathbf{q}_j|$.
strut: Struts can become longer; for all $\{i, j\}$ struts, $|\mathbf{p}_i - \mathbf{p}_j| \leq |\mathbf{q}_i - \mathbf{q}_j|$.

As we did for bar frameworks as in Chapter 2, we say that a tensegrity $(G, \mathbf{p})$ is *rigid* in d-dimensional Euclidean space if any one of the following definitions is satisfied (recalling the definitions of a flex and trivial flex from Chapter 3).

Definition 1: Each continuous flex $\mathbf{p}(t)$ of the tensegrity $(G, \mathbf{p})$ satisfying the tensegrity constraints in $\mathbb{E}^d$ is trivial.

Definition 2: Each analytic flex $\mathbf{p}(t)$ of the tensegrity $(G, \mathbf{p})$ satisfying the tensegrity constraints in $\mathbb{E}^d$ is trivial.

Definition 3: There is an $\epsilon > 0$ such that for every configuration of n labelled vertices $\mathbf{q} = [\mathbf{q}_1; \ldots ; \mathbf{q}_n]$ in $\mathbb{E}^d$ satisfying the tensegrity constraints for the configuration $\mathbf{p}$ and $|\mathbf{p} - \mathbf{q}| < \epsilon$, then $\mathbf{q}$ is congruent to $\mathbf{p}$.

Theorem 4.2.1. *All three of these definitions of rigidity for tensegrities are equivalent.*

Just as with the case for bar frameworks, the proof of this is a part of real algebraic geometry. The inequality tensegrity constraints define what is called a *semi-algebraic set*, and basic results in Bochnak et al. (1998) show how to parametrize sets with analytic functions. For a further discussion of semi-algebraic sets, see Section 8.8. When a tensegrity is not rigid we say it is a *finite mechanism*.

4.3 Infinitesimal Rigidity

Here we will consider infinitesimal displacements $\mathbf{d}$ for the tensegrity. We will be interested in the sign of the extension coefficients for the members; recall from (3.9) that $\tilde{e}_{ij} = (\mathbf{p}_j - \mathbf{p}_i)^{\mathsf{T}}(\mathbf{d}_j - \mathbf{d}_i)$. For a tensegrity $(G, \mathbf{p})$, we say $\mathbf{d} = [\mathbf{d}_1; \ldots ; \mathbf{d}_n]$ is an *infinitesimal flex* if the following *kinematic tensegrity conditions* hold.

cable: For all $\{i, j\}$ cables, $(\mathbf{p}_i - \mathbf{p}_j)^{\mathsf{T}}(\mathbf{d}_i - \mathbf{d}_j) = \tilde{e}_{ij} \leq 0$.
bar: For all $\{i, j\}$ bars, $(\mathbf{p}_i - \mathbf{p}_j)^{\mathsf{T}}(\mathbf{d}_i - \mathbf{d}_j) = \tilde{e}_{ij} = 0$.
strut: For all $\{i, j\}$ struts, $(\mathbf{p}_i - \mathbf{p}_j)^{\mathsf{T}}(\mathbf{d}_i - \mathbf{d}_j) = \tilde{e}_{ij} \geq 0$.

Thus negative extensions are allowed for cables, positive extensions are allowed for struts, and bars must remain the same length. Again, we say that a tensegrity $(G, \mathbf{p})$ *infinitesimally rigid* if every infinitesimal flex $\mathbf{d}$ is a rigid-body motion.

Theorem 4.3.1. *If a tensegrity $(G, \mathbf{p})$ is infinitesimally rigid, then it is rigid.*

Proof. Suppose that $(G, \mathbf{p})$ is not rigid. We use Definition 2, which guarantees that there is an analytic flex $\mathbf{p}(t)$ of $(G, \mathbf{p})$ that satisfies the tensegrity constraints in $\mathbb{E}^d$, but is not a rigid-body motion. We would like simply to take the first derivative of this motion at $t = 0$ and use that for the infinitesimal flex. But there is a complication: if the motion is of the form $\mathbf{p}(t^2)$, for example, the derivative is $\mathbf{0}$, and yet the motion is not a rigid-body motion. We have to consider higher-order derivatives.

For $k = 0, 1, \ldots$, define

$$\left(\frac{d}{dt}\right)^k \mathbf{p}(t)|_{t=0} = [\mathbf{p}_1^{(k)}; \ldots ; \mathbf{p}_n^{(k)}] = \mathbf{p}^{(k)}.$$

(Note that $\mathbf{p}^1 = \mathbf{d}$, but we are now also considering higher derivatives.) We will also make use of higher derivatives of the squares of lengths, and so we define

$$\tilde{e}_{ij}^{(k)} = \frac{1}{2} \frac{d^k (l_{ij}^2)}{dt^k}\bigg|_{t=0} = \frac{1}{2} \left(\frac{d}{dt}\right)^k (\mathbf{p}_i(t) - \mathbf{p}_j(t))^2 \bigg|_{t=0}. \tag{4.1}$$

(Note that $\tilde{e}_{ij}^{(1)} = \tilde{e}_{ij}$, and $\tilde{e}_{ij}^{(k)}$ is defined for all i, j, whether or not $\{i, j\}$ is a member.)

Let $N = 1, 2, \ldots$ be the smallest integer such that there is a rigid-body motion, where $\mathbf{p}^{(k)}$ is its k-th derivative for $k = 1, 2, \ldots, N - 1$, while $\mathbf{p}^{(N)}$ is not the extension to the N-th derivative of any such rigid-body motion. We can show that such an N must exist – if it does not, then for all $k = 1, 2, \ldots$ and all i, j, whether or not $\{i, j\}$ is a member, $\tilde{e}_{ij}^{(k)} = 0$. Since the coordinate functions of $\mathbf{p}(t)$ are analytic, this implies that all the distances are constant for all t. This contradicts the assumption that $\mathbf{p}(t)$ is not a rigid-body motion, and thus N must exist.

Expanding the expression from (4.1), we have, for each member $\{i, j\}$,

$$\sum_{k=0}^{k=N} \binom{N}{k} (\mathbf{p}_i^{(N-k)} - \mathbf{p}_j^{(N-k)})^{\mathrm{T}} (\mathbf{p}_i^{(k)} - \mathbf{p}_j^{(k)}) = 2\tilde{e}_{ij}^{(N)}, \tag{4.2}$$

where $\binom{N}{k} = N!/k!(N - k)!$ is the appropriate binomial coefficient. Furthermore, by composing the motion $\mathbf{p}(t)$ with a rigid-body motion, we can assume that $\mathbf{p}^{(k)} = 0$, for $k = 1, \ldots, N - 1$. Then (4.2) reduces to the following:

$$(\mathbf{p}_i - \mathbf{p}_j) \cdot (\mathbf{p}_i^{(N)} - \mathbf{p}_j^{(N)}) = \tilde{e}_{ij}^{(N)} \begin{cases} \leq 0 & \text{if } \{i, j\} \text{ is a cable,} \\ = 0 & \text{if } \{i, j\} \text{ is a bar,} \\ \geq 0 & \text{if } \{i, j\} \text{ is a strut.} \end{cases} \tag{4.3}$$

Now consider a new infinitesimal flex, $\bar{\mathbf{d}} = \mathbf{p}^{(N)}$. It is clear from (4.3) that this flex satisfies the conditions described at the start of this section. Further, the new flex $\bar{\mathbf{d}}$ cannot be an infinitesimal rigid-body motion: if $\mathbf{d}$ were the first derivative of a rigid-body motion $\mathbf{q}(t)$ at $t = 0$ of $\mathbf{p}$, then $\frac{1}{N!}\left(\frac{d}{dt}\right)^k \mathbf{q}(t^N) = \mathbf{p}^{(k)}$, for $k = 1, \ldots, N$, contradicting the choice of N. Thus, $\bar{\mathbf{d}}$ is a first-order infinitesimal motion of $\mathbf{p}$, which is not the derivative of rigid-body motion, as desired.   □

Note that what we have done here is show that, if there is a finite motion, then even at a "cusp" point (an example will be shown in Section 8.8.3) there must be some derivative that starts the tensegrity moving out of its current position – and that must itself be an infinitesimal flex, even though it is not the first derivative of the motion itself.

The proof of Theorem 4.3.1 is an alternative proof to the equivalent statement for bar frameworks, Theorem 3.8.1.

4.4 Static Rigidity

As in Chapter 2 for bar frameworks, we investigate the statics of a tensegrity. For any tensegrity $(G, \mathbf{p})$ we define the underlying bar framework $\bar{G}(\mathbf{p})$ to be the same graph G and configuration $\mathbf{p}$ but with all members converted to bars. We define a *proper stress* $\boldsymbol{\omega} = [\cdots ; \omega_{ij}; \ldots]$ of $(G, \mathbf{p})$ to be a stress of $\bar{G}(\mathbf{p})$, with the additional *static tensegrity conditions* that

$$\omega_{ij} \begin{cases} \geq 0 & \text{if } \{i, j\} \text{ is a cable,} \\ \leq 0 & \text{if } \{i, j\} \text{ is a strut.} \end{cases} \tag{4.4}$$

A cable can only carry a positive internal force, which indicates tension, and a strut can only support negative internal force that indicates compression – notice that there is no condition on the sign of the internal force on a bar of G. We also define a *proper self-stress* of $\bar{G}(\mathbf{p})$ to be a self-stress of $\bar{G}(\mathbf{p})$ that also satisfies (4.4). We can equivalently apply the proper sign conditions to internal force densities $\omega_{ij} = t_{ij}/l_{ij}$.

Recall from Chapter 3 that a bar framework is statically rigid if every external self-equilibrated applied load can be equilibrated with an internal force, a stress. We say that a tensegrity $(G, \mathbf{p})$ is *statically rigid* if for every external self-equilibrated applied load, $\mathbf{f} = [\mathbf{f}_1; \ldots \mathbf{f}_n]$ can be equilibrated with a proper stress. Recall equation (3.14) – in terms of internal force densities, the condition that a stress satisfies the force equilibrium equation is that, for each node i,

$$\mathbf{f}_i + \sum_j \omega_{ij}(\mathbf{p}_j - \mathbf{p}_i) = 0,$$

where the sum is taken over all vertices j, which may also be taken to be just those nodes adjacent to node i. In this case we say that the external force $\mathbf{f}$ is *resolved* by the internal force coefficients $\boldsymbol{\omega} = [\ldots, \omega_{ij}, \ldots]$. We might also say that $\mathbf{f}$ is resolved by the corresponding internal foces $\mathbf{t}$).

Examples of tensegrities are shown in Figures 4.1 and 4.3. Although both have the same statically rigid underlying bar framework, the tensegrity in Figure 4.1 is statically rigid, while that in Figure 4.3 is not.

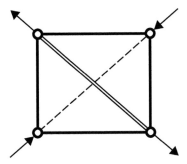

Figure 4.3 The tensegrity is not statically rigid, since it cannot resolve the forces indicated.

Note that this tensegrity definition of static rigidity agrees with our previous definition of static rigidity for bar frameworks. There are two particular cases which will prove useful.

(i) When the configuration is fully dimensional in $\mathbb{E}^d$, a bar framework formed by placing bars between all the nodes is always infinitesimally and statically rigid in $\mathbb{E}^d$.

(ii) The only non-fully-dimensional tensegrity frameworks that are infinitesimally and statically rigid are the simplices formed by placing bars between all k nodes $(1 \leq k \leq d)$ of an affine independent configuration. For the purposes of determining static rigidity, we can restrict to the case when the configuration is fully dimensional.

4.5 Elementary Forces

It is often useful to consider special types of external self-equilibrated applied loads. Suppose that $\mathbf{p}_i$ and $\mathbf{p}_j$ are two nodes in a configuration $\mathbf{p} = [\mathbf{p}_1; \ldots; \mathbf{p}_n]$. Define the load $\mathbf{f}(i, j)$ such that $\mathbf{f}_i = c \cdot (\mathbf{p}_j - \mathbf{p}_i)$, $\mathbf{f}_j = c \cdot (\mathbf{p}_i - \mathbf{p}_j)$, and all the other forces are $\mathbf{0}$, where c is a positive scalar conversion factor from length to force. Notice that the positive sense of $\mathbf{f}(i, j)$ is when the force is acting to push the nodes together, whereas $-\mathbf{f}(i, j)$ is a force acting to pull nodes apart. We call $\pm\mathbf{f}(i, j)$ *elementary forces*. Note that if $\{i, j\}$ is a member of the framework, $\mathbf{f}(i, j)^{\mathsf{T}}/c = \mathbf{f}(j, i)^{\mathsf{T}}/c$ is the row in the rigidity matrix $\mathbf{R}$ that corresponds to this bar.

Lemma 4.5.1. *A tensegrity $(G, \mathbf{p})$, where $\mathbf{p} = [\mathbf{p}_1; \ldots; \mathbf{p}_n]$ is a fully dimensional configuration in $\mathbb{E}^d$, is statically rigid if and only if, for all pairs of nodes $\{i, j\}$, all elementary forces $\pm\mathbf{f}(i, j)$ can be resolved by a proper stress ω.*

Proof. It is clear that if $(G, \mathbf{p})$ is statically rigid, then since elementary forces are self-equilibrated applied loads, they must be resolvable by some proper stress.

To prove the converse we must show that any self-equilibrated applied load can be considered as a sum of elementary forces, with suitable coefficients, i.e. that the vectors $\pm\mathbf{f}(i, j)$ span the space of self-equilibrated applied loads. We can do so by considering an alternative bar framework $\bar{G}_{\text{complete}}(\mathbf{p})$, which has the same nodal positions as $(G, \mathbf{p})$, but whose underlying graph is the complete graph, i.e. there is a bar between every pair of nodes. As $\bar{G}_{complete}(\mathbf{p})$ is known to be statically rigid, the row space of its rigidity matrix $\bar{\mathbf{R}}_{complete}$ must contain all possible self-equilibrated applied loads. But each of the rows of $\bar{\mathbf{R}}_{complete}$ is one of the scaled elementary loads, $\mathbf{f}(i,j)^{\mathrm{T}}/c = \mathbf{f}(j,i)^{\mathrm{T}}/c$, which hence span the row space. Thus $(G,\mathbf{p})$ is statically rigid in $\mathbb{E}^d$. $\qquad\square$

The idea for Lemma 4.5.1 can be extended to the following. Suppose $G_1(\mathbf{p})$ and $G_2(\mathbf{p})$ are two tensegrities with the same configuration of nodes $\mathbf{p}$, and $G_1(\mathbf{p})$ is statically rigid in $\mathbb{E}^d$. Then $G_2(\mathbf{p})$ is also statically rigid in $\mathbb{E}^d$, if and only if, for every cable $\{i, j\}$ of $G_1(\mathbf{p})$, $\mathbf{f}(i, j)$ can resolved by a proper stress in $G_2(\mathbf{p})$, for every strut $\{i, j\}$ of $G_1(\mathbf{p})$, $-\mathbf{f}(i, j)$ can be resolved by a proper stress in $G_2(\mathbf{p})$, and for every bar $\{i, j\}$ of $G_1(\mathbf{p})$, both $\mathbf{f}(i, j)$ and $-\mathbf{f}(i, j)$ can be resolved by a proper stress in $G_2(\mathbf{p})$.

Figure 4.3 can be seen as showing two elementary forces that cannot be resolved by the tensegrity.

4.6 Farkas Alternative

In Chapter 3 we made use of the following basic result from linear algebra:

Theorem 4.6.1 (Theorem of the Alternative). *Let $\mathbf{A}$ be an m-by-n matrix, and $\mathbf{b}$ an n-component vector. Then exactly one of the following occurs:*

$$\mathbf{A}\mathbf{x} = \mathbf{0} \quad \textit{for some n-component column vector } \mathbf{x} \textit{ such that } \mathbf{x}^{\mathrm{T}}\mathbf{b} \neq \mathbf{0}, \qquad (4.5)$$

or

$$\mathbf{A}^{\mathrm{T}}\mathbf{y} = \mathbf{b} \quad \textit{has a solution.} \qquad (4.6)$$

In words, either $\mathbf{b}$ lies in the row space of $\mathbf{A}$, or it has a component in the perpendicular nullspace, the nullspace of $\mathbf{A}$. It is called the Theorem of the Alternative because it is impossible to find both $\mathbf{x}$ and $\mathbf{y}$ to satisfy (4.5) and (4.6) simultaneously.

The current chapter considers inequalities as well as equalities, and hence we consider the extension of Theorem 4.6.1 to the case of inequalities. The result we require is called the Farkas Alternative.

Firstly, we define some notation. If $\mathbf{x}$ is a vector the notation $\mathbf{x} \geq \mathbf{0}$, $\mathbf{x} > \mathbf{0}$, etc. is defined to mean that each coordinate of the vector satisfies the given inequality.

Lemma 4.6.2 (Farkas Alternative). *Let $\mathbf{A}$ be an m-by-n matrix, and $\mathbf{b}$ an n-component vector. Then exactly one of the following occurs:*

$$\mathbf{A}\mathbf{x} \leq \mathbf{0} \quad \textit{for some n-component column vector } \mathbf{x} \textit{ such that } \mathbf{x}^{\mathrm{T}}\mathbf{b} > \mathbf{0}, \qquad (4.7)$$

or

$$\mathbf{A}^{\mathrm{T}}\mathbf{y} = \mathbf{b} \quad \textit{for some m-component vector } \mathbf{y} \textit{ such that } \mathbf{y} \geq \mathbf{0}. \qquad (4.8)$$

Thus we are still considering solutions to $\mathbf{A}^{\mathrm{T}}\mathbf{y} = \mathbf{b}$, but with the added constraint that $\mathbf{y} \geq \mathbf{0}$. Thus the $\mathbf{b}$'s for which (4.8) hold do not fill out a subspace (the rowspace of $\mathbf{A}$), but

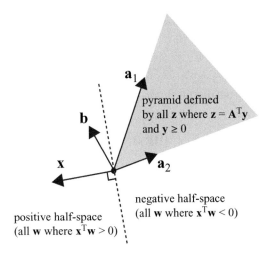

pyramid defined
by all **z** where **z** = **A**^{T}**y**
and **y** ≥ 0

a$_1$

b

x

a$_2$

negative half-space
(all **w** where **x**^{T}**w** < 0)

positive half-space
(all **w** where **x**^{T}**w** > 0)

Figure 4.4 In this case, the vector **b** can be separated from the pyramid of positive linear combinations of the row vectors of the matrix **A** in the Farkas Alternative.

rather describe an open-ended cone with a vertex at the origin. If **b** lies outside this cone, then (4.7) states that there is a hyperplane through the origin (described by its normal, **x**) that separates the space into two half-spaces: in the "positive" half-space lies **b**, where **x**^{T}**b** > **0**; and in the "negative" half-space lies the cone, where **x**^{T}(**A**^{T}**y**) < **0**. A formal proof of the existence of the "separating hyperplane," and hence of Lemma 4.6.2 will be found in, for instance, Gale (1960), a book on linear economic models.

Figure 4.4 shows the separation argument for a simple case in the plane, where **A** is a 2 × 2 matrix whose rows are the vectors **a**$_1$ and **a**$_2$, $A = \left[\mathbf{a}_1^T; \mathbf{a}_2^T\right]$ and where **b** does not satisfy Condition (4.8). The vector **x** is chosen to define a hyperplane (a line in this case) separating **b** from all positive combinations of **a**$_1$ and **a**$_2$. Thus Condition (4.7) is satisfied.

4.7 Equivalence of Static and Infinitesimal Rigidity

In order to apply the Farkas Alternative in the form expressed in (4.7), using a rigidity matrix for the matrix **A**, we need to do some housekeeping on our inequalities to make them consistent.

We can make the tensegrity conditions on extension coefficients consistently one-way if we define a new vector of extension coefficients $\bar{e}$ in terms of $\tilde{e}$ defined in (4.3) with components

cable: For all $\{i, j\}$ cables, $\bar{e}_{ij} = \tilde{e}_{ij}$.
bar: For all $\{i, j\}$ bars, we define two coefficients, $\bar{e}_{ij+} - \bar{e}_{ij-} = \tilde{e}_{ij}$, where either $\bar{e}_{ij+} = 0$
    or $\bar{e}_{ij-} = 0$.
strut: For all $\{i, j\}$ struts, $\bar{e}_{ij} = -\tilde{e}_{ij}$.

We can write the above relations as a matrix $\mathbf{H}$, and the kinematic tensegrity conditions then become

$$\bar{\mathbf{e}} = \mathbf{H}\tilde{\mathbf{e}} \leq \mathbf{0}. \tag{4.9}$$

We can consider $\mathbf{H}$ to carry out row operations on the rigidity matrix to give a revised rigidity matrix $\mathbf{HR}$:

$$(\mathbf{HR})\mathbf{d} \leq \mathbf{0} \tag{4.10}$$

We can also make the static tensegrity conditions one-way by redefining a new vector of force densities $\bar{\boldsymbol{\omega}}$ in terms of $\boldsymbol{\omega}$ as defined in Section 3.3 where

cable: For all $\{i, j\}$ cables, $\bar{\omega}_{ij} = \omega_{ij}$.
bar: For all $\{i, j\}$ bars, we define two force densities, $\bar{\omega}_{ij+} - \bar{\omega}_{ij-} = \omega_{ij}$, where either
$\bar{\omega}_{ij+} = 0$ or $\bar{\omega}_{ij-} = 0$.
strut: For all $\{i, j\}$ struts, $\bar{\omega}_{ij} = \omega_{ij}$.

The above definitions can also be described with the matrix $\mathbf{H}$, and the static tensegrity conditions then become

$$\bar{\boldsymbol{\omega}} = \mathbf{H}\boldsymbol{\omega} \geq \mathbf{0}. \tag{4.11}$$

Note that the new extension coefficient vector $\bar{\mathbf{e}}$ and the new stress vector $\bar{\boldsymbol{\omega}}$ are work-conjugate (compare equation 3.3) only with the additional conditions that for bars, either the positive or negative extension coefficient, and the corresponding force density, are zero. With this in place there will be an equilibrium equation, which, with the static tensegrity conditions, then gives

$$(\mathbf{RH})^{\mathsf{T}}\bar{\boldsymbol{\omega}} = \mathbf{f} \quad \text{for some } \bar{\boldsymbol{\omega}} \text{ such that } \bar{\boldsymbol{\omega}} \geq \mathbf{0}. \tag{4.12}$$

Theorem 4.7.1. *A tensegrity* $(G, \mathbf{p})$ *is infinitesimally rigid in* $\mathbb{E}^d$ *if and only if it is statically rigid.*

Proof. By the remarks Section 4.5 we may restrict to the case when the configuration $\mathbf{p}$ is fully dimensional.

We can apply the Farkas Alternative, Theorem 4.6.2, using the kinematic (4.10) and static (4.12) tensegrity conditions. We consider these equations for each of elementary forces $\pm\mathbf{f}(i, j)^{\mathsf{T}}$. The condition $\mathbf{bx} > 0$ in (4.7) becomes $\mathbf{f}(i, j)^{\mathsf{T}}\mathbf{d} = (\mathbf{p}_i - \mathbf{p}_j)^{\mathsf{T}}(\mathbf{d}_i - \mathbf{d}_j) \neq 0$. Thus Condition (4.7) of the Farkas Alternative becomes the statement that there is an infinitesimal flex $\mathbf{d}$ of $(G, \mathbf{p})$ which is not a rigid-body motion. On the other hand if there is no such $\mathbf{d}$ satisfying Condition (4.7) for all possible choices of $\mathbf{b} = \pm\mathbf{f}(i, j)$, this implies that $(G, \mathbf{p})$ is infinitesimally rigid, by the remarks in Section 4.5.

Condition (4.8) of the Farkas Alternative for the choices above implies that the stress coefficients $\hat{\boldsymbol{\omega}}$ resolve the elementary force $\mathbf{b} = \pm\mathbf{f}(i, j)$. Thus if Condition (4.8) holds for

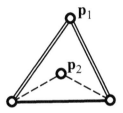

Figure 4.5 An example of a tensegrity that is not rigid; however, although the distance between the two nodes p_1 and p_2 can be increased, it cannot be decreased.

all choices of $\mathbf{b} = \pm\mathbf{f}(i, j)$ for all $\{i, j\}$, that is equivalent to the static rigidity of $(G, \mathbf{p})$. From the observations above and the Farkas Alternative, this is also equivalent to the infinitesimal rigidity of $(G, \mathbf{p})$. $\square$

The proof above actually provides somewhat more than the statement of the equivalence of static and infinitesimal rigidity. Suppose we have a tensegrity $(G, \mathbf{p})$ with a fully dimensional configuration $\mathbf{p}$, and we wish to know if there is an infinitesimal flex $\mathbf{d}$ that moves the nodes $\mathbf{p}_i$ and $\mathbf{p}_j$ either closer together or further apart strictly. Place a cable or strut, respectively, between them. If this new tensegrity has a proper self-stress on that member, it "blocks" the motion, and there is no such infinitesimal flex $\mathbf{d}$. If there is no such proper self-stress, there is such a motion. This is regardless of whether or not $(G, \mathbf{p})$ is infinitesimally rigid. For example, the nodes labelled $\mathbf{p}_1$ and $\mathbf{p}_2$ in the tensegrity in Figure 4.5 are such that their distance cannot decrease to the first-order, but it can increase.

4.8 Roth–Whiteley Criterion for Infinitesimal Rigidity

Often it is the case that it is easy to determine if a bar framework is infinitesimally rigid, but because of the inherent inequalities involved, the infinitesimal rigidity can seem to complicate matters. The following is a simple criterion taken from the paper of Roth and Whiteley (1981) for the infinitesimal rigidity of tensegrities, that can make the task of determining infinitesimal rigidity somewhat easier. In Section 4.10 we show some examples where this criterion does simplify things quite a bit.

Theorem 4.8.1 (Roth–Whiteley). *Let* $(G, \mathbf{p})$ *be a tensegrity in* $\mathbb{E}^d$*, and* $\bar{G}(\mathbf{p})$ *the corresponding bar framework, where all the members of* G *have been replaced by bars. Then* $(G, \mathbf{p})$ *is infinitesimally rigid (and equivalently statically rigid) if and only if the following two conditions are satisfied:*

(i) $\bar{G}(\mathbf{p})$ is infinitesimally rigid in $\mathbb{E}^d$, and
(ii) there is a proper self-stress with force coefficients ω for $(G, \mathbf{p})$, where for each cable and strut $\{i, j\}$ of G, $\omega_{ij} \neq 0$.

Proof. Suppose that $(G, \mathbf{p})$ is statically rigid. Then clearly $\bar{G}(\mathbf{p})$ is also statically rigid, since there are more possibilities for resolving any given external load. So Condition *i* holds. For condition *ii*, initially consider any cable $\{i, j\}$, together with an elementary force $\mathbf{f}(i, j)$. Since $(G, \mathbf{p})$ is statically rigid that elementary force can be resolved by a proper internal stress in $(G, \mathbf{p})$. Now add the elementary force $-\mathbf{f}(i, j)$, resolved by a proper internal stress that is zero on every member except $\{i, j\}$. Combining $\mathbf{f}(i, j)$ and $-\mathbf{f}(i, j)$ gives zero external load, resolved by a proper stress on $(G, \mathbf{p})$ that is non-zero on $\{i, j\}$. This process can be repeated for all cables, and reversed for all struts. The sum of all those proper internal self-stresses is non-zero on all the cables and struts of $(G, \mathbf{p})$. This shows that Condition *ii* holds.

Conversely suppose that Condition *i* and *ii* hold, and let $\mathbf{d}$ be any infinitesimal flex of $(G, \mathbf{p})$. For any cable or strut $\{i, j\}$ of G, if $\tilde{e}_{ij} = (\mathbf{p}_i - \mathbf{p}_j)(\mathbf{d}_i - \mathbf{d}_j) \neq 0$, it would imply that $\omega_{ij}\tilde{e}_{ij} < 0$, and hence $\omega^\mathsf{T} d < 0$, contradicting the equilibrium property of ω given in Equation 3.18. Thus every member has no extension/contraction, and $\mathbf{d}$ is an infinitesimal flex of $\bar{G}(\mathbf{p})$ as well. Hence Condition *i* implies that $\mathbf{d}$ an infinitesimal rigid-body motion and $(G, \mathbf{p})$ is infinitesimally rigid.   □

One very handy corollary is the following, where we count the number of members. The extra member is necessary because of the stress that is required for rigidity for tensegrities.

> **Corollary 4.8.2.** *If a tensegrity $(G, \mathbf{p})$, with n nodes in $\mathbb{E}^d$, b members, and at least one strut or cable, is infinitesimally rigid, then $b \geq nd - d(d+1)/2 + 1$. If there are fixed nodes and n variable nodes, then simply $b \geq nd + 1$.*

4.9 First-Order Stiffness

The Roth–Whiteley Theorem, 4.8.1, has interesting implications for the stiffness of a tensegrity structure. An engineer will typically analyse a structure (see Section 3.7) that includes cables by assuming that they are bars, i.e. by using the corresponding bar framework $\bar{G}(\mathbf{p})$ in Theorem 4.8.1. This is justified by assuming that any cable that might slacken is initially stressed, and if the analysis finds that the associated member will go into compression, in the actual structure this will be accommodated by a reduction in the initial tension. Of course, this relies on Condition *ii* of the proof, that there exists a proper self-stress that leaves every bar (and strut) stressed.

However, Roth–Whiteley have shown that the structure is rigid, i.e. it has positive stiffness for any applied self-equilibrated applied loads, even if the tensegrity is actually unstressed when a load is applied. However, in this case, it will not be possible to define a stiffness matrix, as described for a bar framework in Section 3.6. There will be a discontinuity in the stiffness around the zero load point. For instance, the same load applied with the opposite sign might be carried by a different set of members, with obvious potential implication for the stiffness (one load might be carried by stiff members, another by rather flexible members). A simple 1D example is shown in Figure 4.6.

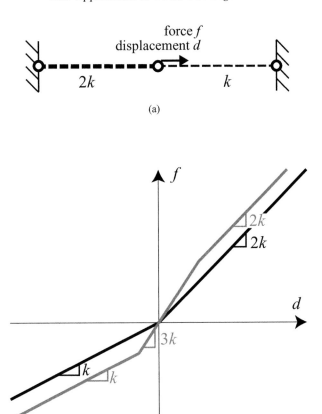

Figure 4.6 (a) A tensegrity consisting of two joined cables attached to rigid foundations. The cables have stiffness k and $2k$, where the stiffness is a ratio of the force carried to the extension. A force f is applied horizontally, and the resultant displacement is d. (b) A plot showing the variation of the force with the displacement: (i) when there is no prestress; (ii) when the cables are initially stressed. For each segment of each plot, the slope (stiffness) is shown. A discontinuity in slopes shows a change in which (if any) cables are slack.

4.10 Application to Circle Packings

Here we give an example of the situation where the first-order results of the present chapter can be used to show the stability of jammed packings of circular disks in a container. This also shows a physical model of a strut that is not just a bar in compression.

Consider a rigid box that contains a collection of non-overlapping frictionless circular disks. This is called a *packing* of disks. Suppose further that the disks are jammed tightly so that they cannot move. For example, Figure 4.7 shows one such configuration. We say that such a packing is *jammed* if the only continuous motion of the packing disks, fixing the

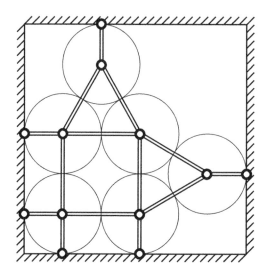

Figure 4.7 An example of a jammed packing of circular disks in a square box. The nodes of the associated tensegrity are the centres of the packing disks; two nodes are connected by a strut if the corresponding packing disks touch on their boundaries; where a disk touches the bounding box, there is a fixed node connected to that disk with a strut.

boundary, the radii of the disks, and not allowing any two packing disks to overlap (except on their common boundary), is the identity motion where nothing moves.

For such a packing in a polygonal box in the plane, there is a natural tensegrity that captures the rigidity properties of the packing quite well. Place a node at the centre of each packing disk and join two nodes by a strut if the corresponding packing disks touch on their boundaries. Also, place a fixed node on the polygonal boundary at every point that is touched by a packing disk, and connect that node to the centre of that packing disk. In this way a tensegrity of all struts and some fixed nodes is created. It is clear that such a packing is jammed if and only if the associated strut tensegrity is rigid.

Theorem 4.10.1. *Any tensegrity $(G, \mathbf{p})$ with all members as struts (connected to fixed nodes) is rigid if and only if it is infinitesimally rigid.*

Proof. If $(G, \mathbf{p})$ is infinitesimally rigid, then it is rigid by Theorem 4.3.1. For the converse, suppose that $\mathbf{d}$ is a non-zero infinitesimal flex of $(G, \mathbf{p})$, where $\mathbf{p} = [\mathbf{p}_1; \ldots; \mathbf{p}_n]$. Since there are fixed nodes, $\mathbf{d}$ being non-zero is equivalent to it being not an infinitesimal rigid-body motion. We wish to show that there is a finite non-fixed motion of the packing disks maintaining them as a packing. For each node i, and $t \geq 0$ define $\mathbf{p}_i(t) = \mathbf{p}_i + t\mathbf{d}_i$. For $\{i, j\}$ strut in G, we calculate (the square of) its distance as a function of the parameter t:

$$|\mathbf{p}_i(t) - \mathbf{p}_j(t)|^2 = ((\mathbf{p}_i - \mathbf{p}_j) + t(\mathbf{d}_i - \mathbf{d}_j))^2$$
$$= (\mathbf{p}_i - \mathbf{p}_j)^2 + 2t(\mathbf{p}_i - \mathbf{p}_j) \cdot (\mathbf{d}_i - \mathbf{d}_j) + t^2(\mathbf{d}_i - \mathbf{d}_j)^2.$$

For small t, the middle term dominates if the inner product is positive by the strut condition. Otherwise the last term dominates, and in either case, the distance increases strictly for small

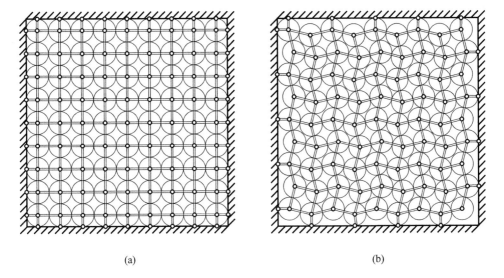

(a)   (b)

Figure 4.8 The packing in (a) is a jammed packing. Yet it is quite fragile in the sense that very small perturbations will allow it to become unjammed. Example (b) shows a small perturbation of the packing of (a). To continuously move from (a) to (b) the walls of the container must expand ever so slightly.

t for some $\{i, j\}$ since not all the $\mathbf{d}_i$ are equal. It is easy to check that the packing disks do not penetrate the boundaries as well. So for sufficiently small $t > 0$ this serves as a motion unjamming the packing. Thus the associated tensegrity must be infinitesimally rigid if the packing is to be jammed.   □

This result together with the Roth–Whiteley Theorem 4.8.1 provide useful tools for determining whether a packing of disks is jammed. For example, Corollary 4.8.2 implies that for a packing with n disks to be jammed in a planar box, it must have at least $2n + 1$ contacts, counting contacts between disks as well as disk–boundary contacts.

As another example, consider the packing in Figure 4.8(a). It is easy to check that the underlying bar framework is infinitesimally rigid by building it from the corners of the square using Type I Henneberg operations. It is also easy to see that there is a proper self-stress, non-zero on all the struts, by adding stresses along horizontal and vertical lines. So the Roth–Whiteley criterion implies that this packing is jammed.

This points to some of the advantages and drawbacks of this method and the definition of rigidity. These methods can make it easy (in some cases) to determine whether a packing is jammed. For example, one can use linear programming methods to determine the existence of a stress reasonably efficiently. One the other hand, the hard distance constraints may not be physically realistic as with the square packing of Figure 4.8(a) with a large number of packing disks. Small perturbations of the packing configuration as with 4.8(b), not obtainable by a continuous deformation through packings, can lead to nearby packings that will not be jammed, and will essentially "fall apart" as the number of packing disks gets large.

The idea of using the lack of infinitesimal rigidity to "unjam" packings was used in Danzer (1963), and the idea was systematically applied in Tarnai and Gáspár (1983) to packings of circular disks on a sphere and later in many other situations. Ironically, when the ambient space is a sphere, the guarantees of Theorem 4.10.1 do not always apply. Nevertheless, the idea still seemed to work well regardless. These ideas have also been applied to unjam computer models of granular materials in Donev et al. (2004) and (2007).

4.11 Exercises

1. Which sets of three of elementary loads (all non-zero, none along the bars) in the regular hexagon, with nodes at the vertices and bars along the edges, are resolvable by an internal stress?

2. Suppose that you have an infinitesimally rigid tensegrity in $\mathbb{E}^d$ with n nodes, b bars, and $c \geq 1$ cables with no struts. Corollary 4.8.2 implies that $c \geq nd - d(d+1)/2 - b + 1$. Suppose that no cable is redundant. In other words, after the removal of any cable, the resulting tensegrity is not infinitesimally rigid. Show that $2(nd - d(d+1)/2 - b) \geq c$.

3. Consider the unit cube in $\mathbb{E}^3$. Place a node at each vertex of the cube and a bar along each edge. The resulting framework is not rigid. What is the smallest number of cables one can add so that the resulting tensegrity is infinitesimally rigid, and what is such a collection of cables that is minimal?

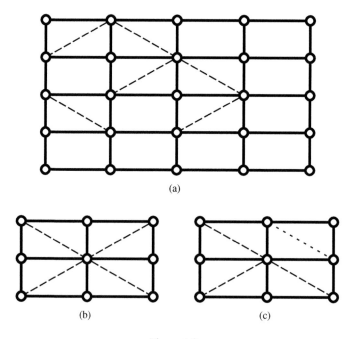

(a)

(b)   (c)

Figure 4.9

4. Suppose that one has an $(n + 1)$-by-$(m + 1)$ rectangular grid of bars and nodes, as in Problem 3.11 of Chapter 3, but where some of the small rectangles are braced with cables or struts instead of bars as shown in Figure 4.9. Note, however, that when a strut is used for one diagonal of a small rectangle, it is equivalent, for the rigidity, to using a cable for the other diagonal. There are n rows and m columns of small rectangles.

   a. Show that, in general, $n + m$ cables and struts are necessary for rigidity.
   b. Show that the rigidity of the braced grid does not change when a pair of rows or a pair of columns are interchanged.
   c. Show that the 2-by-2 grid in Figure 4.9(b) is rigid, and thus if the upper-right cable is missing, it is equivalent to having an implied cable as in Figure 4.9(c)
   d. Show that in Figure 4.9(a) the upper-left 3-by-3 is rigid with six cables, but there is no way to rigidify the 4-by-4 grid by adding two more cables in the small rectangles.

5

Energy Functions and the Stress Matrix

5.1 Introduction

In Chapters, 3 and 4, a local first-order analysis was presented which allowed us to address problems of infinitesimal and static rigidity. In the present chapter, we introduce the effects of having a stressed structure, which we explore by defining energy functions for the stored strain energy in the system, and using stress matrices. A consequence of this is that, under certain conditions, we can consider global properties of the framework, in particular global rigidity, introduced in Section 2.4.5, and universal rigidity, introduced in Section 2.4.6. Examples of the simple tensegrities that we can study with this approach are shown in Figure 5.1

Chapter 6 will go on to use energy functions and stress matrices to look at the prestress stability of tensegrities.

5.2 Energy Functions and Rigidity

A very natural point of view to determine the stability of a structure is to introduce the concept of *energy functions*. We assume that the structure is *elastic*, which implies that the *strain energy* stored in a tensegrity is purely a function of its configuration. Each member has a well-defined energy function that is solely dependent on the member's length. Thus, for example, we are not interested here in plastic behaviour or friction, where energy is dissipated from the tensegrity as it changes its configuration. The total energy for a given configuration is the sum of the energies of all of the member lengths.

Although we have not previously used the concept of strain energy, it should be noted that the concept of equilibrium, as introduced in Section 3.3, is based on the idea of balancing the change of internal energy with the work done by applied loads. Heyman (1998) explores the history of the use of energy methods in structural analysis; Castigliano (1879) gave a full formulation of the use of energy principles, but the ideas clearly have earlier antecedents.

From an engineering perspective, if we wish to understand and estimate the deflection of a structure due to load, it is important that we use an energy function that is a realistic representation of the materials used, and we will explore that further in Section 5.2.1. However, from a geometric perspective, the key result is given later, in Theorem 6.2.1. This

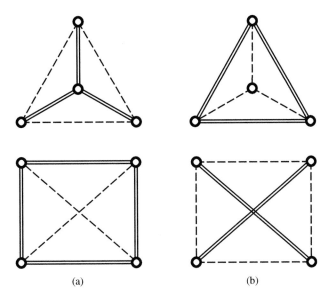

Figure 5.1 The tensegrities in (a) are not globally rigid in the plane, although they are rigid in the plane. The tensegrities in (b) are globally rigid in any Euclidian space that contains the plane.

shows that, if a structure can be shown to be rigid by using any energy function, then that structure is rigid when hard geometric constraints are introduced. Note that the converse is not true: if a structure is not shown to be rigid for a wide class of energy functions, it may nonetheless be rigid for hard geometric constraints. Theorem 6.2.1 allows freedom in the choice of energy function, and in the present chapter, we introduce a simple quadratic energy function in Section 5.3. In Chapter 6 we will go on to introduce a fairly general form of energy function for making further progress – and physical scientists will be cheered to see that a sensible choice of energy function to make progress with geometrical questions is also a physically reasonable choice.

5.2.1 Physical Energy Function

We cannot directly measure the energy stored in an elastic member, but we can measure the force applied to cause the member to change its length, and that force is the rate of change of the stored energy with respect to the length. Typically, the force will be measured as a function of the change of length of the member, but it can also be considered as a function of the length of the member. Consider the situation shown in Figure 5.2 which shows a plot of internal force against length for configurations $\mathbf{q}$. Local to the current configuration $\mathbf{p}$, we approximate the internal force as

$$t_{ij}(\mathbf{q}) = t_{ij}(\mathbf{p}) + k_{ij}(\mathbf{p})(l_{ij}(\mathbf{q}) - l_{ij}(\mathbf{p})), \tag{5.1}$$

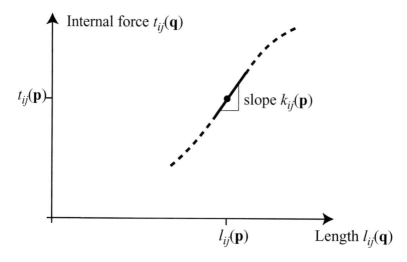

Figure 5.2 A plot showing the relationship between the internal force $t_{ij}(\mathbf{q})$ and the length $l_{ij}(\mathbf{q})$ for a member $\{i, j\}$ at a varying configuration $\mathbf{q}$. The graph is shown for configurations close to the configuration of interest, $\mathbf{p}$. Locally the relationship has a slope $k_{ij}(\mathbf{p})$. Note, the situation shown is for a cable, or a bar in tension, where $t_{ij}(\mathbf{q})$ is positive; for a strut $t_{ij}(\mathbf{q})$ would be negative. Commonly in engineering, it is assumed that $k_{ij}(\mathbf{q})$ is a constant, equal to $\frac{AE}{l_0}$, where A is the cross-sectional area of the member, E is the Young's modulus of the material, and l_0 is the rest length, when the internal force is zero.

i.e. we assume that the force–extension plot is a straight line, and hence the internal strain energy E_f (with an unknown constant E_c) is given as

$$E_f = \frac{k_{ij}(\mathbf{p})}{2}(l_{ij}(\mathbf{q}) - l_{ij}(\mathbf{p}))^2 + t_{ij}(\mathbf{p})(l_{ij}(\mathbf{q}) - l_{ij}(\mathbf{p})) + E_c, \qquad (5.2)$$

which can be rewritten in terms of length $l_{ij}(\mathbf{q})$ rather then extension $l_{ij}(\mathbf{q}) - l_{ij}(\mathbf{p})$ as

$$E_f = \frac{k_{ij}(\mathbf{p})}{2}l_{ij}(\mathbf{q})^2 + (t_{ij}(\mathbf{p}) - k_{ij}(\mathbf{p})l_{ij}(\mathbf{p}))l_{ij}(\mathbf{q}) + E_0, \qquad (5.3)$$

where $E_0 = E_c + k_{ij}(\mathbf{p})l_{ij}^2(\mathbf{q})/2 - t_{ij}(\mathbf{p})l_{ij}(\mathbf{p})$ is a constant term.

Chapter 6 will return to more general energy functions, but first we will consider a simple form where only the quadratic term in (5.3) is considered. In fact, this is not as unphysical as it might at first sight appear. In the field of 'static balancing' of mechanisms (Herder, 2001), where, for instance, Anglepoise or Luxo lamps are held in equilibrium across a range of configurations (French and Widden, 2000), zero-free-length springs, as introduced by Carwardine (1935) are an essential component. These are springs where, within the working range in tension, $k_{ij}(\mathbf{q})$ is a constant, the apparent rest length is zero, and hence the term $t_{ij}(\mathbf{p}) - k_{ij}(\mathbf{p})l_{ij}(\mathbf{p})$ in (5.3) equals zero. These springs have been used in a tensegrity context as cables within a "zero stiffness tensegrity" by Schenk et al. (2007). By contrast, the quadratic form does appear to be rather unphysical for struts, but nonetheless this term makes up an important component of more general energy functions.

5.3 Quadratic Energy Function

We wish to explore a tensegrity with a tensegrity graph G and a configuration $\mathbf{p}$: to do this, we define a energy function based on a general configuration $\mathbf{p}$, and here for initial simplicity we will assume that this function is quadratic.

For each member of $\{i, j\}$ of G, we define a scalar (stiffness) k_{ij} with the conditions that

(i) $k_{ij} \geq 0$ when $\{i, j\}$ is a cable;
(ii) $k_{ij} \leq 0$ when $\{i, j\}$ is a strut.

We do not set a condition when $\{i, j\}$ is a bar, although note that setting $k_{ij} = 0$ is equivalent to deleting the member $\{i, j\}$. We define the energy function, or the *potential function* $E_\omega : \mathbb{E}^{dn} \to \mathbb{E}^1$ to be

$$E_\omega(\mathbf{q}) = \sum_{\{ij\}\in M(G)} k_{ij} \left(\frac{|\mathbf{q}_i - \mathbf{q}_j|^2}{2} \right), \qquad (5.4)$$

where $M(G)$ are the members of G. There is one term for each member of G, and equivalently this can be written as

$$E_\omega(\mathbf{q}) = \sum_{\{ij\}\in M(G)} k_{ij} \left(\frac{l_{ij}^2}{2} \right), \qquad (5.5)$$

where l_{ij} is the length of member $\{i, j\}$. Notice that E_ω is a quadratic function in the coordinates of $\mathbb{E}^{nd}$, and in case there are no fixed nodes, is a quadratic form. In other words, as a function in its coordinates, E_ω is a polynomial, where each term is either a square, or the product of two different coordinates of the configuration $\mathbf{q}$ multiplied by the constant force density. In case there are fixed nodes, their coordinates act as constants as well. The theory of quadratic forms comes into play in a very useful way in the discussion later.

5.4 Equilibrium

In order to consider equilibrium of the tensegrity we are interested in those configurations of the tensegrity that correspond to critical points for the energy function. We will be particularly concerned with critical points that are minima of the stress–energy function, but we look at the general situation with any critical point first. Essentially we will initially consider the "weak" statement of equilibrium described in Section 3.3.2.

Recall that a critical point for a differentiable real-valued function of several real variables is a point in the domain of the function, where all the directional derivatives are zero. In other words, if the domain of the function is restricted to any straight line through a critical point, the resulting real-valued function of one variable will have its first derivative zero at the critical point.

In order to test for critical points, consider a particular configuration $\mathbf{p}$ in $\mathbb{E}^d$ with n labelled points. Let $\mathbf{p} + t\mathbf{d}$, where $\mathbf{d} = [\mathbf{d}_1; \ldots; \mathbf{d}_n]$, denote another configuration in $\mathbb{E}^d$. We think of $\mathbf{d}$ as a direction or velocity in the configuration space $\mathbb{E}^{nd}$, or equivalently, a set of directions one for each node of the configuration $\mathbf{p}$. (This is a sort of discrete vector field and is closely related to infinitesimal flexes, which came up in Chapter 2.) We will evaluate the energy potential function on the line $\mathbf{p} + t\mathbf{d}$, $s \in \mathbb{E}^1$. However, we introduce a restriction that when the tensegrity graph G has fixed nodes, we never consider configurations where the fixed nodes move. So we say that $\mathbf{d}$ is a *permissible displacement* if $d_i = 0$ for all fixed nodes i. When the first derivative of $E_\omega(\mathbf{p} + s\mathbf{d})$ is zero at $s = 0$ for all permissible $\mathbf{d} \in \mathbb{E}^{nd}$, then $\mathbf{d}$ is a *critical point* for E_ω.

We can also consider a "strong" nodal statement of equilibrium by considering the force density in each member. The strain energy in each bar $\{i, j\}$ is $E_{ij} = k_{ij}l_{ij}^2/2$, and thus the internal force is

$$t_{ij} = \frac{dE_{ij}}{dl_{ij}} = k_{ij}l_{ij}, \tag{5.6}$$

and for the quadratic energy function, the force density is

$$\omega_{ij} = \frac{t_{ij}}{l_{ij}} = k_{ij}, \tag{5.7}$$

i.e. is a constant for any length of the bar. Note that we can equivalently write $\omega_{ij} = dE_{ij}/d(l_{ij}^2/2) = k_{ij}$. Following Chapter 3, the stress is $\omega = (\ldots, \omega_{ij}, \ldots)$. We say that ω is a *proper stress* if

(i) $\omega_{ij} \geq 0$ when $\{i, j\}$ is a cable;
(ii) $\omega_{ij} \leq 0$ when $\{i, j\}$ is a strut.

There is no condition when $\{i, j\}$ is a bar. We say that ω is a *self-stress* for the configuration $\mathbf{p}$ in $\mathbb{E}^d$ for the framework $(G, \mathbf{p})$ if, for each node i that is not fixed,

$$\sum_j \omega_{ij}(\mathbf{p}_j - \mathbf{p}_i) = 0, \tag{5.8}$$

where the sum is taken over all nodes j, but due to our convention about non-edges having zero stress, it is equivalent to taking the sum over only nodes j of $V(G)$ that have an edge in common with the node i. Note that there is no equilibrium condition on the nodes that are fixed.

Figure 5.3 revisits the framework originally seen in Figure 3.2(c), now shown as a tensegrity. It turns out that a proper self-stress for this tensegrity has all four cable force densities as $+1$ and, and both strut force densities as -1. Figure 3.9 graphically shows the strong nodal equilibrium at one node: note that a negative force density reverses the directed line segment from $\mathbf{p}_i$ to $\mathbf{p}_j$, that represents the force, and, of course, the value of each force density rescales the vector in the vector sum in (5.8).

Next, we revisit the equivalence of the "strong" and "weak" forms of equilibrium originally addressed in Section 3.3.3, specifically for the quadratic energy function.

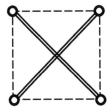

cables   struts
$$\omega = [\,\overbrace{1,1,1,1}\,,\,\overbrace{-1,-1}\,]$$

Figure 5.3 A tensegrity in the plane with nodes forming a square

Theorem 5.4.1 (Equilibrium). *A stress* $\omega = [\ldots, \omega_{ij}, \ldots]$ *is a self-stress for a configuration* $\mathbf{p} = [\mathbf{p}_1; \ldots; \mathbf{p}_n]$ *if and only if* $\mathbf{p}$ *is a critical point for the quadratic function* E_ω.

Proof. Let $\mathbf{p} + t\mathbf{d}$, where $\mathbf{d} = [\mathbf{d}_1; \ldots; \mathbf{d}_n]$, denote another configuration in $\mathbb{E}^d$, each $\mathbf{d}_i \in \mathbb{E}^d$. Then we expand each term of $E_\omega(\mathbf{p} + t\mathbf{p}')$:

$$|\mathbf{p}_i + t\mathbf{d}_i' - (\mathbf{p}_j + t\mathbf{d}_j)|^2 = |(\mathbf{p}_i - \mathbf{p}_j) + t(\mathbf{d}_i - \mathbf{d}_j)|^2$$
$$= |\mathbf{p}_i - \mathbf{p}_j|^2 + 2t(\mathbf{p}_i - \mathbf{p}_j) \cdot (\mathbf{d}_i - \mathbf{d}_j) + t^2|\mathbf{d}_i - \mathbf{d}_j|^2,$$

where we are using the standard dot product of vectors. Then from (5.4), using $k_{ij} = \omega_{ij}$ from (5.7),

$$E_\omega(\mathbf{p} + t\mathbf{d}) = \sum_{\{i,j\} \in M(G)} \omega_{ij}\big(|(\mathbf{p}_i - \mathbf{p}_j)|^2 + 2t(\mathbf{p}_i - \mathbf{p}_j) \cdot (\mathbf{d}_i - \mathbf{d}_j) + t^2|\mathbf{d}_i - \mathbf{d}_j|^2\big).$$

$$(5.9)$$

From this, it is clear that at $t = 0$ the first derivative of $E_\omega(\mathbf{p} + t\mathbf{d})$ with respect to t is 0 if and only if

$$\sum_{\{i,j\} \in M(G)} \omega_{ij}(\mathbf{p}_i - \mathbf{p}_j) \cdot (\mathbf{d}_i - \mathbf{d}_j) = 0. \qquad (5.10)$$

So $\mathbf{p}$ will be a critical point for $E_\omega(\mathbf{p} + t\mathbf{p}')$ if and only if (5.10) holds for all permissible $\mathbf{d}$.

Let j be any node of G that is not fixed, and consider only displacements $\mathbf{d}$ where $d_i = \mathbf{0}$ for all $i \neq j$. Then (5.10) reduces to

$$\sum_{\{i,j\} \in M(G)} \omega_{ij}(\mathbf{p}_i - \mathbf{p}_j) \cdot (-\mathbf{d}_j) = 0, \qquad (5.11)$$

and this must hold for all $\mathbf{d}_j \in \mathbb{E}^d$. But if we take $\mathbf{d}_j = -\sum_{\{i,j\} \in M(G)} \omega_{ij}(\mathbf{p}_i - \mathbf{p}_j)$, this implies

$$\sum_{\{i,j\} \in M(G)} \omega_{ij}(\mathbf{p}_i - \mathbf{p}_j) = \mathbf{0}, \qquad (5.12)$$

which shows that if $\mathbf{p}$ is a critical point for E_ω, then it is an equilibrium configuration for the stress ω.

Conversely suppose that the equilibrium condition (5.10) holds for all nodes j of G that are not fixed. Define $\hat{\mathbf{d}}_j = [\mathbf{0}; \ldots; \mathbf{0}; \mathbf{d}_j; \mathbf{0}; \ldots, \mathbf{0}]$, so $\mathbf{d} = \sum_j \hat{\mathbf{d}}_j$. Since the equilibrium condition (5.10) holds for each j, (5.10) holds for each $\hat{\mathbf{d}}_j$. But the expression on the left in (5.10) is linear in the coordinates of $\mathbf{d}$. This gives us the equality in (5.8) and finishes the proof. □

> **Remark 5.4.2.** *The conclusion of the theorem that critical points have a self-stress is quite fundamental and holds in a much greater generality. However, in this chapter we just need it for the quadratic case.*

5.5 The Principle of Least Energy

One of the simplest ways to show that a tensegrity framework is rigid and especially to show that it is globally rigid is to use energy functions, the most basic of which was defined in Section 2.4.5. The basic principle is to look for configurations where the energy is a local minimum, and is closely related to Castigliano's principle in the engineering literature. It has been used throughout mathematics at least since the advent of calculus. Here we start with a very special situation, that is nevertheless quite useful and representative of the more general cases that we will discuss later.

> **Theorem 5.5.1.** *Let ω be a proper self-stress for a tensegrity graph G (necessarily with fixed nodes) such that the configuration $\mathbf{p}$ in $\mathbb{E}^d$ is the unique minimum for the associated energy function E_ω. Then the tensegrity framework $(G, \mathbf{p})$ is globally rigid in $\mathbb{E}^d$.*

Proof. Suppose that the configuration $\mathbf{q} = (\cdots ; \mathbf{q}_i; \ldots)$ satisfies the tensegrity constraints for $(G, \mathbf{p})$ of Chapter 4, where $\mathbf{p} = (\cdots ; \mathbf{p}_i; \ldots)$. Since $\omega = (\ldots, \omega_{ij}, \ldots)$ is a proper self-stress for G which doesn't vary with configuration, for all i, j we have that

$$\omega_{ij}|\mathbf{q}_i - \mathbf{q}_j| \le \omega_{ij}|\mathbf{p}_i - \mathbf{p}_j|.$$

Hence we have

$$E_\omega(\mathbf{q}) = \sum_{\{i, j\} \in M(G)} \omega_{ij}|\mathbf{q}_i - \mathbf{q}_j|^2 \le \sum_{\{i, j\} \in M(G)} \omega_{ij}|\mathbf{p}_i - \mathbf{p}_j|^2 = E_\omega(\mathbf{p}).$$

Thus the configuration $\mathbf{q}$ is a minimum for the energy function E_ω. Since $\mathbf{p}$ is the unique minimum for E_ω, $\mathbf{p} = \mathbf{q}$, as desired. □

We call such a ω, such as in the hypothesis of Theorem 5.5.1, a *rigidifying self-stress* for $(G, \mathbf{p})$ in $\mathbb{E}^d$. Note that a rigidifying self-stress is necessarily a self-stress by Theorem 5.4.1.

It is certainly not the case that Theorem 5.5.1 applies to all tensegrities, even those that may be locally rigid. For example, shown in Figure 5.4 is a tensegrity with a proper self-stress whose configuration is a minimum for the quadratic energy function, but this configuration is not a unique minimum. There is another configuration that satisfies the tensegrity constraints, but is not congruent to the original.

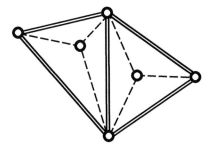

Figure 5.4 A tensegrity in $\mathbb{E}^2$ that is not globally rigid; the right half can be folded over the left half, for example. For this tensegrity, the stress matrix, defined in Section 5.6, is positive semi-definite, but does not have maximal rank.

One of the most natural applications of the principle of least work is to "spider webs", which have a natural correspondence with engineered cable-net structures, such as those developed by Frei Otto. We say that a tensegrity graph, usually containing some fixed nodes, is a *spider web* or a *spider web graph* if all its members are cables. The ideas developed here for spider webs are closely related to those in the famous paper, "How to draw a graph" by Tutte (1963).

Proposition 5.5.2. *Any proper self-stress ω for a spider web graph G that is non-zero (i.e. positive) on each member of G and such that every node is connected to a fixed node by members of G, has a unique configuration $\mathbf{p}$ such that ω is a self-stress for $(G, \mathbf{p})$, and $\mathbf{p}$ is the minimum point for the associated quadratic energy function E_ω. Thus $(G, \mathbf{p})$ is universally globally rigid.*

Proof. Since each of the terms of E_ω are non-negative, it is clear that $E_\omega(\mathbf{q}) \geq 0$ for all configurations $\mathbf{q}$. Choose some given configuration $\mathbf{q}(0)$. The connectivity condition ensures that there a constant $C > 0$ such that when any node $|\mathbf{q}_i| \geq C$, then $E_\omega(\mathbf{q}) \geq E_\omega(\mathbf{q}(0))$. Thus the function E_ω has at least one minimum point, say at $\mathbf{p}$.

Let $\mathbf{q}$ be any configuration that is a critical point for E_ω. Define $\mathbf{d} = \mathbf{p} - \mathbf{q}$. Then for $0 \leq t \leq 1$,

$$E_\omega(\mathbf{q} + t\mathbf{d}) = E_\omega(\mathbf{q}) + 2t \left[\sum_{\{ij\} \in M(G)} \omega_{ij}(\mathbf{q}_i - \mathbf{q}_j) \cdot (\mathbf{d}_i - \mathbf{d}_j) \right]$$

$$+ t^2 \left[\sum_{\{ij\} \in M(G)} \omega_{ij}(\mathbf{d}_i - \mathbf{d}_j)^2 \right].$$

But the middle term is 0, and the last term is strictly positive for $t > 0$, unless $\mathbf{d}_i = \mathbf{d}_j$ for all cables $\{i, j\}$. So $\mathbf{p}_i - \mathbf{q}_i = \mathbf{p}_j - \mathbf{q}_j$, and $\mathbf{p}_i - \mathbf{p}_j = \mathbf{q}_i - \mathbf{q}_j$ for all cables $\{i, j\}$. So by the connectivity hypothesis, we inductively show that $\mathbf{p}_i = \mathbf{q}_i$ for all nodes i. Thus $\mathbf{p} = \mathbf{q}$ as desired.   □

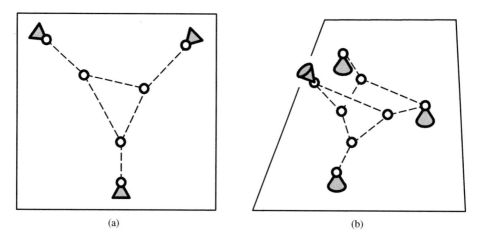

(a)   (b)

Figure 5.5 Two globally rigid spider web graphs with fixed nodes. In (a), all nodes must lie in the plane defined by the three fixed points. In (b), the support points may be non-planar, and hence the spider-web may be non-planar

Figure 5.5 shows some examples of spider web graphs.

Although spider webs are informative and interesting, we will next consider more general forms that also include struts.

5.6 The Stress Matrix

In this section we consider the case when the energy function E_ω is a quadratic form. In other words, there are no linear or constant terms in E_ω, which is the case when there are no fixed nodes – but we do in general have both struts and cables. We can compute the matrix of the quadratic form in terms of the coordinates of the nodes of the configuration $\mathbf{p} = [\mathbf{p}_1; \dots; \mathbf{p}_n]$.

Let $\{i, j\}$ be a member of a tensegrity graph G, and define the stress

$$\omega(\{i, j\}) = (0, \dots, 0, 1, 0, \dots, 0),$$

where all the coefficients are 0, except for $\omega_{ij} = 1$. Then for any configuration $\mathbf{p}$ in $\mathbb{E}^d$,

$$E_{\omega(\{i, j\})}(\mathbf{p}) = \frac{1}{2}|\mathbf{p}_i - \mathbf{p}_j|^2 = \frac{(x_i - x_j)^2}{2} + \frac{(y_i - y_j)^2}{2} + \cdots,$$

where $\mathbf{p}_i = (x_i; y_i; \dots)$ and $\mathbf{p}_j = (x_j; y_j; \dots) \in \mathbb{E}^d$. Observe that

$$(x_i - x_j)^2 = x_i^2 - 2x_i x_j + x_j^2 = \begin{bmatrix} x_i & x_j \end{bmatrix} \begin{bmatrix} 1 & -1 \\ -1 & 1 \end{bmatrix} \begin{bmatrix} x_i \\ x_j \end{bmatrix},$$

so we define an n-by-n symmetric matrix $\mathbf{\Omega}_{ij}$, where all the entries are 0 except for the entries (i, i) and (j, j), which are 1, and the (i, j) and (j, i) entries, which are -1. Then

$$\begin{bmatrix} x_1 & \cdots & x_n \end{bmatrix} \mathbf{\Omega}_{ij} \begin{bmatrix} x_1 \\ \vdots \\ x_n \end{bmatrix} = (x_i - x_j)^2,$$

and similarly for the other coordinates

$$\begin{bmatrix} y_1 & \cdots & y_n \end{bmatrix} \mathbf{\Omega}_{ij} \begin{bmatrix} y_1 \\ \vdots \\ y_n \end{bmatrix} = (y_i - y_j)^2,$$

and so on. So for an arbitrary stress $\omega = (\ldots, \omega_{ij}, \ldots)$ we have

$$\sum_{\{i,j\} \in M(G)} \omega_{ij}(x_i - x_j)^2 = \sum_{\{i,j\} \in M(G)} \omega_{ij} \begin{bmatrix} x_1 & \cdots & x_n \end{bmatrix} \mathbf{\Omega}_{ij} \begin{bmatrix} x_1 \\ \vdots \\ x_n \end{bmatrix}$$

$$= \begin{bmatrix} x_1 & \cdots & x_n \end{bmatrix} \left(\sum_{\{i,j\} \in M(G)} \omega_{ij} \mathbf{\Omega}_{ij} \right) \begin{bmatrix} x_1 \\ \vdots \\ x_n \end{bmatrix},$$

and similarly for the other coordinates. We define

$$\mathbf{\Omega} = \sum_{\{i,j\} \in M(G)} \omega_{ij} \mathbf{\Omega}_{ij}$$

as the *stress matrix* corresponding to the stress $\omega = (\ldots, \omega_{ij}, \ldots)$. It is easy to see that $\mathbf{\Omega}$ is a symmetric n-by-n matrix and that

(i) When $i \neq j$, the (ij) entry of $\mathbf{\Omega}$ is $-\omega_{ij}$.
(ii) The sum of the row and column entries of $\mathbf{\Omega}$ is 0.

Note that these conditions determine $\mathbf{\Omega}$ directly, and for some configuration $\mathbf{p} = [\mathbf{p}_1; \ldots; \mathbf{p}_n]$ where $\mathbf{p}_i = [x_i; y_i; \ldots)]$,

$$E_\omega(\mathbf{p}) = \frac{1}{2} \begin{bmatrix} x_1 & \cdots & x_n \end{bmatrix} \mathbf{\Omega} \begin{bmatrix} x_1 \\ \vdots \\ x_n \end{bmatrix} + \frac{1}{2} \begin{bmatrix} y_1 & \cdots & y_n \end{bmatrix} \mathbf{\Omega} \begin{bmatrix} y_1 \\ \vdots \\ y_n \end{bmatrix} + \cdots. \tag{5.13}$$

We can compact the notation further using the Kronecker product. If $\mathbf{A} = (a_{ij})$ and $\mathbf{B} = (b_{ij})$ are two matrices, the *Kronecker product* of $\mathbf{A}$ and $\mathbf{B}$ is defined as

$$\mathbf{A} \otimes \mathbf{B} = \begin{bmatrix} a_{11}\mathbf{B} & a_{12}\mathbf{B} & \cdots \\ a_{21}\mathbf{B} & a_{22}\mathbf{B} & \\ \vdots & & \ddots \end{bmatrix}.$$

(See James and Liebeck, 2001, Chapter 19, for a discussion of this operation on matrices.)
Now we can rewrite (5.13) as

$$E_\omega(\mathbf{p}) = \frac{1}{2}\mathbf{p}^{\mathrm{T}}(\mathbf{\Omega} \otimes \mathbf{I}^d)\mathbf{p},$$

where $\mathbf{I}^d$ is the d-by-d identity matrix and $(\dots)^{\mathrm{T}}$ represents the transpose operation on matrices. The matrix $\mathbf{\Omega} \otimes \mathbf{I}^d$ is sometimes called the *large stress matrix*. (Note that $\mathbf{\Omega} \otimes \mathbf{I}^d$ and $\mathbf{I}^d \otimes \mathbf{\Omega}$ differ only in a permutations of the rows and columns.)

In the case of a spider web with fixed nodes, the energy function had a strict minimum when each node was connected, by a sequence of edges in the graph G, to a fixed node. But in this case, with no fixed nodes, that cannot happen. For example, condition *ii* for a stress matrix (the sum of each row or column equals zero) implies that $\mathbf{\Omega}$ has the vector of all 1's in its kernel. But if there is a configuration vector $\mathbf{q} \in \mathbb{E}^{nd}$ such that $E_\omega(\mathbf{q}) = (1/2)\mathbf{q}^{\mathrm{T}}(\mathbf{\Omega} \otimes \mathbf{I}^d)\mathbf{q} < 0$, then $\lambda\mathbf{q}^{\mathrm{T}}(\mathbf{\Omega} \otimes \mathbf{I}^d)\lambda\mathbf{q} = \lambda^2\mathbf{q}^{\mathrm{T}}(\mathbf{\Omega} \otimes \mathbf{I}^d)\mathbf{q} \to \infty$ as $\lambda \to \infty$, and there is no minimum for E_ω. In the language of quadratic forms, E_ω is not positive semi-definite. (Recall that E_ω is *positive semi-definite* if for all $\mathbf{p} \in \mathbb{E}^{nd}$, $E_\omega(\mathbf{p}) \geq 0$.) In any case, it can never turn out that any E_ω is positive definite, since there is always the vector of all 1's in the kernel of $\mathbf{\Omega}$.

The stress matrix given here has an identical engineering antecedent, introduced by Schek (1974) for the form finding of a "cable net" (finding the configuration of a spider web, in the terminology used here). The stress matrix is sometimes known in the engineering literature as the "force density matrix," and its connection to the stress matrix was commented on by Tibert and Pellegrino (2003).

5.6.1 Equilibrium

In Chapter 3, the equations of equilibrium (3.16, 3.17) were written in terms of a matrix (the rigidity or equilibrium matrix) that depended on the configuration, multiplying a vector that contained the forces or force densities. Here we will reverse that situation using the stress matrix.

> **Proposition 5.6.1.** *A stress ω is in equilibrium with a nodal force vector $\mathbf{f} = [\mathbf{f}_1^T, \dots, \mathbf{f}_n^T]$ for the configuration $\mathbf{p}$ if and only if*
>
> $$(\mathbf{\Omega} \otimes \mathbf{I}^d)\mathbf{p} = \mathbf{f}.$$

Proof. Calculate

$$\mathbf{\Omega}_{ij}\begin{bmatrix} x_1 \\ \vdots \\ x_n \end{bmatrix} = \begin{bmatrix} 0 \\ \vdots \\ 0 \\ x_i - x_j \\ 0 \\ x_j - x_i \\ \vdots \\ 0 \end{bmatrix}$$

and then

$$\Omega \begin{bmatrix} x_1 \\ \vdots \\ x_n \end{bmatrix} = \begin{bmatrix} \vdots \\ \sum_j \omega_{ij}(x_i - x_j) \\ \vdots \end{bmatrix},$$

and this implies the equilibrium condition (3.14) by looking at each coordinate at a time. □

Suppose that a configuration $\mathbf{p}$ in $\mathbb{E}^d$ is a critical point for an energy function E_ω, for some stress ω. Then by Theorem 5.4.1, ω is self-stress for $\mathbf{p}$.

Proposition 5.6.2. *A stress ω is a self-stress for the configuration $\mathbf{p}$ if and only if*

$$(\Omega \otimes I^d)\mathbf{p} = \mathbf{0}.$$

Proof. This follows directly from Proposition 5.6.1; alternatively, for any symmetric matrix $\mathbf{Q}$, the gradient of quadratic form $\mathbf{p} \to \mathbf{p}^T \mathbf{Q}\mathbf{p}$ is the function $\mathbf{p} \to 2\mathbf{Q}\mathbf{p}$. So the result follows from Theorem 5.4.1. □

5.7 The Configuration Matrix

It is very helpful, for the calculations to come, to be able to rewrite the equilibrium equations directly in terms of the stress matrix Ω without having to use the Kronecker product. For any configuration $\mathbf{p} = [\mathbf{p}_1; \ldots; \mathbf{p}_n]$, we regard each point $\mathbf{p}_i \in \mathbb{E}^d$ as a column vector. We then assemble these into a single d-by-n matrix

$$\mathbf{P} = \begin{bmatrix} \mathbf{p}_1 & \mathbf{p}_2 & \cdots & \mathbf{p}_n \end{bmatrix},$$

which we call the *configuration matrix* for the configuration $\mathbf{p}$. Note that we still regard each $\mathbf{p}_i$ as a column vector. Furthermore it is convenient to define the following $(d + 1)$-by-n matrix:

$$\hat{\mathbf{P}} = \begin{bmatrix} \mathbf{p}_1 & \mathbf{p}_2 & \cdots & \mathbf{p}_n \\ 1 & 1 & \cdots & 1 \end{bmatrix}, \tag{5.14}$$

which we call the *augmented configuration matrix* $\hat{\mathbf{P}}$. The first row of $\mathbf{P}$ and $\hat{\mathbf{P}}$ is the row of x-coordinates of the points of the configuration; the second row of $\mathbf{P}$ and $\hat{\mathbf{P}}$ is the row of y-coordinates, and so on. The only difference between $\mathbf{P}$ and $\hat{\mathbf{P}}$ is the additional row of ones in in $\hat{\mathbf{P}}$. With this notation it is clear that if Ω is the stress matrix corresponding to a self-stress ω, then $\mathbf{P}\Omega = 0$ and $\Omega\mathbf{P}^T = 0$ are equivalent to the equilibrium conditions. Similarly $\hat{\mathbf{P}}\Omega = 0$ and $\Omega\hat{\mathbf{P}}^T = 0$ are equivalent to the equilibrium conditions as well. Indeed any n-by-n symmetric matrix Ω that satisfies $\hat{\mathbf{P}}\Omega = 0$ for some augmented configuration matrix will correspond to a self-stress for the corresponding configuration $\mathbf{p}$. Of course, we are often interested in the case when certain of the off-diagonal entries of Ω are 0, and when the signs of other entries are determined.

We next investigate the relation between the affine properties of the configuration **p** and the augmented configuration matrix **p**. Recall that the *affine span* of the vectors $\mathbf{p}_1, \ldots, \mathbf{p}_n$ is

$$< \mathbf{p}_1, \ldots, \mathbf{p}_n > = \{\mathbf{p}_0 \mid \mathbf{p}_0 = \lambda_1 \mathbf{p}_1 + \cdots \lambda_n \mathbf{p}_n, \ \lambda_1 + \cdots \lambda_n = 1\}.$$

From this the following is clear.

Proposition 5.7.1. *The affine span of the nodes of the configuration* $\mathbf{p} = [\mathbf{p}_1; \ldots; \mathbf{p}_n]$ *is the same as the linear span of the columns of the augmented configuration matrix* $\hat{\mathbf{P}}$ *intersected with those vectors whose last coordinate is* 1, *thus the* $dim(< \mathbf{p}_1, \ldots, \mathbf{p}_n >) + 1 = rank(\hat{\mathbf{P}})$.

The notation for the configuration matrix helps to understand the effect of an affine transformation. Suppose that **V** is a d-by-d matrix and $\mathbf{w} \in \mathbb{E}^d$, so that an affine transformation is defined as in Section 3.3.6. Then the augmented configuration matrix of the affine image is given by

$$\begin{bmatrix} \mathbf{Vp}_1 + \mathbf{w} & \mathbf{Vp}_2 + \mathbf{w} & \cdots & \mathbf{Vp}_n + \mathbf{w} \\ 1 & 1 & \cdots & 1 \end{bmatrix} = \begin{bmatrix} \mathbf{V} & \mathbf{w} \\ 0 & 1 \end{bmatrix} \begin{bmatrix} \mathbf{p}_1 & \mathbf{p}_2 & \cdots & \mathbf{p}_n \\ 1 & 1 & \cdots & 1 \end{bmatrix}. \tag{5.15}$$

Proposition 5.7.2. *Suppose that the columns of the* $(d + 1)$-by-n *augmented configuration matrix* $\hat{\mathbf{P}}$ *as defined in* (5.14) *span* $\mathbb{E}^{d+1}$. *Let* **q** *be any configuration whose augmented configuration matrix* $\hat{\mathbf{Q}}$ *is such that the row span of* $\hat{\mathbf{Q}}$ *is contained in the row span of* $\hat{\mathbf{P}}$. *Then* $\hat{\mathbf{Q}}$ *is given by* (5.15).

Proof. It is clear that $\hat{\mathbf{Q}}$ is given by

$$\begin{bmatrix} \mathbf{V} & \mathbf{w} \\ \mathbf{c} & d \end{bmatrix} \begin{bmatrix} \mathbf{p}_1 & \mathbf{p}_2 & \cdots & \mathbf{p}_n \\ 1 & 1 & \cdots & 1 \end{bmatrix} = \hat{\mathbf{Q}},$$

where **c** is a 1-by-d row, d is a 1-by-1 scalar, and **V** and **w** are as in Section 3.3.6. But for the last coordinate of each column of $\hat{\mathbf{Q}}$ to be 1 we must have for $i = 1, \ldots, n$,

$$\begin{bmatrix} \mathbf{c} & d \end{bmatrix} \begin{bmatrix} \mathbf{p}_i \\ 1 \end{bmatrix} = \begin{bmatrix} 1 \end{bmatrix}. \tag{5.16}$$

Since the vectors $\begin{bmatrix} \mathbf{p}_i \\ 1 \end{bmatrix}$ span $\mathbb{E}^{d+1}$, the only solution to $\begin{bmatrix} \mathbf{c} & d \end{bmatrix} \begin{bmatrix} \mathbf{p}_i \\ 1 \end{bmatrix} = \begin{bmatrix} 0 \end{bmatrix}$ is the zero vector. So any solution to (5.16) is unique, and so $\begin{bmatrix} \mathbf{c} & d \end{bmatrix} = \begin{bmatrix} 0 & 1 \end{bmatrix}$. This is what is to be proved. $\square$

Notice that Equation (5.15) gives another proof of Theorem 3.3.2 that an affine map preserves the equilibrium condition for a self-stress.

5.8 Universal Configurations Exist

We are now in a position to define universal configurations. We say that a configuration **p** is *universal with respect to a self-stress* ω if any configuration **q** that is in self-equilibrium with respect to ω is an affine image of **p**.

Theorem 5.8.1. *Let* Ω *be an n-by-n stress matrix such that the configuration* $\mathbf{p} = (\mathbf{p}_1, \ldots, \mathbf{p}_n)$ *in* $\mathbb{E}^d$ *is in self-equilibrium with respect to the corresponding stress* ω, *and the affine span of* $\mathbf{p}$ *is all of* $\mathbb{E}^d$. *Then*

$$rank(\Omega) \leq n - d - 1, \tag{5.17}$$

and $\mathbf{p}$ *is universal with respect to* ω *if and only if (5.17) holds with equality. Furthermore, there always is a universal configuration* $\tilde{\mathbf{p}}$ *in* $\mathbb{E}^k \supset \mathbb{E}^d, k \geq d$ *which projects orthogonally onto* $\mathbf{p}$.

Proof. Let $\hat{\mathbf{P}} = \begin{bmatrix} \mathbf{p}_1 & \mathbf{p}_2 & \cdots & \mathbf{p}_n \\ 1 & 1 & \cdots & 1 \end{bmatrix}$ be the augmented configuration matrix corresponding to the configuration $\mathbf{p}$. The rows of $\hat{\mathbf{P}}$ are in the cokernel of Ω, by the equilibrium condition. In other words, $\begin{bmatrix} x_1 & \cdots & x_n \end{bmatrix} \Omega = \mathbf{0}$ for the x-coordinates of the configuration, and similarly for the other coordinates and the row of 1's. Since the affine span of the nodes of the configuration is all of $\mathbb{E}^d$, by Proposition 5.7.1, the rank of $\hat{\mathbf{P}}$ is $d + 1$ and the rows are independent. Thus the dimension of the cokernel of Ω is at least $d + 1$. This implies the inequality (5.17).

It is also clear that the cokernel of Ω is the linear span of the entire row space of $\hat{\mathbf{P}}$ if and only if (5.17) is an equality. When the row space of $\hat{\mathbf{P}}$ is the whole cokernel of Ω, then the row space of any other augmented configuration matrix, $\hat{\mathbf{Q}}$, corresponding to a configuration $\mathbf{q}$ in equilibrium with respect to Ω, will be contained in the row space of $\hat{\mathbf{P}}$. Then Proposition 5.7.2 implies that $\mathbf{q}$ is an affine image of $\mathbf{p}$. Hence $\mathbf{p}$ is universal.

If $\mathbf{p}$ is not universal, then the rows of $\hat{\mathbf{P}}$ do not span the cokernel of Ω, but it is always possible to add rows to $\hat{\mathbf{P}}$ so that (5.17) does hold. This corresponds to the universal configuration $\tilde{\mathbf{p}}$ which projects orthogonally onto $\mathbf{p}$. $\qquad\square$

Theorem 5.8.1 encapsulates almost everything that we want to show about global rigidity. However, the result leaves open the question what to do with configurations that might be affine images of the starting configuration $\mathbf{p}$. This will be addressed later. However, even if we do have configurations that are affine images of $\mathbf{p}$ to contend with, or even if the configuration is not universal with respect to the self-stress ω, we can still get a lot of information when the stress matrix is positive semi-definite.

We say that a tensegrity $(G, \mathbf{p})$, $\mathbf{p} = [\mathbf{p}_1, \ldots, \mathbf{p}_n]$ is *unyielding* if any other configuration $\mathbf{q} = [\mathbf{q}_1, \ldots, \mathbf{q}_n]$, satisfying the constraints of Chapter 3, must have all those constraints satisfied as equalities. In other words, for all members $\{i, j\}$ of G, (not just the bars), break $|\mathbf{q}_i - \mathbf{q}_j| = |\mathbf{p}_i - \mathbf{p}_j|$.

Theorem 5.8.2. *If a tensegrity* $(G, \mathbf{p})$, $\mathbf{p} = [\mathbf{p}_1, \ldots, \mathbf{p}_n]$ *has a self-stress* ω *with a positive semi-definite stress matrix* Ω *and all the members* G *with a non-zero force density, then* $(G, \mathbf{p})$ *is unyielding.*

Proof. We use a variation of the principle of least energy. If the configuration $\mathbf{q} = (\mathbf{q}_1, \ldots, \mathbf{q}_n)$ satisfies the tensegrity constraints of Chapter 3, then

$$E_\omega(\mathbf{q}) = \sum_{\{ij\} \in M(G)} \omega_{ij} |\mathbf{q}_i - \mathbf{q}_j|^2 \leq \sum_{\{ij\} \in M(G)} \omega_{ij} |\mathbf{p}_i - \mathbf{p}_j|^2 = E_\omega(\mathbf{p}),$$

where the inequality is strict if any of the tensegrity constraints are strict. But since Ω, and therefore E_ω, is positive semi-definite, we see that $(G, \mathbf{p})$ is unyielding, as desired.   □

5.9 Projective Invariance

Recall the definition of a projective transformation in Section 3.10, and that a non-singular projective transformation of a bar framework preserves its infinitesimal rigidity in Section 3.10.2. The projective representation of a point $\mathbf{p}_i$ is where one extra coordinate is attached (at the end, say) of $\mathbf{p}_i$. This gives the augmented configuration matrix as in Section 5.7. In terms of the notation in Section 5.7 the projective transformation is given by the $(d + 1)$-by-$(d + 1)$ matrix $\mathbf{V}$ operating on the configuration matrix $\hat{\mathbf{P}}$. But, of course, we must take care to adjust the last coordinate. Regarding any multiple of $\hat{\mathbf{p}}_i$ as representing the point in projective space, a $(d + 1)$-by-$(d + 1)$ matrix $\mathbf{V}$ describes the image of $\mathbf{p}$ as

$$\mathbf{V}\begin{bmatrix} \mathbf{p}_1 & \mathbf{p}_2 & \cdots & \mathbf{p}_n \\ 1 & 1 & \cdots & 1 \end{bmatrix} = \begin{bmatrix} \mathbf{q}_1 & \mathbf{q}_2 & \cdots & \mathbf{q}_n \\ \lambda_1 & \lambda_2 & \cdots & \lambda_n \end{bmatrix}.$$

But we want to project these back into the standard representation of the configuration with the last coordinates equal to 1. This can be accomplished by multiplying on the right by a diagonal matrix. This gives

$$\mathbf{V}\begin{bmatrix} \mathbf{p}_1 & \mathbf{p}_2 & \cdots & \mathbf{p}_n \\ 1 & 1 & \cdots & 1 \end{bmatrix}\begin{bmatrix} 1/\lambda_1 & & \\ & \ddots & \\ & & 1/\lambda_n \end{bmatrix} = \begin{bmatrix} \mathbf{q}_1 & \mathbf{q}_2 & \cdots & \mathbf{q}_n \\ \lambda_1 & \lambda_2 & \cdots & \lambda_n \end{bmatrix}\begin{bmatrix} 1/\lambda_1 & & \\ & \ddots & \\ & & 1/\lambda_n \end{bmatrix}$$

$$= \begin{bmatrix} \mathbf{q}_1/\lambda_1 & \mathbf{q}_2/\lambda_2 & \cdots & \mathbf{q}_n/\lambda_n \\ 1 & 1 & \cdots & 1 \end{bmatrix} = \hat{\mathbf{Q}}, \tag{5.18}$$

where $\hat{\mathbf{Q}}$ is the configuration matrix of the projective image of $\hat{\mathbf{P}}$. In order for this image configuration be defined, we ensure that all the $\lambda_i \neq 0$, for $i = 1, \ldots, n$. Here we show the following.

Theorem 5.9.1. *The image of a configuration* $\mathbf{p}$ *under a non-singular projective transformation has a corresponding stress matrix of the same rank and number of positive eigenvalues as the stress matrix of* $\mathbf{p}$.

Proof. Suppose the stress matrix of $\mathbf{p}$ with augmented configuration matrix $\hat{\mathbf{P}}$ is Ω. Then the corresponding stress matrix for the projective image, given by (5.18), is

$$\begin{bmatrix} \lambda_1 & & \\ & \ddots & \\ & & \lambda_n \end{bmatrix}\Omega\begin{bmatrix} \lambda_1 & & \\ & \ddots & \\ & & \lambda_n \end{bmatrix} = \Omega_q. \tag{5.19}$$

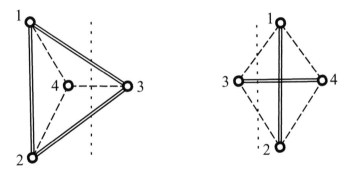

Figure 5.6 The two tensegrities here are projective images of each other. The line that is sent to infinity is indicated by a dotted line in each case. When this line crosses a member, it changes the member from a strut to a cable, or vice versa.

To see this, observe that $\hat{P}\Omega = 0$ if and only if

$$\hat{Q}\Omega_q = V \begin{bmatrix} \mathbf{p}_1 & \mathbf{p}_2 & \cdots & \mathbf{p}_n \\ 1 & 1 & \cdots & 1 \end{bmatrix} \begin{bmatrix} 1/\lambda_1 & & \\ & \ddots & \\ & & 1/\lambda_n \end{bmatrix} \begin{bmatrix} \lambda_1 & & \\ & \ddots & \\ & & \lambda_n \end{bmatrix} \Omega \begin{bmatrix} \lambda_1 & & \\ & \ddots & \\ & & \lambda_n \end{bmatrix}$$

$$= V\hat{P}\Omega \begin{bmatrix} \lambda_1 & & \\ & \ddots & \\ & & \lambda_n \end{bmatrix} = 0,$$

which shows that Ω and Ω_q have the same rank and number of positive eigenvalues. □

Note that the stress density ω_{ij} on member $\{i, j\}$ corresponding to the configuration $\mathbf{p}$ is transformed to $\lambda_i \lambda_j \omega_{ij}$ from (5.19).

We say that a projective point in homogeneous coordinates $(x_1; \ldots; z; \lambda)$ is *at infinity* if $\lambda = 0$. If we think of moving the projective transformation continuously from the identity to the given transformation, defined by (5.18), the only time the sign of a stress changes is when a node moves across the points at infinity, which are a subspace one dimension lower than the ambient space. For example, in the plane, the points that are transformed to the points at infinity form another line in the projective plane. When that line intersects the line segment between $\mathbf{p}_i$ and $\mathbf{p}_j$ the stress density ω_{ij} changes sign. This allows us to create several other tensegrities with a proper self-stress from ones with a seemingly different pattern of cables and struts. The roles of cables and struts switch when the line at infinity intersects their interior. Figure 5.6 shows an example of this behaviour.

Note that this gives another proof of the projective invariance of infinitesimal rigidity of bar frameworks as given in Section 3.10. This also provides a proof of the projective invariance of the infinitesimal rigidity of tensegrities, with the understanding that a cable will be changed to a strut, and vice versa, when the line that goes to infinity intersects the member. The dimension of the space of stresses is seen to be the same since the correspondence from one space to the other is linear for fixed projective transformation.

5.10 Unyielding and Globally Rigid Examples

We apply the results of the previous sections to some illustrative examples. Consider the configuration of Figure 5.3. We label the nodes cyclically around the square, and use the force densities as indicated. Then the stress matrix Ω is:

$$\Omega = \begin{bmatrix} 1 & -1 & 1 & -1 \\ -1 & 1 & -1 & 1 \\ 1 & -1 & 1 & -1 \\ -1 & 1 & -1 & 1 \end{bmatrix}.$$

Since $n = 4$, and $d = 2$, Theorem 5.8.1 implies that the rank of Ω is at most 1. But clearly Ω is not the $\mathbf{0}$ matrix with rank 0. So Ω is of rank 1, and has only one non-zero eigenvector. The trace of Ω, the sum of its diagonal values, is four, which is positive. Thus the only non-zero eigenvalue of Ω is four, Ω is positive semi-definite, and Theorem 5.8.2 implies that $(G, \mathbf{p})$ is unyielding. Since every pair of nodes of G has a cable or strut between them, $(G, \mathbf{p})$ is universally globally rigid.

Note also that $(G, \mathbf{p})$ is universal with respect to the stress matrix Ω. But Theorem 5.8.1, by itself, does not preclude the possibility that there might be another configuration that is a non-congruent affine image of $\mathbf{p}$ with corresponding members of equal length. This is clearly not the case here, and, indeed, affine transformations are often not a problem; universality of the configuration $\mathbf{p}$, and Ω being positive semi-definite are usually the most relevant considerations.

It is possible that there can be a tensegrity that is unyielding, and yet is not globally rigid. Consider the tensegrity of Figure 5.4, which consists of two globally rigid tensegrities of Figure 5.1 identified along an edge. We can add the self-stress in each half, each of which has a positive semi-definite stress matrix, such that we get a self-stress for the entire structure whose stress matrix is positive semi-definite. So this tensegrity is unyielding, but it is also clear that it is not globally rigid, since it is possible to fold one of triangles on top of the other in the plane. This folding map is not the restriction of an affine transformation in the plane, so we see that the configuration must not be universal with respect to the positive definite stress matrix indicated (or any other self-stress). But what is the universal configuration? Each triangle is determined up to congruence by the first example, so the only possibility is that the universal configuration for the whole tensegrity is when the affine span of the two triangles is 3-dimensional, like two leaves of a book. Note that although this framework is universal and the stress matrix is positive semi-definite, it is still not globally rigid. It is not even rigid in three-space. There is a finite motion, where one plane rotates about the other in three-space, that is an affine motion.

It is also possible to combine tensegrities, and appropriately chosen self-stresses for each, in such a way that on one (or more) of the members, the sum of the force densities vanish, and yet we still obtain an unyielding tensegrity. For example, the tensegrity of Figure 5.7 combines the self-stress for each of the square tensegrity and the triangle tenseg-rity of Figure 5.1 to give a self-stress with a positive semi-definite stress matrix. Since we can choose the force density on a strut of one tensegrity to be -1 and the corresponding force density on the cable of the other tensegrity to be $+1$, the force densities cancel. Thus we

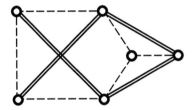

Figure 5.7 This is an example of forming a new tensegrity, along with its self-stress, by the combination of two other tensegrities and their self-stresses, in this case taken from Figure 5.1. Since the force densities along the two members that overlap can be chosen to cancel, we can eliminate that member and still have an unyielding tensegrity.

can remove that member and we will still have an unyielding tensegrity for the combination. Note that this combined tensegrity in the plane is still not universal for any self-stress, since its affine span is not 3-dimensional, but when, say the triangle tensegrity is rotated into three-space, then the configuration is universal for the combined self-stress.

5.11 Universal Tensegrities

Recall from Subsection 2.4.6 that a universally rigid framework is rigid in all Euclidean spaces containing the given framework $(G, \mathbf{p})$ and super stable frameworks, which are defined in Section 5.14, are examples of universally rigid frameworks.

Super stable tensegrities have the property that they are universally rigid (or universally globally rigid), but there are others, such as those in Figure 5.5(b), that are universally rigid, but not super stable. Part of the graph is universally rigid; that part can be added to the pinned nodes, and then the process continued. The stress does not extend to the whole graph since one of the members in the second graph must have a zero stress in the whole graph if it is to be in equilibrium. In Connelly and Gortler (2015) a complete characterization is given to determine when a framework or tensegrity is universally (globally) rigid.

5.12 Small Unyielding Tensegrities

There is a class of unyielding tensegrities that often turn out to be useful. These are examples of tensegrities where there are just $d + 2$ nodes in $\mathbb{E}^d$. Recall that an n-dimensional *simplex* σ is the convex hull of $n + 1$ points $\mathbf{p}$ in $\mathbb{E}^d$ such that they are affine independent – in other words, no $k + 2$ of the points lie in a k-dimensional hyperplane in $\mathbb{E}^d$.

> **Proposition 5.12.1.** *Suppose that an a-dimensional simplex σ_1 and a b-dimensional simplex σ_2 have a point that is in the relative interior of both simplices. Create a configuration consisting of the nodes of σ_1 and σ_2, and a tensegrity graph G consisting of struts corresponding to all the edges of σ_1 and all the edges of σ_2 and cables connecting each node of σ_1 to each of σ_2. This tensegrity $(G, \mathbf{p})$ has a proper self-stress, non-zero on each cable and strut, such that $\mathbf{p}$ is a minimum point for the associated quadratic form for the self-stress ω. Thus $(G, \mathbf{p})$ is unyielding.*

Proof. Let $(\mathbf{p}_1, \ldots, \mathbf{p}_{a+1})$ be the nodes of σ_1, and let $(\mathbf{p}_{(a+1)+1}, \ldots, \mathbf{p}_{(a+1)+(b+1)})$ be the nodes of σ_2. Since they share a point each in their relative interiors, there are scalars, all positive, $\lambda_1, \lambda_2, \ldots, \lambda_{(a+1)+(b+1)}$ such that

$$\sum_{i=1}^{a+1} \lambda_i \mathbf{p}_i = \sum_{i=(a+1)+1}^{(a+1)+(b+1)} \lambda_i \mathbf{p}_i, \tag{5.20}$$

and

$$\sum_{i=1}^{a+1} \lambda_i = 1 = \sum_{i=(a+1)+1}^{(a+1)+(b+1)} \lambda_i.$$

The configuration of the nodes of both simplices is $\mathbf{p} = (\mathbf{p}_1, \ldots, \mathbf{p}_{(a+1)+(b+1)})$. Define a stress matrix as

$$\begin{bmatrix} \lambda_1 \\ \lambda_2 \\ \vdots \\ \lambda_{a+1} \\ -\lambda_{(a+1)+1} \\ \vdots \\ -\lambda_{(a+1)+(b+1)} \end{bmatrix} \begin{bmatrix} \lambda_1 & \lambda_2 & \cdots & \lambda_{a+1} & -\lambda_{(a+1)+1} & \cdots & -\lambda_{(a+1)+(b+1)} \end{bmatrix} = \mathbf{\Omega}.$$

From this we see that the quadratic form corresponding to $\mathbf{\Omega}$ is positive semi-definite and that the force densities $\omega_{ij} = \lambda_i \lambda_j$ if i and j are nodes of the same simplex, and $\omega_{ij} = -\lambda_i \lambda_j$ if i and j are nodes of different simplices. (Recall that the off-diagonal entries of $\mathbf{\Omega}$ are the corresponding force densities but with the opposite sign by Condition i in Section 5.6.) It is clear that $\mathbf{\Omega}$ is a symmetric matrix, so by Condition ii for a stress matrix, we only need to check that row and column sums are 0. The row sum can be calculated by multiplying $\mathbf{\Omega}$ on the right by column vector of all ones. But

$$\begin{bmatrix} \lambda_1 & \lambda_2 & \cdots & \lambda_{a+1} & -\lambda_{(a+1)+1} & \cdots & -\lambda_{(a+1)+(b+1)} \end{bmatrix} \begin{bmatrix} 1 \\ 1 \\ \vdots \\ 1 \end{bmatrix}$$

$$= \sum_{i=1}^{a+1} \lambda_i - \sum_{i=(a+1)+1}^{(a+1)+(b+1)} \lambda_i = 1 - 1 = 0. \tag{5.21}$$

Hence $\mathbf{\Omega}$ is a stress matrix for ω by Conditions i and ii in Section 5.6. To show that $\mathbf{p}$ is a minimum point for E_ω, again we calculate $E_\omega(\mathbf{p})$. We already know that the quadratic form for $\mathbf{\Omega}$ is positive semi-definite, and thus E_ω is positive semi-definite. Hence $\mathbf{p}$ is a minimum

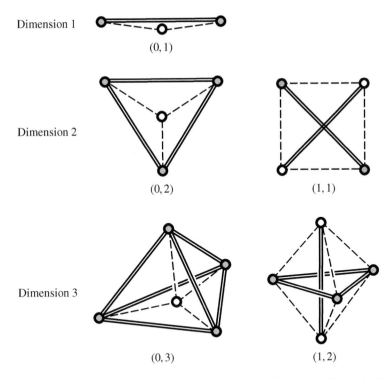

Figure 5.8 Examples of (a,b)-tensegrities up to dimension 3; the nodes of the a-simplex are not shaded, while those of the b-simplex are. Note that we have already seen examples of $(0,2)$- and $(1,1)$-tensegrities in Figure 5.1(b).

point for E_ω if and only if $E_\omega(\mathbf{p}) = 0$. Let $(x_1, x_2, \ldots, x_{a+b+2})$ be the first coordinates of each point of $\mathbf{p}$. Then from (5.20) we have

$$
\begin{bmatrix} \lambda_1 & \lambda_2 & \cdots & \lambda_{a+1} & -\lambda_{(a+1)+1} & \cdots & -\lambda_{(a+1)+(b+1)} \end{bmatrix}
\begin{bmatrix} x_1 \\ x_2 \\ \mathbf{c} \cdots \\ x_{a+b+2} \end{bmatrix}
$$

$$
= \sum_{i=1}^{a+1} \lambda_i x_i - \sum_{i=(a+1)+1}^{(a+1)+(b+1)} \lambda_i x_i = 0. \tag{5.22}
$$

Applying a similar argument to the other coordinates we see that $E_\omega(\mathbf{p}) = 0$. Thus $\mathbf{p}$ is a minimum point for E_ω, and $(G, \mathbf{p})$ is unyielding by Theorem 5.8.2. $\qquad\square$

With Proposition 5.12.1 in mind we say that a tensegrity, constructed as above with an a-dimensional simplex and a b-dimensional simplex intersecting in their relative interiors, is an *(a,b)-tensegrity*. Note that any (a,b)-tensegrity is globally rigid since all of the members are either a cable or strut and the tensegrity is unyielding. Figure 5.8 shows some examples of such unyielding (a,b)-tensegrities in dimensions 1, 2, and 3 with the (a,b) designation for each.

5.13 Affine Motions Revisited

We have still not completely dealt with affine motions that arise even when we have a universal configuration for a positive semi-definite stress matrix. We need to understand the nature of affine motions with regard to which pairs of distances are increasing or decreasing. Suppose that we have an affine transformation of $\mathbb{E}^d$ given by $\mathbf{p}_i \rightarrow \mathbf{V}\mathbf{p}_i + \mathbf{w}$, where $\mathbf{V}$ is a d-by-d matrix and $\mathbf{w} \in \mathbb{E}^d$ for each $\mathbf{p}_i \in \mathbb{E}^d$. We want to determine when the distance between $\mathbf{p}_i$ and $\mathbf{p}_j$ increases, decreases, or stays the same under such a transformation. We do this by calculating the squares of the distances involved and subtracting:

$$|(\mathbf{V}\mathbf{p}_i + \mathbf{w} - (\mathbf{V}\mathbf{p}_j + \mathbf{w})|^2 - |\mathbf{p}_i - \mathbf{p}_j|^2 = (\mathbf{V}\mathbf{p}_i - \mathbf{V}\mathbf{p}_j)^2 - (\mathbf{p}_i - \mathbf{p}_j)^2$$

$$= [\mathbf{V}(\mathbf{p}_i - \mathbf{p}_j)]^2 - (\mathbf{p}_i - \mathbf{p}_j)^2$$

$$= (\mathbf{p}_i - \mathbf{p}_j)^T \mathbf{V}^T \mathbf{V}(\mathbf{p}_i - \mathbf{p}_j)$$

$$- (\mathbf{p}_i - \mathbf{p}_j)^T \mathbf{I}^d(\mathbf{p}_i - \mathbf{p}_j)$$

$$= (\mathbf{p}_i - \mathbf{p}_j)^T [\mathbf{V}^T \mathbf{V} - \mathbf{I}^d](\mathbf{p}_i - \mathbf{p}_j), \qquad (5.23)$$

where $\mathbf{I}^d$ denotes the d-by-d identity matrix, the squaring operation refers to the dot product, and $(\ldots)^T$ is the transpose operation. From this calculation we see that the symmetric matrix $\mathbf{V}^T \mathbf{V} - \mathbf{I}^d$ and its associated quadratic form determine when distances increase, decrease, or stay the same.

It is quite natural and helpful if we think in terms of the projective plane (or projective space in dimensions greater than 3) that is defined in terms of lines through the origin in $\mathbb{E}^d$. So if we have a configuration $\mathbf{p}$ in $\mathbb{E}^d$ we say that a *member direction* for the member $\{i, j\}$ is the equivalence class determined by $\mathbf{p}_i - \mathbf{p}_j$, where two directions are equivalent if they are scalar multiples of each other. If $\mathbf{p}_i = \mathbf{p}_j$ we will say that they do not determine a direction.

In the case $d = 3$, the directions are a standard model for the points of the real projective plane. Let $\mathbf{Q}$ be any d-by-d non-zero symmetric matrix. We say that the set of directions defined by

$$C = \{\mathbf{v} \in \mathbb{E}^d \mid \mathbf{v}^T \mathbf{Q}\mathbf{v} = 0\}$$

is a *conic at infinity*. It is clear that C is well-defined since scalar multiples of a vector satisfy the same quadratic equation defining C. So we can say whether a direction or a set of directions lies on a conic at infinity. Note that when $\mathbf{Q}$ is definite (positive or negative) the corresponding conic at infinity is empty. In three-space, it is also possible that $\mathbf{Q}$ could determine a single plane through the origin, which is regarded as a projective line in the projective plane of directions, or it could determine two distinct planes through the origin, which is regarded as two projective lines in the projective plane of directions. But generally one would expect that C would be the set of lines from the origin to the points of an ellipse, say, in some plane not through the origin.

For an affine transformation given by $\mathbf{p}_i \rightarrow \mathbf{V}\mathbf{p}_i + \mathbf{w}$, we say that a direction given by a vector $\mathbf{v} \in \mathbb{E}^d$ is *length preserving* if the terms in Equation (5.23) are 0 with $\mathbf{p}_i - \mathbf{p}_j = \mathbf{v}$.

Proposition 5.13.1. *Suppose that D is a set of directions in $\mathbb{E}^d$. There is an affine transformation that is not a congruence and preserves lengths in all the directions in D if and only if the directions in D lie on a conic at infinity. Furthermore when the directions D do lie on a conic at infinity, there is a finite continuous motion of all of $\mathbb{E}^d$ that preserves the lengths of the members initially along the directions D.*

Proof. Equation (5.23) shows that an affine transformation determines a conic where $\mathbf{Q} = \mathbf{V}^\mathrm{T}\mathbf{V} - \mathbf{I}^d$. It is clear that $\mathbf{Q} = \mathbf{0}$ if and only if $\mathbf{V}^\mathrm{T}\mathbf{V} = \mathbf{I}^d$ which holds if and only if $\mathbf{V}$ is an orthogonal matrix. So for an affine transformation that is not a congruence, the directions that are length preserving lie on a conic at infinity.

Conversely suppose that the non-zero symmetric matrix $\mathbf{Q}$ determines a conic at infinity. Then by the spectral theorem (eigenvector eigenvalue decomposition) in linear algebra, there is a d-by-d orthogonal matrix $\mathbf{X}$, where $\mathbf{X}^\mathrm{T}\mathbf{X} = \mathbf{I}^d$, such that

$$\mathbf{Q} = \mathbf{X}^\mathrm{T}\begin{bmatrix} \lambda_1 & & & \\ & \lambda_2 & & \\ & & \ddots & \\ & & & \lambda_d \end{bmatrix}\mathbf{X},$$

where the middle matrix is diagonal. Then define

$$\mathbf{V}_t = \begin{bmatrix} \sqrt{1+t\lambda_1} & & & \\ & \sqrt{1+t\lambda_2} & & \\ & & \ddots & \\ & & & \sqrt{1+t\lambda_d} \end{bmatrix}\mathbf{X},$$

for $t\lambda_i \geq -1$ and for all $i = 1,\ldots,d$. Then for those t,

$$\mathbf{V}_t^\mathrm{T}\mathbf{V}_t - \mathbf{I}^d = \mathbf{X}^\mathrm{T}\begin{bmatrix} 1+t\lambda_1 & & & \\ & 1+t\lambda_2 & & \\ & & \ddots & \\ & & & 1+t\lambda_d \end{bmatrix}\mathbf{X} = \mathbf{I}^d + t\mathbf{Q} - \mathbf{I}^d = t\mathbf{Q}.$$

Reading Equation (5.23) from the other direction we see that $\mathbf{V}_t$ provides the affine transformation corresponding to the same conic at infinity. Since $\mathbf{V}_0 = \mathbf{I}^d$, then $\mathbf{V}_t\mathbf{p}_i$ provides a continuous motion of the nodes of the original configuration $\mathbf{p}$. $\qquad\square$

Note that the motion that is described in the proof of Proposition 5.13.1 continues until the matrix $\mathbf{V}_t$ is singular for the nearest values of t to 0, where $t\lambda_i = -1$. These configurations will necessarily have an affine span of dimension less than d.

An example when Proposition 5.13.1 applies when the square and triangle of Figure 5.7 lie each in a separate plane in three-space. All the members lie in two planes, so the directions of the members lie on two lines at infinity, a degenerate conic. This accounts for the affine flex in that case.

A more interesting example is shown in Figure 5.9. Start with a single line segment parallel to the y-z plane but which intersects the unit circle in the x-y plane at the point $(1,0,0)$. Rotate this segment about the z-axis to get several other disjoint line segments as in Figure 5.9(a). Reflect all the struts about the x-z plane to get another set of lines (These lines are part of two rulings of lines on a hyperboloid of revolution.) We wish to explore any motion that may be possible keeping these lines straight, but with any crossings remaining in contact.

To explore this motion we consider the assembly as a tensegrity. Place a node at each intersection point, and join subsequent nodes along lines with struts. Connect all the nodes that lie on one line with cables to the end nodes of the line, which ensures that the struts along the line do not change in length, and remain straight to give an unyielding tensegrity (although note that this line construction is not infinitesimally rigid). However, it is possible for the struts along two distinct lines to rotate around their common node, but note that the stressed directions of this tensegrity lie on a circle at infinity. So the entire construction is a finite mechanism keeping all the struts and cables at a fixed length. (This example is described in the book, Geometry and the Imagination, Hilbert and Cohn-Vossen (1981), pages 16–17.)

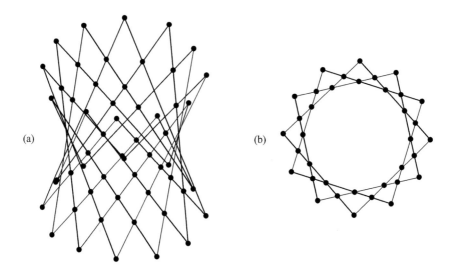

(a)   (b)

Figure 5.9 Figure (a) represents part of a finite mechanism, where the continuous bars lie in a hyperboloid of revolution $x^2 + y^2 - z^2 = 1$ in three-space. In this case, each bar is rigid, and the point of intersection between bars does not allow each individual bar to bend, but allows the bars to hinge relative to each other while ensuring that they do not move apart (hence the usual symbol for a node is not used in this figure). The motion is such that, in one direction, the whole mechanism becomes planar as in (b), and in the other direction it becomes collinear. This follows from the analysis in Theorem 5.13.1, because when the bars of (a) are translated to intersect at the same point, the lines through the translated bars form a circular cone $x^2 + y^2 - z^2 = 0$, say. In other words, they lie on a circle at infinity.

5.14 The Fundamental Theorem of Tensegrity Structures

We now put the information we have together to state the basic theorem that allows us to determine rigidity and global rigidity of tensegrities.

Theorem 5.14.1. *Suppose a tensegrity* $(G, \mathbf{p})$, $\mathbf{p} = [\mathbf{p}_1; \dots; \mathbf{p}_n]$ *in* $\mathbb{E}^d$, *with no fixed nodes has a self-stress* ω *with a stress matrix* Ω *such that*

1. *The matrix* Ω *is positive semi-definite,*
2. *The configuration* $\mathbf{p}$ *is universal with respect to* ω *and has a d-dimensional affine span; in other words, the rank of* Ω *is* $n - d - 1$,
3. *The directions of the members that have a non-zero force density, or are bars, do not lie on a conic at infinity.*

Then $(G, \mathbf{p})$ *universally globally rigid.*

Proof. We may assume without loss of generality that all the members of $(G, \mathbf{p})$ have a non-zero force density. By Theorem 5.8.2, $(G, \mathbf{p})$ is unyielding. Since by Condition 2, $\mathbf{p}$ is universal with respect to ω, if any other configuration $\mathbf{q} = [\mathbf{q}_1; \dots; \mathbf{q}_n]$ is such that $(G, \mathbf{q})$ satisfies the tensegrity constraints for $\mathbf{p}$, then $\mathbf{q}$ must by an affine image of $\mathbf{p}$. But by Proposition 5.13.1 every affine image of $\mathbf{p}$ that satisfies the equality distance constraints, must be a congruence. The only conics at infinity for $\mathbb{E}^2$ are two points, i.e. two directions. Thus $(G, \mathbf{p})$ is globally rigid in all dimensions, as was to be shown. □

The conditions (1), (2), (3) are so important that we say that any tensegrity $(G, \mathbf{p})$ that satisfies them is called *super stable*, a word coined by Alex Tsow, who was a student at Cornell. See also the discussion about affine motions in this context in Connelly et al. (2018).

We can apply Theorem 5.14.1 to show that several tensegrities are super stable. For example, we can combine tensegrities of Section 5.10, which are super stable themselves, to get several others that are super stable. Figure 5.10 shows how to combine $(1, 1)$ tensegrities to get a tensegrity on five nodes that is super stable.

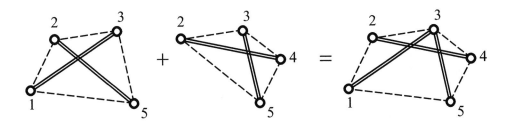

Figure 5.10 Two $(1, 1)$ tensegrities on the left are combined to give a tensegrity with five nodes on the right. The force densities are adjusted so that the force density on the $\{2, 5\}$ member cancels. In order for there to be equilibrium at the 3 node, the $\{3, 5\}$ member must have a negative force density, and so it must be a strut.

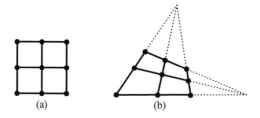

(a)   (b)

Figure 5.11 Tensegrity (b) has its member directions not on two points at infinity, and thus satisfies Property (3) of Theorem 5.14.1. But there is a projective transformation that takes it to Tensegrity (a) whose member directions do lie on a conic at infinity, which consists of two points.

5.14.1 *Projective Invariance of Super Stability*

Note that Properties (1) and (2) of Theorem 5.14.1 are preserved under projective transformation by Theorem 5.9.1. But Property (3) is not always preserved. For example, the tensegrity in Figure 5.11 consists of six line segments each with three nodes and three colinear members, a long strut connected to an inner node by two smaller cables. All members in this configuration have a non-zero stress. Since it consists of two sets of parallel lines, the member directions lie on at conic at infinity, in this case two points. So Property (3) is violated.

On the other hand there is a positive semi-definite stress matrix for both Tensegrity (a) and (b) of Figure 5.11, since there is a positive semi-definite stress on each of the six line segments. But for Tensegrity (a), and therefore Tensegrity (b), the stress matrix is of rank 5 since there is a realization of the grid in $\mathbb{E}^3$ on the quadric surface $z = xy$ that projects orthogonally onto Tensegrity (a). So this 3-dimensional configuration is the universal configuration for the positive definite stresses that are non-zero on each member of Tensegrity (a). This has the effect of implying that if a tensegrity is superstable in the plane, then so is a non-singular projective image, since counterexamples for the projective invariance of Property (3) of Theorem 5.14.1 cannot occur in the presence of Properties (1) and (2).

Another condition that seems to help with Property (3) is when the configuration is in *general position* in $\mathbb{E}^d$, which means that every subset of k nodes has a $(k-1)$-dimensional affine span when $k \le d+1$. Specifically, a theorem of Alfakih and Ye (2013) states that if Properties (1) and (2) of Theorem 5.14.1 for a self-stress of $\mathbf{p}$, and $\mathbf{p}$ is in general position, then Property (3) holds.

It is also interesting to compare the discussion here and in Section 5.9 with the lifting of weavings in Whiteley (1989).

5.14.2 *Applications of the Fundamental Theorem: Cauchy Polygons*

Suppose that $\mathbf{p} = [\mathbf{p}_1; \dots; \mathbf{p}_n]$ are the nodes in cyclic order of a convex polygon in the plane. Let $\{i, i+1\}$, $i = 1, \dots, n$, indices modulo n, be the cables, and let $\{i, i+2\}$, $i = 1, \dots, n-2$ be the struts. We call this tensegrity a *Cauchy polygon*, $C_n(\mathbf{p})$.

Proposition 5.14.2. *All Cauchy polygons are super stable.*

Proof. We proceed by induction on the number of nodes n in the configuration $\mathbf{p} = [\mathbf{p}_1; \ldots; \mathbf{p}_n]$ of the Cauchy polygon, starting with $n = 4$, $C_4(\mathbf{p})$. When $n = 4$ the Cauchy polygon $C_4(\mathbf{p})$ is a $(1, 1)$ polygon, and from the discussion in Section 5.10, it has a self-stress with positive semi-definite stress matrix of rank $1 = 4 - 2 - 1$. In the plane, a conic at infinity is just one or two points, and the $(1, 1)$ polygon has at least four distinct stressed directions. Thus a $(1, 1)$ polygon, which is the same as a Cauchy polygon $C_4(\mathbf{p})$, is super stable.

We now assume that any Cauchy polygon $C_n(\mathbf{p})$, for some $n \geq 4$, is super stable, and we wish to show that any Cauchy polygon $C_{n+1}(\mathbf{p})$ is super stable. Recall that the configuration for $C_{n+1}(\mathbf{p})$ is $\mathbf{p} = [\mathbf{p}_1; \ldots; \mathbf{p}_{n+1}]$. Remove $\mathbf{p}_n$ from $\mathbf{p}$ to give $\mathbf{q} = [\mathbf{p}_1; \ldots; \mathbf{p}_{n-1}; \mathbf{p}_{n+1}]$, and apply the inductive hypothesis to this Cauchy polygon $C_n(\mathbf{q})$. Let $\omega(C_n) = [\ldots, \omega_{ij}(C_n), \ldots]$, be a proper, non-zero, self-stress for $C_n(\mathbf{q})$. Let $C_4(\mathbf{r})$ be a Cauchy polygon on the four nodes $\mathbf{r} = [\mathbf{p}_1; \mathbf{p}_{n+1}; \mathbf{p}_n; \mathbf{p}_{n-1}]$, and let $\omega(C_4) = [\ldots, \omega_{ij}(C_4), \ldots]$ be the corresponding proper, non-zero, self-stress for $C_4(\mathbf{r})$. Note that $\{n - 2, n + 1\}$ is a strut in C_n and it is a cable in C_4. So we can rescale one of the self-stresses so that $\omega_{n-2,n+1}(C_n) = -\omega_{n-2,n+1}(C_4)$. We add these two self-stresses to give a self-stress $\omega(C_{n+1})$, where $\omega_{ij}(C_{n+1}) = \omega_{ij}(C_n) + \omega_{ij}(C_4)$, when the member $\{i, j\}$ lies in both C_n and C_4. When $\{i, j\}$ lies in just one of C_n or C_4, then $\omega_{ij}(C_{n+1})$ is just the force density for the graph it lies in.

We see that the quadratic form corresponding to the stress matrix $\mathbf{\Omega}(C_{n+1})$ corresponding to the self-stress $\omega(C_{n+1})$ is clearly positive semi-definite, since it is the sum of two other semi-definite forms corresponding to the stress matrices for the self-stresses $\omega(C_n)$ and $\omega(C_4)$.

We now show that the rank of $\mathbf{\Omega}(C_{n+1})$ is $n + 1 - 3 = n + 2$. In other words we need to show that a universal configuration for $\omega(C_{n+1})$ is $C_n(\mathbf{p})$. Since the quadratic form for $\mathbf{\Omega}(C_{n+1})$ is the sum of two positive semi-definite quadratic forms, each of the those forms must themselves be $\mathbf{0}$ for the configuration $\mathbf{p}$. Thus both universal configurations for $\mathbf{\Omega}(C_n)$ and $\mathbf{\Omega}(C_4)$ must have a 2-dimensional affine span. The nodes of the configurations C_n and C_4 overlap on the points, $\mathbf{p}_{n-2}, \mathbf{p}_{n-1}, \mathbf{p}_{n+1}$, whose affine span is 2-dimensional. Thus the span of the whole universal configuration corresponding to $\omega(C_{n+1})$ is 2-dimensional.

Lastly, we must show that the self-stress $\omega(C_{n+1})$ is proper. In other words the sign of the force densities on the members must be positive for the cables, negative for the struts, and 0 elsewhere. We have arranged that $\omega_{n-2,n+1} = 0$, as desired. All the other members of C_{n+1} are the sum of force densities of the same sign as desired, except for $\omega_{n-1,n+1}$, which is the sum of a positive $\omega_{n-1,n+1}(C_n)$ and a negative $\omega_{n-1,n+1}(C_4)$. But notice that we have equilibrium at the point $\mathbf{p}_{n+1}$, and that there are only three members incident to the node $\mathbf{p}_{n+1}$, coming from the nodes $\mathbf{p}_1$, $\mathbf{p}_{n-1}$, and $\mathbf{p}_n$. The members $\{1, n + 1\}$ and $\{n, n + 1\}$ are cables and have a positive force density. If $\omega_{n-1,n+1} \geq 0$, by the convexity of the Cauchy polygon at $\mathbf{p}_{n+1}$, equilibrium could not hold. Thus $\omega_{n-1,n+1} < 0$, as desired. This finishes the proof that Cauchy polygons are super stable. See Figure 5.10 for a graphic example of this process. □

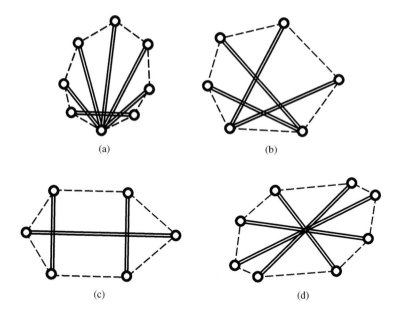

Figure 5.12 Examples of super stable planar tensegrities, where the cables form a convex polygon, and the struts form some of the diagonals. Example (a), suggested by Branko Grünbaum, is obtained by connecting one node by a strut to all the others except for the two adjacent nodes on the polygon, which are themselves connected by a strut. Example (b), suggested by Ben Roth, generalizes the examples of (a) by having two nodes such that all the other nodes are connected by struts to one or the other of these two nodes. Example (c) has two struts and two cables forming a rectangle, with the third strut perpendicular to both the other struts. Example (d) is symmetric about a central point with antipodal nodes connected by struts. Examples (c) and (d) depend on the particular position of the nodes for there to be the required self-stress for super stability, while examples (a) and (b) are always super stable as long as the nodes form a convex polygon as indicated.

In Figure 5.12 are some other examples of polygons in the plane with cables along the edges and struts along the interior that are super stable.

5.14.3 An Application to Cauchy's Theorem

It is often interesting to consider a tensegrity where we add further restrictions. For example, if a cable is replaced by a bar, the cable constraint still remains; the additional constraint may not be required for the property of being globally rigid, but the property remains useful, as will be demonstrated in the proof of Cauchy's "Arm Lemma" in the following section.

Suppose we restrict Cauchy polygons to the case when all of the outer cables except one are replaced with bars, as in Figure 5.13. This leads to the following lemma due to Cauchy (originally published, in a slightly different form, with some fixable errors in the proof) that has come to be known as the "Arm Lemma" because the structure resembles an opening arm, and is a special case of the universal global rigidity of tensegrity Cauchy polygons, Theorem 5.14.2

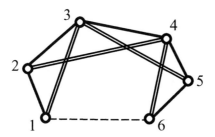

Figure 5.13 The n outside edges of this convex polygon are bars, except the $\{1, n\}$ member, which is a cable. The internal edges are struts as with the Cauchy polygon as described in Section 5.14.2. So the strut condition is equivalent to the angles at $\mathbf{p}_2, \ldots, \mathbf{p}_{n-1}$ not being allowed to decrease.

Lemma 5.14.3. *Suppose that* $\mathbf{p}_1, \ldots, \mathbf{p}_n$ *is a convex polygonal path in* $\mathbb{E}^2$ *and* $\mathbf{q}_1, \ldots, \mathbf{q}_n$ *is another path in* $\mathbb{E}^d$ *(not necessarily convex) with* $d \geq 2$ *such that* $|\mathbf{p}_i - \mathbf{p}_{i+1}| = |\mathbf{q}_i - \mathbf{q}_{i+1}|$ *for* $i = 1, \ldots, n - 1$, *and the internal angle at* $\mathbf{p}_i$ *is less than or equal to the angle at* $\mathbf{q}_i$ *for* $i = 2, \ldots, n - 1$. *Then* $|\mathbf{p}_1 - \mathbf{p}_n| \leq |\mathbf{q}_1 - \mathbf{q}_n|$.

Proof. Form a Cauchy polygon from the points on the path, and replace all the cables by bars, except for $\{1, n\}$ which remains a cable. If $\mathbf{p}$ is the original convex configuration, and $\mathbf{q}$ is another configuration that satisfies all the bar constraints, then the strut constraints can be regarded as saying that the internal angle at each point $\mathbf{q}_i$ is no smaller than the internal angle at $\mathbf{p}_i$ for $i = 2, \ldots, n - 1$. So if $\mathbf{q}$ is not congruent to $\mathbf{p}$, then the cable constraint must be violated. In other words the distance from $\mathbf{p}_1$ to $\mathbf{p}_n$ must be less than the distance from $\mathbf{q}_1$ to $\mathbf{q}_n$.   □

We now use the Arm Lemma 5.14.3 to prove Cauchy's theorem.

Proof of Cauchy's Theorem 3.9.1. As the convex polytope P is repositioned to the convex polytope Q, label each edge $\{i, j\}$ of P with a $+$, $-$, or nothing, depending whether the dihedral angle at $\{i, j\}$ increases, decreases, or stays the same, respectively. If there are at least four changes in sign (or none) as one proceeds around $\mathbf{p}_i$, for each vertex $\mathbf{p}_i$, then the argument in Section 3.9.1 shows that, indeed, all the angles do not change and P is congruent to Q.

To do this, consider a given vertex $\mathbf{p}_i$ and its adjacent vertices and faces. This is part of a convex cone with $\mathbf{p}_i$ at its vertex. We have to eliminate the possibility that there are exactly two changes in sign on the edges around $\mathbf{p}_i$. Slice the cone with a plane near $\mathbf{p}_i$ to get a convex polygon X where the vertices of X correspond to the edges of the cone. Choose two points $\mathbf{a}$ and $\mathbf{b}$ on X that separate the $+$ points from the $-$ points, as in Figure 5.14, and choose the path from $\mathbf{a}$ to $\mathbf{b}$ containing the $+$ vertices. Let $\mathbf{a}'$ and $\mathbf{b}'$ be the corresponding points in Q. The Arm Lemma 5.14.3 implies that $|\mathbf{a}' - \mathbf{b}'| > |\mathbf{a} - \mathbf{b}|$. Reversing the roles of P and Q implies $|\mathbf{a} - \mathbf{b}| > |\mathbf{a}' - \mathbf{b}'|$, a contradiction unless there are no changes in sign for the dihedral angles to start with.   □

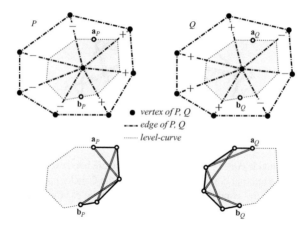

Figure 5.14 This shows two corresponding vertices of the convex polytopes P and Q.

Needless to say, Cauchy did not think of his Arm Lemma this way, but his proof made a mistake when he tried to do it by induction. See Exercise 5.15 for a counter-example to using induction on the number of vertices. See also Chern (1967) and Schoenberg and Zaremba (1967) for a self-contained elementary proof of the Arm Lemma 5.14.3.

5.14.4 Convex Planar Super Stable Polygons

In addition to the examples of Cauchy polygons of Figure 5.10 and the polygons of Figure 5.12, we can characterize all such super stable polygons with cables forming a convex polygon and struts forming some of the diagonals.

> **Theorem 5.14.4.** *(Connelly, 1982) Let a tensegrity $(G, \mathbf{p})$ be such that the nodes $\mathbf{p} = [\mathbf{p}_1; \ldots ; \mathbf{p}_n]$ form a convex polygon, where $\{i, i+1\}$, for $i = 1, \ldots, n$ (indices modulo n) are all the cables, the struts are some of the (internal) diagonals, and there is a non-zero proper self-stress ω for $(G, \mathbf{p})$. Then $(G, \mathbf{p})$ is super stable for the corresponding stress matrix Ω.*

Proof. We need to show the three conditions for super stability. Condition (3) is the easiest, since there are at least three distinct stressed directions.

Now consider condition (2). Assume that Ω is positive semi-definite, but is not maximal rank, i.e. the rank $< n - 3$. Then, by the discussion in Section 5.8, there is a configuration $\mathbf{q} = (\mathbf{q}_1, \ldots, \mathbf{q}_n)$ in $\mathbb{E}^3$, whose affine span is 3-dimensional, with a self-stress corresponding to Ω, and each $\mathbf{q}_i$ projects orthogonally onto $\mathbf{p}_i$ for $i = 1, \ldots, n$. Consider the convex hull H of the nodes of the lift $\mathbf{q}$ in $\mathbb{E}^3$. There must be an internal edge e of two of the top faces of H which connects some pair of nodes of $\mathbf{q}$ and separates the other nodes of $\mathbf{q}$. If some strut e' with a non-zero force density corresponding to $\Omega(t_0)$ crosses under e, then it is clear that there is contradiction since a rotation along e will change the length of e' to the first-order, contradicting the equilibrium condition for nodes containing e'. If there is no such e', then there can be no force density even on e in its projection onto $\mathbf{p}$ in the

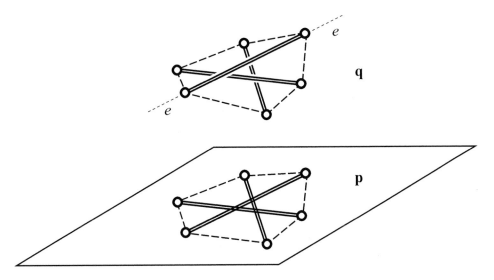

Figure 5.15 The planar configuration **p** is lifted to a configuration **q** in $\mathbb{E}^3$. The two nodes along line e have been raised further than the other nodes, so the other struts pass beneath the strut that lies along line e.

plane, since again there is an infinitesimal motion in the plane that lengthens e and all the internal struts to the first-order. Thus there is no such configuration **q** in equilibrium with a 3-dimensional affine span. So, if Ω is positive semi-definite, it must have 3-dimensional kernel. See Figure 5.15 for a picture of this.

Finally, consider both condition (1) and (2). Let Ω_C be the stress matrix for the corresponding Cauchy polygon for the configuration **p**, and define $\Omega(t) = (1 - t)\Omega_C + t\Omega$, for $0 \le t \le 1$. By Proposition 5.14.2, $\Omega(0) = \Omega_C$ satisfies (1) and (2) for super stability, namely it is positive semi-definite and has rank $n - 3$. Suppose that $\Omega = \Omega(1)$ does not satisfy either (1) or (2). We wish to find a contradiction. Since the set of eigenvalues of $\Omega(t)$ vary continuously with t, and the kernel of $\Omega(t)$ always contains the kernel of $\Omega(0)$, if either (1) or (2) fails, there is a (smallest) $0 < t_0 \le 1$, where $\Omega(t_0) \ne 0$ has rank less than $n - 3$ and its kernel is at least 4-dimensional. But we have already shown that this cannot be the case above. So $\Omega = \Omega(1)$ is positive semi-definite with a 3-dimensional kernel, and thus $(G, \mathbf{p})$ is super stable. □

5.15 Exercises

1. In Figure 5.5(a) the cable lines from the fixed nodes intersect in a point. Show that this tensegrity has a proper self-stress that is non-zero on all the cables and thus is necessarily (uniformly) globally rigid.
2. Suppose that a tensegrity $(G, \mathbf{p})$ such as the one Figure 5.5(b), where the underlying graph G is a tree with the end points as fixed nodes (possibly overlapping), no zero length members, and each non-fixed node is contained in the relative interior of the

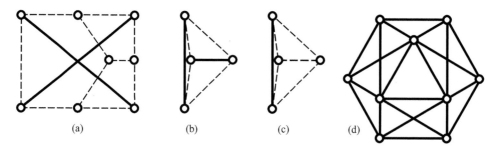

(a)   (b)   (c)   (d)

Figure 5.16 Which of these tensegrities is universally globally rigid? Here the solid lines are bars, and the three nodes in the vertical bar are meant to be collinear in Figures (b) and (c).

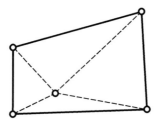

Figure 5.17 This is any convex quadrilateral, with a fifth node in its interior connected by cables to the four exterior nodes.

convex hull of its neighbouring nodes. Show that $(G, \mathbf{p})$ has a self-stress that is non-zero on all its cables and hence is universally globally rigid.

3. Determine which of the tensegrities in Figure 5.16 are universally globally rigid.

4. Show that all the tensegrities in Figure 5.16 are rigid in $\mathbb{E}^3$.

5. Show that all the tensegrities of Figure 5.12 have a non-zero proper self-stress, and are thus universally globally rigid.

6. Consider the bar framework $(G, \mathbf{p})$ in the line $\mathbb{E}^1$ with six distinct nodes, three with a positive coordinate, the other three with a negative coordinate, where the bars connect all positive nodes with all negative nodes. Show that $(G, \mathbf{p})$ is globally rigid in $\mathbb{E}^2$ but is a finite mechanism in $\mathbb{E}^3$. See Bang et al. (1993) for a hint.

7. Show that if a tensegrity graph G without bars, just struts and cables, is such that the subgraph of cables is not connected, while it is connected when the struts are included, then any non-zero stress matrix Ω, corresponding a proper self-stress, must have a negative eigenvalue.

8. Show that the tensegrity $(G, \mathbf{p})$ in Figure 5.17, with struts or bars on the outside is such that any non-zero stress matrix Ω, corresponding a proper self-stress, must have exactly one negative eigenvalue, one positive eigenvalue, and three zero eigenvalues.

9. Suppose P is a convex polytope in $\mathbb{E}^3$. Create a tensegrity $(G, \mathbf{p})$ by putting a node at each vertex of P, cables along each edge of P, and struts from a node in the interior of P to each vertex of P. In Lovász (2001) it is shown that any proper self-stress

for $(G, \mathbf{p})$ has a stress matrix Ω with one negative eigenvalue, four zero eigenvalues, and all other eigenvalues positive. Use this to show that any centrally symmetric polytope with cables on the edges and struts joining antipodal nodes is super stable. (See Bezdek and Connelly (2006) for more details.)

10. Suppose $(G, [\mathbf{p}_1; \mathbf{p}_2; \mathbf{p}_3; \mathbf{p}_4])$ is a convex arm and $(G, [\mathbf{q}_1; \mathbf{q}_2; \mathbf{q}_3; \mathbf{q}_4])$ is another arm, where $\mathbf{q}_2 = \mathbf{p}_2$, $\mathbf{q}_3 = \mathbf{p}_3$, $|\mathbf{p}_1 - \mathbf{p}_2| = |\mathbf{q}_1 - \mathbf{q}_2|$, $|\mathbf{p}_3 - \mathbf{p}_4| = |\mathbf{q}_3 - \mathbf{q}_4|$.

    a. Find an example where $(G, [\mathbf{q}_1; \mathbf{q}_2; \mathbf{q}_3; \mathbf{q}_4])$ is another convex arm, and the angles at $\mathbf{q}_2$ and $\mathbf{q}_3$ are not smaller than the angles at $\mathbf{p}_2$ and $\mathbf{p}_3$, respectively, but moving $\mathbf{p}_1$ to $\mathbf{q}_1$ or $\mathbf{p}_4$ to $\mathbf{q}_4$ separately does not give a convex arm.

    b. Find another example where the angles at $\mathbf{q}_2$ and $\mathbf{q}_3$ are not larger than the angles at $\mathbf{p}_2$ and $\mathbf{p}_3$, respectively, but $|\mathbf{q}_1 - \mathbf{q}_4| > |\mathbf{p}_1 - \mathbf{p}_4|$.

11. (Barvinok, 1995) Show that if a framework has $b < d(d+1)/2$ bars in $\mathbb{E}^d$, then it is a finite mechanism, given by affine motions, ending in $\mathbb{E}^{d-1}$.

12. Show that the frameworks of Figures 5.9 and 5.11, where all the nodes on each line are connected to each other, satisfy conditions (1) and (2) of Theorem 5.14.1 for $d = 3$, but not condition (3). The critical point is to show that the universal configuration is 3-dimensional.

6

Prestress Stability

6.1 Introduction

Here we start by adding ideas from Chapter 3 concerning first-order stiffness to the quadratic energy function of Chapter 5. We define a more general energy function that allows us to define when a structure is prestress stable. We shall see that in many structural contexts, prestress stiffness is actually what we want, allowing us to understand the structural response of both infinitesimally rigid structures and also those stiff structures that are underbraced, such as Snelson's tensegrity sculptures (Snelson, 2009).

6.2 A General Energy Function

To cover the widest range of situations, both physical and geometric, we here introduce a very general energy function. The central assumption is that the strain energy in a member is purely a function of the member's length l_{ij}, and that the total strain energy is found by summing the strain energy stored in the members. In fact, to make the subsequent development clearer, we choose to make the strain energy in a member a function of $l_{ij}^2/2$. We also assume that the function is appropriately differentiable.

Suppose that $(G, \mathbf{p})$ is a tensegrity, where $\mathbf{p} = [\mathbf{p}_1; \ldots; \mathbf{p}_n]$ is a configuration in $\mathbb{E}^d$. For each member $\{i, j\}$ of G, define an *energy function* E_{ij} that is a twice continuously differentiable, real-valued function defined for all non-negative real numbers. For any configuration $\mathbf{q} = [\mathbf{q}_1; \ldots; \mathbf{q}_n]$ in $\mathbb{E}^d$ define the *total energy* to be

$$E(\mathbf{q}) = \sum_{ij} E_{ij} \left(\frac{|\mathbf{q}_i - \mathbf{q}_j|^2}{2} \right), \tag{6.1}$$

where the sum is taken over all members $\{i, j\}$ of G. The energy of any given member $\{i, j\}$ at the configuration $\mathbf{q}$ is $E_{ij}(|\mathbf{q}_i - \mathbf{q}_j|^2/2)$.

As in Section 5.4, the derivative of E_{ij} with respect to the length of the member is the internal force carried by the member, and the derivative with respect to $l_{ij}^2/2$ is the force density.

$$\frac{dE_{ij}}{dl_{ij}} = t_{ij} \tag{6.2}$$

$$E'_{ij} = \frac{dE_{ij}}{d(l_{ij}^2/2)} = \omega_{ij} = \frac{t_{ij}}{l_{ij}} \tag{6.3}$$

Note that, unlike in Section 5.4, the force density in a member, and hence the stress ω, may vary with configuration.

We say that an energy function E for a tensegrity $(G, \mathbf{p})$ is *proper* if the first derivative of E_{ij}, i.e. the force density, is strictly positive when $\{i, j\}$ is a cable, strictly negative when $\{i, j\}$ is a strut, and at a unique minimum when $\{i, j\}$ is a bar. This means that when a cable, strut, or bar constraint is violated, there is a cost in terms of energy that has to be paid. Sometimes we may wish that these derivative conditions hold for all real values of each E_{ij}, but for most purposes it is enough simply to assume that they hold close to $\mathbf{p}$.

Consider the second derivative of E_{ij}, which we denote as b_{ij},

$$b_{ij} = E''_{ij} = \frac{d^2 E_{ij}}{d(l_{ij}^2/2)^2}. \tag{6.4}$$

We shall also assume that b_{ij} is strictly positive for all values in its domain, or at least that this holds locally to the configuration of interest. The effect of this is to ensure that, as a cable, strut, or bar constraint is increasingly violated, there is an increasing magnitude of force density in the member. Note that this was not true for the quadratic energy function introduced in 5.3. We shall see that having a positive second derivative effectively adds the first-order stiffness from Chapter 3 to the properties given by the quadratic energy function in Chapter 5.

We now introduce the key theorem that describes why energy functions can give rigidity results, through the "principle of least energy".

Theorem 6.2.1. *Suppose that E is a proper energy function for the tensegrity $(G, \mathbf{p})$ in $\mathbb{E}^d$ and that $\mathbf{p}$ is a strict local minimum point for E up to congruent copies. (In other words, if $E(\mathbf{q}) = E(\mathbf{p})$ and $\mathbf{q}$ is close enough to $\mathbf{p}$ as a configuration in $\mathbb{E}^d$, $\mathbf{q}$ and $\mathbf{p}$ are congruent.) Then $(G, \mathbf{p})$ is rigid in $\mathbb{E}^d$.*

Proof. The derivative conditions on each E_{ij} imply that they are strictly monotone increasing for a cable, strictly monotone decreasing for a strut, and at a strict minimum for a bar. So if $\mathbf{q}$ is close enough to $\mathbf{p}$ as a configuration in $\mathbb{E}^d$ and it satisfies the cable, strut, and bar constraints for a tensegrity, then each $E_{ij}(|\mathbf{q}_i - \mathbf{q}_j|^2/2) \leq E_{ij}(|\mathbf{p}_i - \mathbf{p}_j|^2/2)$, with strict inequality if the cable or strut condition is slack. So $E(\mathbf{q}) \leq E(\mathbf{p})$. Since $\mathbf{p}$ is a strict local minimum point for E up to congruent copies, we know that $\mathbf{q}$ and $\mathbf{p}$ are congruent. Thus $(G, \mathbf{p})$ is rigid in $\mathbb{E}^d$. □

6.2.1 The Physical Energy Function Revisited

In Section 5.2.1 we described an energy function for a member in terms of, at the current configuration $\mathbf{p}$, the tension carried by the member $t_{ij}(\mathbf{p})$, length $l_{ij}(\mathbf{p})$, and member stiffness $k_{ij}(\mathbf{p})$, as shown in Figure 5.2. The energy was described in (5.3):

$$E_f = \frac{k_{ij}(\mathbf{p})}{2} l_{ij}(\mathbf{q})^2 + \left(t_{ij}(\mathbf{p}) - k_{ij}(\mathbf{p})l_{ij}(\mathbf{p})\right) l_{ij}(\mathbf{q}) + E_0$$

If we write $\alpha = l_{ij}^2/2$, then for this case the energy function in (6.1) is

$$E_{ij}(\alpha) = k_{ij}(\mathbf{p})\alpha + \sqrt{2}(t_{ij}(\mathbf{p}) - k_{ij}(\mathbf{p})l_{ij}(\mathbf{p}))\alpha^{\frac{1}{2}},$$

and the first and second derivatives for a general local configuration $\mathbf{q}$ are

$$E'_{ij}(\alpha) = \frac{dE_{ij}}{d\alpha} = k_{ij}(\mathbf{p}) + \frac{\sqrt{2}}{2}(t_{ij}(\mathbf{p}) - k_{ij}(\mathbf{p})l_{ij}(\mathbf{p}))\alpha^{-\frac{1}{2}},$$

$$E''_{ij}(\alpha) = \frac{d^2 E_{ij}}{d\alpha^2} = -\frac{\sqrt{2}}{4}(t_{ij}(\mathbf{p}) - k_{ij}(\mathbf{p})l_{ij}(\mathbf{p}))\alpha^{-\frac{3}{2}},$$

which, evaluated at the current configuration $\mathbf{p}$, $\alpha = l_{ij}^2(\mathbf{p})/2$, $\alpha^{-\frac{1}{2}} = \sqrt{2}/l_{ij}(\mathbf{p})$, $\alpha^{-\frac{3}{2}} = 2\sqrt{2}/l_{ij}^3(\mathbf{p})$ gives

$$E'_{ij}(\alpha(\mathbf{p})) = \frac{t_{ij}(\mathbf{p})}{l_{ij}(\mathbf{p})} = \omega_{ij}(\mathbf{p}) \quad ; \quad E''_{ij}(\alpha(\mathbf{p})) = (k_{ij}(\mathbf{p}) - \omega_{ij}(\mathbf{p}))\frac{1}{l_{ij}^2(\mathbf{p})} = b_{ij}.$$

Thus, as noted above, the first derivative is the force density carried by the member. For the second derivative to be positive, we must have $k_{ij}(\mathbf{p}) > \omega_{ij}(\mathbf{p})$.

6.2.2 Equilibrium and Critical Configuration Condition

We now consider conditions on our proper energy function that will imply that E is at a strict local minimum at the configuration $\mathbf{p}$. First we consider the First Derivative Test as in multivariable calculus. We say that $\mathbf{p}$ is a *critical point* for the energy function E if $\nabla E_{\mathbf{p}} = \mathbf{0}$, where $\nabla E_{\mathbf{p}}$ is the gradient of E at the configuration $\mathbf{p}$. It is possible to calculate ∇E in terms of the standard coordinates in $\mathbb{E}^d$ for each of the nodes. But it is instructive to do the calculation of directional derivatives written as scalar multiples of infinitesimal displacements $\mathbf{d} = [\mathbf{d}_1; \ldots; \mathbf{d}_n]$, where each $\mathbf{d}_i$ is in $\mathbb{E}^d$. From basic calculus we know that $\nabla E_{\mathbf{p}}(\mathbf{d}) = \frac{d}{dt} E(\mathbf{p} + t\mathbf{d})|_{t=0}$ is the directional derivative of E in the direction $\mathbf{d}$ at the configuration $\mathbf{p}$. In the following calculation, the product of vectors is the inner product:

$$\frac{d}{dt} E(\mathbf{p} + t\mathbf{d}) = \sum_{ij} \frac{d}{dt} E_{ij} \left(\frac{((\mathbf{p}_i - \mathbf{p}_j) + t(\mathbf{d}_i - \mathbf{d}_j))^2}{2} \right)$$

$$= \sum_{ij} \frac{d}{dt} E_{ij} \left(\frac{(\mathbf{p}_i - \mathbf{p}_j)^2 + 2t(\mathbf{p}_i - \mathbf{p}_j)(\mathbf{d}_i - \mathbf{d}_j) + t^2(\mathbf{d}_i - \mathbf{d}_j)^2}{2} \right) \quad (6.5)$$

$$= \sum_{ij} E'_{ij} \left(\frac{(\mathbf{p}_i - \mathbf{p}_j)^2 + 2t(\mathbf{p}_i - \mathbf{p}_j)(\mathbf{d}_i - \mathbf{d}_j)}{2} + t^2(\mathbf{d}_i - \mathbf{d}_j)^2 \right)$$

$$\times \left((\mathbf{p}_i - \mathbf{p}_j)(\mathbf{d}_i - \mathbf{d}_j) + t(\mathbf{d}_i - \mathbf{d}_j)^2 \right),$$

where E'_{ij} is the derivative of E_{ij}. When we restrict to $t = 0$ this gives

$$\frac{d}{dt} E(\mathbf{p} + t\mathbf{d}) \bigg|_{t=0} = \sum_{ij} E'_{ij} \left(\frac{(\mathbf{p}_i - \mathbf{p}_j)^2}{2} \right) (\mathbf{p}_i - \mathbf{p}_j)(\mathbf{d}_i - \mathbf{d}_j)$$

$$= \sum_{ij} \omega_{ij} (\mathbf{p}_i - \mathbf{p}_j)(\mathbf{d}_i - \mathbf{d}_j) \quad (6.6)$$

$$= \omega R(\mathbf{p})\mathbf{d},$$

where $\omega_{ij} = E'_{ij}((\mathbf{p}_i - \mathbf{p}_j)^2)$ is the force density in member $\{i, j\}$, $\omega = [\ldots, \omega_{ij}, \ldots]$ is the stress, and $R = R(\mathbf{p})$ is the rigidity matrix defined in (3.10). Thus the configuration $\mathbf{p}$ is a critical point for the energy function E if $\omega = [\ldots, \omega_{ij}, \ldots]$ is a self-stress for $(G, \mathbf{p})$.

Note that 6.6 is only a necessary condition for the configuration to be a local minimum for the energy function E. The question remains as to whether E will be at a local minimum at $\mathbf{p}$. A natural condition to impose is given by second derivatives, which is as follows.

6.2.3 Second Derivatives

The idea here is to approximate the multivariable function E by a quadratic function. The constant part of the function is arbitrary and unimportant, and the linear parts are essentially given by Section 6.2.2. What remains is to calculate the function given by the quadratic terms, which is known as a *quadratic form*. With respect to the Euclidean coordinates, this is given by a symmetric matrix, the *Hessian* of E, often called the *tangent stiffness matrix*. The entries of this Hessian are the mixed partial derivatives of E with respect to the Euclidean coordinates. It is possible to calculate this Hessian directly, but we will continue our more coordinate-free method in terms of second derivatives. We choose the same infinitesimal displacements $\mathbf{d} = [\mathbf{d}_1; \ldots; \mathbf{d}_n]$, where each $\mathbf{d}_i$ is a vector in $\mathbb{E}^d$ as before. This time we continue the calculation of (6.5) to find the second derivative of the energy function E at the configuration $\mathbf{p}$:

$$\frac{d^2}{dt^2} E(\mathbf{p}+t\mathbf{d})|_{t=0} = \sum_{ij} E''_{ij} \left(\frac{(\mathbf{p}_i - \mathbf{p}_j)^2}{2} \right) [(\mathbf{p}_i - \mathbf{p}_j)(\mathbf{d}_i - \mathbf{d}_j)]^2$$

$$+ E'_{ij} \left(\frac{(\mathbf{p}_i - \mathbf{p}_j)^2}{2} \right) (\mathbf{d}_i - \mathbf{d}_j)^2 \qquad (6.7)$$

$$= \sum_{ij} b_{ij} [(\mathbf{p}_i - \mathbf{p}_j)(\mathbf{d}_i - \mathbf{d}_j)]^2 + \omega_{ij} (\mathbf{d}_i - \mathbf{d}_j)^2$$

We can write this in matrix form. If member $\{i, j\}$ is numbered as k (this dual labeling was mentioned in Section 3.2), and there are b members in G, define the matrix $\mathbf{B}$ as a b-by-b diagonal matrix whose kth diagonal entry is b_{ij}. Then (6.7) can be written as

$$\frac{d^2}{dt^2} E(\mathbf{p}+t\mathbf{d})|_{t=0} = \mathbf{d}^T \mathbf{R}(\mathbf{p})^T \mathbf{B} \mathbf{R}(\mathbf{p})\mathbf{d} + \mathbf{d}^T \mathbf{\Omega} \otimes \mathbf{I}^d \mathbf{d}$$

$$= \mathbf{d}^T \left(\mathbf{R}(\mathbf{p})^T \mathbf{B} \mathbf{R}(\mathbf{p}) + \mathbf{\Omega} \otimes \mathbf{I}^d \right) \mathbf{d} \qquad (6.8)$$

$$= \mathbf{d}^T \mathbf{K} \mathbf{d},$$

where

$$\mathbf{K} = \mathbf{R}(\mathbf{p})^T \mathbf{B} \mathbf{R}(\mathbf{p}) + \mathbf{\Omega} \otimes \mathbf{I}^d \qquad (6.9)$$

is the *tangent stiffness matrix* or *total stiffness matrix*. The matrix $\mathbf{\Omega} \otimes \mathbf{I}^d$ is the large stress matrix explained in Chapter 5.

The tangent stiffness matrix defined in (6.8) is the Hessian of the energy function E, which leads to the following result.

Theorem 6.2.2. *If a tensegrity $(G, \mathbf{p})$ in $\mathbb{E}^d$ has a proper energy function E such that the Hessian $\mathbf{K} = \mathbf{R}(\mathbf{p})^T \mathbf{B} \mathbf{R}(\mathbf{p}) + \mathbf{\Omega} \otimes \mathbf{I}^d$ is positive semi-definite with only rigid-body motions in its kernel, then $(G, \mathbf{p})$ is rigid in $\mathbb{E}^d$.*

Proof. By the Principle of Least Energy Theorem 6.2.1, it is enough to show that the energy function E is at a local minimum, modulo rigid-body motions, at the configuration $\mathbf{p}$. Since the space of configurations that are rigid-body motions form a smooth manifold, if there is a motion, satisfying the cable and strut constraints, that is not in that manifold, the motion can be taken initially to be orthogonal to it. This defines a displacement $\mathbf{d}$ that is not infinitesimally a rigid-body motion. The second derivative of E in that direction will be positive, contradicting the minimality of E near the configuration $\mathbf{p}$. Thus $(G, \mathbf{p})$ is rigid in $\mathbb{E}^d$. □

Note that the matrix $\mathbf{R}(\mathbf{p})^T \mathbf{B} \mathbf{R}(\mathbf{p})$ is always positive semi-definite, since we have assumed that the coefficients $b_{ij} > 0$. On the other hand the large stress matrix does not have to be positive semi-definite. However, although the large stress matrix could very well have some negative eigenvalues, full tangent stiffness matrix could still be positive semi-definite, with only the infinitesimal rigid-body motions in its kernel. With this in mind, we say that a tensegrity $(G, \mathbf{p})$ is *prestress stable* for an energy function E if $\mathbf{p}$ is a critical configuration for E and the Hessian $\mathbf{K} = \mathbf{R}(\mathbf{p})^\mathrm{T}\mathbf{B}\mathbf{R}(\mathbf{p}) + \mathbf{\Omega} \otimes \mathbf{I}^d$ is positive semi-definite with only rigid-body motions in its kernel.

From the geometric point of view, we say that a tensegrity $(G, \mathbf{p})$ is *prestressably stable* if there exists a proper energy function E such that $(G, \mathbf{p})$ is prestress stable with respect to that energy function. The point here is that the geometric tensegrity $(G, \mathbf{p})$ is given, and we must find a proper energy function that makes $(G, \mathbf{p})$ prestress stable. It is clear from the condition of Theorem 6.2.2 that we can choose the stiffness coefficients b_{ij} as large as we please, in the definition of the E_{ij} functions, but we have to limit the stress ω to be a proper self-stress.

Note that these definitions imply that if a tensegrity $(G, \mathbf{p})$ is infinitesimally rigid, then it is prestressably stable. The idea is that we can take any proper self-stress but scale it down sufficiently small so that it does not destroy the maximal positive definite property of the stiffness matrix. So the stress part and the stiffness part are positive definite on the space orthogonal to the infinitesimal rigid-body motions.

6.2.4 Stiffness Formulations

It is instructive to compare the formulation for the tangent stiffness matrix in (6.8) with the first-order stiffness described in Section 3.6. To do so, we will first reformulate the tangent stiffness matrix to match the description given by Guest (2006), which was derived from direct differentiation of equilibrium expressions at nodes. First, we write the rigidity matrix in terms of the equilibrium matrix, from (3.4):

$$\mathbf{A}\mathrm{diag}(\mathbf{l}) = \mathbf{R}^\mathrm{T},$$

(remembering that $\mathrm{diag}(\mathbf{l})$ is a diagonal matrix with member lengths l_{ij} along the diagonal) and substitute this into (6.9) to give

$$\mathbf{K} = \mathbf{A}^\mathrm{T}\mathrm{diag}(\mathbf{l}(\mathbf{p}))\mathbf{B}\mathrm{diag}(\mathbf{l}(\mathbf{p}))\mathbf{A} + \mathbf{\Omega} \otimes \mathbf{I}^d,$$

which is more concisely written

$$\mathbf{K} = \mathbf{A}^\mathrm{T}\hat{\mathbf{G}}\mathbf{A} + \mathbf{\Omega} \otimes \mathbf{I}^d, \tag{6.10}$$

where $\hat{\mathbf{G}}$ is a diagonal matrix with entries $b_{ij}l_{ij}^2$, and all matrices are implicitly defined at the configuration $\mathbf{p}$. In terms of the physical energy function, the diagonal terms of $\hat{\mathbf{G}}$ are given by $(k_{ij}(\mathbf{p}) - \omega_{ij}(\mathbf{p}))$.

Consider a bar framework that is unstressed (we here consider a bar framework rather than a tensegrity framework, because of the complexities introduced by the discontinuity in the slope of the internal force vs. length plot at $t_{ij} = \omega_{ij} = 0$ for a cable or a strut, as shown in Figure 4.2). In this case, $\mathbf{\Omega}$ is a zero matrix, and $\hat{\mathbf{G}} = \mathbf{G}$, the matrix defined in (3.6). Thus the stiffness matrix $\mathbf{K}$ equals the first-order stiffness matrix $\mathbf{K_m}$ defined in (3.24) when the matrix is unstressed.

Consider now if stress is gradually added to an unstressed framework (physically, this might be done by, for instance, tightening a turnbuckle). The first point to note is that we

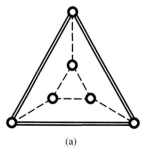
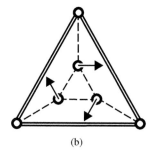

(a)   (b)

Figure 6.1 An under-braced tensegrity with one stress and one infinitesimal motion that is nevertheless prestress stable with any non-zero proper self-stress. A non-trivial infinitesimal flex is indicated in (b).

now have no difficulty working with a tensegrity framework if we have a proper stress that is non-zero for all struts and cables, as we move away from the discontinuity in the internal force vs. length behaviour. If we have $\omega_{ij} \ll k_{ij}$, then the matrix $\mathbf{B}$ (equivalently, $\hat{\mathbf{G}}$) will be little affected – but the behaviour of the structure (described by $\mathbf{K}$) may be significantly changed by the addition of the $\boldsymbol{\Omega} \otimes \mathbf{I}^d$ terms in (6.10) or (6.9). This change of behaviour for some classic tensegrities is described in Calladine (1978), and is described as being due to "product forces" in Pellegrino and Calladine (1986).

As the level of stress in a tensegrity changes, the stiffness varies with the key term being the dimensionless ratio ω_{ik}/k_{ij}. This is described for cable-nets (spider-webs) by Calladine (1982), and is explored for two example tensegrities in Guest (2011).

6.2.5 Examples

An Under-Braced Tensegrity

Consider the tensegrity shown in Figure 6.1(a), which is not infinitesimally rigid: the inner triangle can undergo the infinitesimal rotation $\mathbf{d}$ shown in Figure 6.1(b) without any member changing in length to first order. There is a self-stress for this configuration when the three lines connecting the inner triangle to the outer triangle meet at a point, with positive force densities for the interior cables, and negative for densities for the struts of the exterior triangle.

Since $\mathbf{d}$ is an infinitesimal motion, we know that the $\mathbf{R}^T\mathbf{BR}$ contribution to the stiffness in (6.9) is zero (the terms $b_{ij}[(\mathbf{p}_i - \mathbf{p}_j)(\mathbf{d}_i - \mathbf{d}_j)]^2$ in (6.7) are all zero for all $\{i, j\}$). Thus we look at the contribution from the large stress matrix $\boldsymbol{\Omega} \otimes \mathbf{I}^d$ (the terms $\omega_{ij}(\mathbf{d}_i - \mathbf{d}_j)^2$ in (6.7)). The terms $\omega_{ij}(\mathbf{d}_i - \mathbf{d}_j)^2$ are positive for each cable, whereas because the infinitesimal displacements $\mathbf{d}_i = \mathbf{0}$ for the outer triangle, the terms $\omega_{ij}(\mathbf{d}_i - \mathbf{d}_j)^2$ are zero for each strut. Thus the summation is positive, and $\mathbf{d}^T\mathbf{Kd} > 0$.

We can see that this calculation is enough to conclude that this tensegrity is prestress stable, because the infinitesimal motion $\mathbf{d}$ that we calculated is a basis for a complementary space to the infinitesimal motions of $(G, \mathbf{p})$. This is explained in Section 6.4.

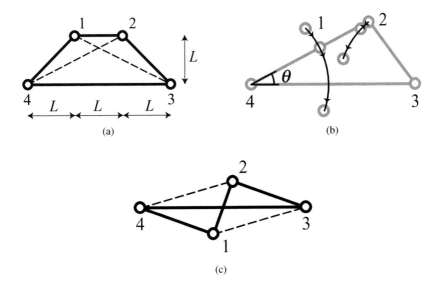

Figure 6.2 A tensegrity in $\mathbb{E}^2$ that forms a four bar-linkage made of bars with hard length constraints, braced with cables that are able to stretch. Three critical configurations are displayed. (a) The initial configuration. (b) The nodal path enforced by the bars, parameterized by the angle θ. The cables are not shown. (c) The final configuration. The energy E stored in the cables as a function of θ is shown in Figure 6.3.

A Braced Four-Bar Linkage

Consider the tensegrity shown in Figure 6.2(a). Clearly, with geometrically hard constraints, this tensegrity is rigid, and we can choose an energy function that will show that to be the case. Here, however, we will think about the structure in a physical sense, and consider how the behaviour can vary with different prestressing.

We will consider that the four outer bars of the tensegrity are inflexible. These four-bars alone form a "four-bar linkage", or a "four bar mechanism," a system with a single motion commonly used in mechanical engineering. The path that the nodes take for this motion are shown in Figure 6.2(b), parameterized by the angle θ shown. We then add two flexible cables, each of which have the "ideal" behaviour shown in Figure 4.2, giving an energy function E_{ij} shown in Figure 6.2(a). The resultant energy function for the entire structure E is shown in Figure 6.3(b) for two different cable choices, which have two different rest lengths.

For the choice, $l_0 = 2L$, E is at a minimum at the configuration shown in Figure 6.2(a). Thus the structure is stable here (and, from Theorem 6.2.1 is hence rigid for geometrically hard constraints). As the structure is displaced along the path shown in Figure 6.3(b), the energy reaches a peak, and then both cables become slack.

By contrast, for the choice, $l_0 = L$, E is at a maximum at the configuration shown in Figure 6.2(a). The energy decreases along the path shown Figure 6.2(b), and reaches a minimum for the configuration shown in Figure 6.2(c). Thus this configuration is rigid

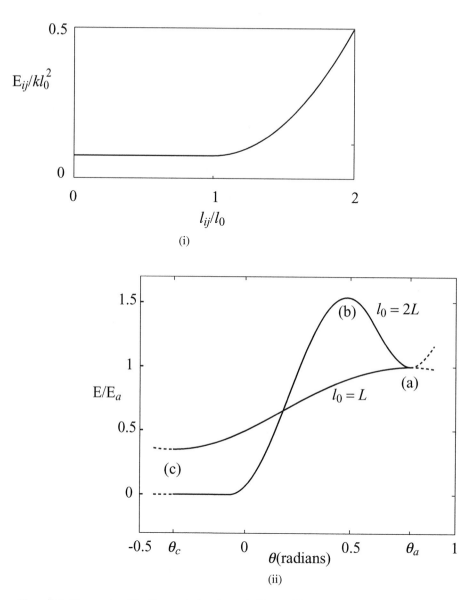

Figure 6.3 Energy stored by the tensegrity shown in Figure 6.2 as the configuration (defined by θ) changes. (i) The energy in a single cable, E_{ij} as a function of length l_{ij}. The "engineering" behavior shown for a cable in Figure 4.2 is assumed, and hence we have $E_{ij} = 0$ for $l_{ij} \leq l_0$ (cable is slack) and $E_{ij} = k(l - l_0)^2/2$ for $l_{ij} \geq l_0$ (cable follows Hooke's Law, with stiffness k). (ii) The total energy E stored in both cables, normalized with respect to the energy in the initial configuration E_a. Two graphs are shown, one with both cables having $l_0 = L$, the other with both having $l_0 = 2L$. For the case $l_0 = 2L$, cable $\{1, 3\}$ goes slack at a configuration between (a) and (b), and both cables go slack at a configuration between (b) and (c).

for geometrically hard constraints, and indeed is super stable by Proposition 5.14.2, as the tensegrity is now a Cauchy polygon.

It is instructive to consider the stiffness formulation described in Section 6.2.4 and equation (6.10) for the configuration shown in Figure 6.2(a). As we consider the outer bars to be inflexible, we only have to consider the (scalar) stiffness associated with a non-trivial flex of the four outer bars – for which we choose the particular deformation described by a unit vector $\mathbf{d}_m$ tangential to the path shown in Figure 6.2(b), where nodes 3 and 4 are fixed, initially node 1 moves down and right, and node 2 moves up and right. Consider the stiffness,

$$k_m = d_m^{\mathrm{T}}(\mathbf{A}^{\mathrm{T}}\hat{\mathbf{G}}\mathbf{A} + \mathbf{\Omega} \otimes \mathbf{I}^d)d_m. \tag{6.11}$$

If we assume the energy function shown in Figure 6.3, then the tension in the cables is given by $t_{ij}dE_{ij}/dL_{ij} = k(l - l_0)$ for $l_{ij} \geq l_0$, and the diagonal entry in the matrix $\hat{\mathbf{G}}$ associated with each cable has the value $\hat{g}_{ij} = k - \omega_{ij} = k - t_{ij}/l_{ij} = kl_0/l_{ij}$. For this particular structure, working through (6.11) shows that the structure has positive stiffness, i.e. is stable, for $l_0/L > 30\sqrt{5}/39 (\approx 1.72)$.

6.3 Quadratic Forms

We will need some simple basic facts about quadratic forms that come from symmetric matrices. This will greatly simplify the calculations that are involved in determining whether a tensegrity is prestressable. If $\mathbf{A}$ is a symmetric square N-by-N matrix, the *quadratic form* associated to $\mathbf{A}$ is Q_A, where $Q_A(\mathbf{p}) = \mathbf{p}^{\mathrm{T}}\mathbf{A}\mathbf{p}$ and $\mathbf{p}$ is an N-dimensional vector. So Q_A is a real-valued continuous function defined on all N-dimensional real vectors. If the symmetric matrix $\mathbf{A}$ is positive definite, this means that $Q_A(\mathbf{p}) > 0$ for all $\mathbf{p} \neq \mathbf{0}$, and of course $Q_A(\mathbf{0}) = 0$.

When $\mathbf{A}$ is positive semi-definite it means that $Q_A(\mathbf{p}) \geq 0$ for all $\mathbf{p}$ an N-dimensional vector. So this allows some non-zero vectors to have 0 value. But note that when $\mathbf{A}$ is positive semi-definite, the *zero set* of Q_A, $\{\mathbf{p} \mid Q_A(\mathbf{p}) = 0\} = \{\mathbf{p} \mid \mathbf{A}\mathbf{p} = 0\}$. In other words, the zero set of Q_A is the same as the nullspace of $\mathbf{A}$, which is a linear subspace of the space of N-dimensional vectors. This can be seen by using a coordinate system where $\mathbf{A}$ is a diagonal matrix. The zero set is seen to be an eigenspace corresponding to the eigenvalue 0. But beware that the zero set for an indefinite quadratic form, i.e. with positive and negative eigenvalues, is not a linear subspace.

The following lemma is helpful in determining when a given tensegrity $(G, \mathbf{p})$ prestressably stable. This is essentially a theorem using the continuity of quadratic forms.

Lemma 6.3.1. *Let $\mathbf{A}$ be a positive semi-definite symmetric square matrix and $\mathbf{B}$ a symmetric square matrix of the same size. There is a $t > 0$ such that $\mathbf{A} + t\mathbf{B}$ is positive definite if and only if $\mathbf{B}$ is positive definite when restricted to K, the nullspace of $\mathbf{A}$.*

Proof. Clearly, $\mathbf{B}$ must be positive definite on K, else we will have $Q_B(\mathbf{p}) \leq 0$ for some $\mathbf{p}$ in K, and $Q_{A+tB}(\mathbf{p}) = Q_A(\mathbf{p}) + tQ_B(\mathbf{p}) \leq 0$ for any $t > 0$.

Suppose that $\mathbf{B}$ is positive definite when restricted to K, which means that $Q_B(\mathbf{p}) > 0$ for every $\mathbf{p} \neq \mathbf{0}$ in K. Consider the unit sphere $\mathbb{S} = \{\mathbf{p} \mid \mathbf{p}^2 = 1\}$. Let N_δ for $\delta > 0$ be the set of vectors in $\mathbb{S}$ that are a distance less than δ to some point in $\mathbb{S} \cap K$. Since $\mathbb{S} \cap K$ is compact and by the continuity of the function Q_B, if δ is chosen small enough $Q_B(\mathbf{p}) > 0$ for every $\mathbf{p}$ in N_δ. The set of points in $\mathbb{S}$ but not in N_δ, $\mathbb{S} - N_\delta$, is also compact. So the function Q_B has its minimum value greater than $-c$ in $\mathbb{S} - N_\delta$, where $c > 0$. Also Q_A has its minimum value greater than $\epsilon > 0$ in $\mathbb{S} - N_\delta$. To recap:

if $\mathbf{p}$ is in N_δ, $Q_A(\mathbf{p}) \geq 0$ and $Q_B(\mathbf{p}) > 0$;
if $\mathbf{p}$ is in $\mathbb{S} - N_\delta$, $Q_A(\mathbf{p}) > \epsilon > 0$ and $Q_B(\mathbf{p}) \geq -c$.

We choose $t = \epsilon/c > 0$:

if $\mathbf{p}$ is in N_δ, $Q_A(\mathbf{p}) + t Q_B(\mathbf{p}) > 0$;
if $\mathbf{p}$ is in $\mathbb{S} - N_\delta$, $Q_A(\mathbf{p}) + t Q_B(\mathbf{p}) > \epsilon + (\epsilon/c)(-c) = 0$.

Thus $Q_{A+tB}(\mathbf{p}) = Q_A(\mathbf{p}) + t Q_B(\mathbf{p}) > 0$ for all $\mathbf{p}$ is in $\mathbb{S}$.

If $\mathbf{p} \neq \mathbf{0}$, then $\mathbf{p}/|\mathbf{p}|$ is in $\mathbb{S}$, and $(\mathbf{p}^T/|\mathbf{p}|)(A + tB)(\mathbf{p}/|\mathbf{p}|) = (1/\mathbf{p}^2)\mathbf{p}^T(A + tB)\mathbf{p} > 0$. Thus $\mathbf{p}^T(A + t\mathbf{B})\mathbf{p} > 0$ and $\mathbf{A} + t\mathbf{B}$ is positive definite. $\square$

We need to adapt Lemma 6.3.1 when the matrix $\mathbf{B}$ is not positive definite but only positive semi-definite when restricted to the kernel of the matrix $\mathbf{A}$.

Lemma 6.3.2. *Let $\mathbf{A}$ be a positive semi-definite symmetric square matrix with nullspace K_A and $\mathbf{B}$ a symmetric square matrix of the same size with nullspace K_B. Suppose there is a linear subspace $K_0 \subset K_B \cap K_A$. There is a $t > 0$ such that $\mathbf{A} + t\mathbf{B}$ is positive semi-definite with nullspace K_0 if and only if $\mathbf{B}$ is positive definite when restricted to any complementary subspace of K_0 in K_A.*

Proof. Note that any matrix is positive semi-definite with kernel K_0 if and only if it is positive definite when restricted to any subspace complementary to K_0. We can then take a basis for K_0, complete it to a basis for K_A using any basis for the complementary subspace of K_0 in K_A and then complete that basis to a basis for the whole space. We then apply Lemma 6.3.1 to $\mathbf{A}$ and $\mathbf{B}$ restricted to space spanned by this basis excluding the basis for K_0. $\square$

We provide here the definition of a complementary space. Consider two linear subspaces U and V of a linear space W. We say that U is *complementary* to V in W if, for every $\mathbf{w} \in W$, there is a unique $\mathbf{u} \in U$ and unique $\mathbf{v} \in V$ such that $\mathbf{u} + \mathbf{v} = \mathbf{w}$. Note that the definition implies that the only vector in both U and V is the zero vector. Note also that $\mathbf{u}$ and $\mathbf{v}$ do not have to be orthogonal. If we also require that every $\mathbf{u} \in U$ and every $\mathbf{v} \in V$ are orthogonal, then V is the unique *orthogonal complement* to U in W.

As an example, the possible complementary spaces to the x-y plane in $\mathbb{E}^3$ are any line through the origin that does not lie in the x-y plane.

6.4 Reducing the Calculation

Suppose we are given a tensegrity $(G, \mathbf{p})$ and we wish to determine whether there is a self-stress that stabilizes it. There are two things that are needed to determine this: the possible infinitesimal displacements K that might describe the destabilizing motion and the possible states of self-stress; and finding both of these was explained in Chapter 3.

Let K_0 be the rigid-body infinitesimal motions of $\mathbf{p}$, while K_1 is complementary subspace of K_0 in K so that $K = K_0 + K_1$ as a direct product. But how do you determine which self-stress stabilizes all possible infinitesimal motions? Suppose that $\Omega_1 \otimes \mathbf{I}^d, \dots, \Omega_k \otimes \mathbf{I}^d$ is a basis for the large stress matrices corresponding to the space of all equilibrium self-stresses of $(G, \mathbf{p})$. Restricting each $\Omega_i \otimes \mathbf{I}^d$ to K_1, we need to know if there is a positive linear combination $\sum_{i=1}^{k} s_i \Omega_i \otimes \mathbf{I}^d$, $s_i \geq 0$ that is positive definite on K_1. This shows the following.

Theorem 6.4.1. *For a tensegrity $(G, \mathbf{p})$ in $\mathbb{E}^d$, let K_1 be a linear subspace of the space of infinitesimal motions of $(G, \mathbf{p})$ complementary to the space of infinitesimal rigid-body motions. Then $(G, \mathbf{p})$ is prestress stable with respect to a proper equilibrium self-stress $\boldsymbol{\omega}$ if and only if $\mathbf{d}^T \Omega \mathbf{d} > 0$ for all $\mathbf{d} \neq \mathbf{0}$ in K_1, where Ω is the stress matrix corresponding to $\boldsymbol{\omega}$.*

By Lemma 6.3.2 we can choose any such convenient K_1, and there will be a stress $\boldsymbol{\omega}$ that stabilizes $(G, \mathbf{p})$ for positive stiffness coefficients if and only if there is such a linear combination.

The question is how hard is it to find a linear combination of quadratic forms (equivalently symmetric matrices) that is positive definite? If the size of the quadratic forms, the dimension of the space of infinitesimal flexes in our case, is one, then the calculation is easy. If the dimension of the quadratic forms, the dimension of the space of equilibrium stresses in our case, is one, then the calculation is easy. Even in the case when both dimensions are two, it is easy. The example in Figure 6.1 has only one rigid-body motion and only one stress. So the calculation that was done for that figure is sufficient to be able to conclude that it is prestress stable by Lemma 6.3.2. There is a literature that addresses the problem of finding a positive *semi-definite* (PSD) linear combination of quadratic forms numerically in Ramana (1997) and Vandenberghe and Boyd (1996), and very often it finds positive definite, not just PSD, linear combinations, using the ellipsoid method in convex programming. So it is possible that these methods could help determine whether a given framework is prestress stable. Note that there is the question of just how the configuration for the framework is defined, and this can muddy the waters as to the yes or no question of rigidity, not to mention whether the framework is prestress stable.

Note also that even when a rigidifying stress is found that provides the self-stress needed, the proof of the existence of the t in Lemma 6.3.1 is definitely non-constructive. Other techniques need to be applied to determine how small t has to be.

6.5 Second-Order Rigidity

For most structures we might wish to be rigid, prestress stability is probably what is needed: these are structures that can have a positive-definite stiffness matrix when correctly prestressed. However, there is another condition for rigidity that we will explore here; although it may not be a useful condition for a structure, it can provide information about when a structure is not rigid with the hard distance constraints that are part of the definition of a tensegrity.

Initially, consider an analytical $\mathbf{p}(t)$, where

$$\mathbf{p}(t) = \mathbf{p} + t\mathbf{p}' + \frac{t^2}{2}\mathbf{p}''$$

(we will see later in Section 6.5.2 that the $\mathbf{p}'$ and $\mathbf{p}''$ can be considered more generally, but the present assumption provides useful motivation for what follows). Note that we are here writing infinitesimal displacements as $\mathbf{p}' = [\mathbf{p}'_1; \ldots; \mathbf{p}'_n]$ instead of $\mathbf{d} = [\mathbf{d}_1; \ldots; \mathbf{d}_n]$, where each $\mathbf{p}'_i$ is a vector in $\mathbb{E}^d$.

We say that $\mathbf{p}'$ is a *first-order motion* of a tensegrity $(G, \mathbf{p})$ in $\mathbb{E}^d$ if it is an infinitesimal motion of $(G, \mathbf{p})$ in $\mathbb{E}^d$ as defined in Chapter 4. Suppose in addition that we have vectors $\mathbf{p}'' = [\mathbf{p}''_1; \ldots; \mathbf{p}''_n]$, where each $\mathbf{p}''_i$ is a vector in $\mathbb{E}^d$. We say that $[\mathbf{p}'; \mathbf{p}'']$ is a *second-order motion* of a tensegrity $(G, \mathbf{p})$ in $\mathbb{E}^d$ if $\mathbf{p}'$ is a first-order motion and if for every member $\{i, j\}$, where $(\mathbf{p}_i - \mathbf{p}_j) \cdot (\mathbf{p}'_i - \mathbf{p}'_j) = 0$ the following holds:

cable: For all $\{i, j\}$ cables, $(\mathbf{p}'_i - \mathbf{p}'_j)^2 + (\mathbf{p}_i - \mathbf{p}_j) \cdot (\mathbf{p}''_i - \mathbf{p}''_j) \leq 0$.
bar: For all $\{i, j\}$ bars, $(\mathbf{p}'_i - \mathbf{p}'_j)^2 + (\mathbf{p}_i - \mathbf{p}_j) \cdot (\mathbf{p}''_i - \mathbf{p}''_j) = 0$.
strut: For all $\{i, j\}$ struts, $(\mathbf{p}'_i - \mathbf{p}'_j)^2 + (\mathbf{p}_i - \mathbf{p}_j) \cdot (\mathbf{p}''_i - \mathbf{p}''_j) \geq 0$.

Notice that the inequalities in the definition of a second-order motion formally come from taking the second derivative of the squared distance formula with the appropriate inequalities on member lengths from Chapter 4.

If we have an analytic rigid-body motion $\mathbf{p}(t)$, for $t \geq 0$, of a configuration $\mathbf{p}$, where $\mathbf{p}(0) = \mathbf{p}$, we call $\frac{d}{dt}\mathbf{p}(t)|_{t=0} = \mathbf{p}'$ a *trivial first-order motion* of $\mathbf{p}$ in $\mathbb{E}^d$, and $[\mathbf{p}'; \mathbf{p}'']$ a *trivial second-order motion* of $\mathbf{p}$ in $\mathbb{E}^d$ if $\mathbf{p}'$ is a trivial first-order motion of $\mathbf{p}$ and $\frac{d^2}{dt^2}\mathbf{p}(t)|_{t=0} = \mathbf{p}''$.

We have to be careful with the definition of what it means to be non-trivial as a second-order motion. If we simply say that $[\mathbf{p}'; \mathbf{p}'']$ are not the first and second derivatives of a rigid-body motion, we have the problem that when $\mathbf{p}' = \mathbf{0}$, the conditions for $[\mathbf{p}'; \mathbf{p}'']$ to be a second-order motion of $(G, \mathbf{p})$ are essentially the same as the conditions for a first-order motion. Thus we say that a tensegrity $(G, \mathbf{p})$ in $\mathbb{E}^d$ is *second-order rigid* if every non-trivial first-order motion $\mathbf{p}'$ of $(G, \mathbf{p})$ in $\mathbb{E}^d$ does not extend to a second-order motion $[\mathbf{p}'; \mathbf{p}'']$ in $\mathbb{E}^d$. If there are no non-trivial first-order motions of $(G, \mathbf{p})$ in $\mathbb{E}^d$, then $(G, \mathbf{p})$ is first-order rigid in $\mathbb{E}^d$, and we simply say that, a fortiori, $(G, \mathbf{p})$ is second-order rigid in E^d as well.

6.5.1 Example of a Second-Order Rigid Framework

Figure 6.4 is an example $(G, \mathbf{p})$ of a bar framework that is second-order rigid. We can see that there is a first-order motion $\mathbf{p}'_1 = [0; 1]$. Since the framework not including node 1 is infinitesimally rigid, we will see later that this means that we can assume that $\mathbf{p}'_i = \mathbf{p}''_i = \mathbf{0}$ for all the nodes except 1. So we see that $\mathbf{p}'$ is non-trivial as a first-order motion of $(G, \mathbf{p})$

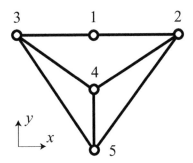

Figure 6.4 The framework of Figure 3.15, shown with a node-numbering scheme.

in the plane. But $\mathbf{p}'$ does not extend to a second-order motion, because of the second-order condition for the two bars at node 1. These condition are

$$0 = (\mathbf{p}'_1 - \mathbf{p}'_2)^2 + (\mathbf{p}_1 - \mathbf{p}_2)(\mathbf{p}''_1 - \mathbf{p}''_2) = 1 + (\mathbf{p}_1 - \mathbf{p}_2)\mathbf{p}''_1$$
$$0 = (\mathbf{p}'_1 - \mathbf{p}'_3)^2 + (\mathbf{p}_1 - \mathbf{p}_3)(\mathbf{p}''_1 - \mathbf{p}''_3) = 1 + (\mathbf{p}_1 - \mathbf{p}_3)\mathbf{p}''_1,$$

but the vectors $(\mathbf{p}_1 - \mathbf{p}_2)$ and $(\mathbf{p}_1 - \mathbf{p}_3)$ are collinear in opposite directions, so the only solution to these equations is when $\mathbf{p}''_1 = \mathbf{0}$, which means $\mathbf{p}'' = \mathbf{0}$. Thus $(G, \mathbf{p})$ is second-order rigid.

Note that it is also straightforward to show that this example if prestress stable.

6.5.2 Second-Order Rigidity Implies Rigidity

It is important to know that, when a tensegrity is second-order rigid, it is not a finite mechanism. So we show the following.

Theorem 6.5.1. *If a tensegrity $(G, \mathbf{p})$ is second-order rigid in $\mathbb{E}^d$, then it is rigid in $\mathbb{E}^d$.*

Proof. We use the analytic definition of rigidity. We need to show that if $\mathbf{p}(t)$ is an analytic motion of configurations in $\mathbb{E}^d$, where $\mathbf{p}(0) = \mathbf{p}$ and $\mathbf{p}(t)$ for $0 \le t \le 1$ satisfies the cable, bar, and strut constraints of Section 6.5, then there is a non-trivial first-order motion $\mathbf{p}'$ that extends to a second-order motion $[\mathbf{p}', \mathbf{p}'']$ of $(G, \mathbf{p})$. We would like to simply let $\mathbf{p}'$ and $\mathbf{p}''$ be the first and second derivatives of $\mathbf{p}(t)$ at $t = 0$, but we have the complication that the first derivative may be trivial, or even $\mathbf{0}$. On the other hand, some derivative of $\mathbf{p}(t)$ must be non-trivial since $\mathbf{p}(t)$ is a rigid-body motion and $\mathbf{p}(t)$ is analytic.

Consider that $\mathbf{p}(t)$ is an analytic motion, with derivatives

$$\mathbf{p}^{(k)} = \frac{d^k}{dt^k}\mathbf{p}(t)|_{t=0}, \text{ for } k = 1, 2, \ldots \text{ in } \mathbb{E}^{nd}.$$

If $\mathbf{p}^{(k)}$ is tangent to the space of rigid-body motions of $\mathbf{p}$ for k, that means it satisfies the conditions for being a trivial first-order motion. If it is tangent for all $k = 1, 2, \ldots$, since $\mathbf{p}(t)$ is analytic, that means that $\mathbf{p}(t)$ actually lies in the space of rigid-body motions for all real t. Thus we know that there is some smallest $k_0 = 1, 2, \ldots$ such that $\mathbf{p}^{(k_0)}$ is non-trivial as a first-order motion. By composing $\mathbf{p}(t)$ with a rigid-body motion of $\mathbb{E}^d$, can assume that $\mathbf{p}^{(k)} = \mathbf{0}$ for $k = 1, \ldots, k_0 - 1$. We define $\mathbf{p}' = \mathbf{p}^{(k_0)}$ and $\mathbf{p}'' = (2/\binom{2k_0}{k_0})\mathbf{p}^{(2k_0)}$.

By our construction $\mathbf{p}'$ is non-trivial as a first-order motion. We need to show that $\mathbf{p}''$ satisfies the conditions of Section 6.5. The Leibnitz Rule for differentiation gives

$$L_{ij}^{(k)} = \frac{d^k}{dt^k}(\mathbf{p}_i(t) - \mathbf{p}_j(t))^2|_{t=0} = \sum_{a=0}^{k} \binom{k}{a}(\mathbf{p}_i^{(a)} - \mathbf{p}_j^{(a)})(\mathbf{p}_i^{(k-a)} - \mathbf{p}_j^{(k-a)}),$$

for $k = 1, \ldots, 2k_0$, and $\{i, j\}$ a member of G. We know that $(\mathbf{p}_i^{(a)} - \mathbf{p}_j^{(a)}) = \mathbf{0}$, for $a = 1, \ldots, k_0 - 1$, and so all the terms of $L_{ij}^{(k)}$ are 0 for $k = k_0 + 1, \ldots, 2k_0 - 1$ and all $\{i, j\}$. If $\{i, j\}$ is a member of G, and $L_{ij}^{(k_0)}$ is not 0, it must be of the correct sign for the cable and strut conditions by the corresponding cable and strut conditions on $\mathbf{p}(t)$, and we do not have to check the condition for $\mathbf{p}''$. Conversely, if $L_{ij}^{(k_0)}$ is 0 and $\{i, j\}$ is a member of G, then there are only three non-zero terms in $L_{ij}^{(2k_0)}$. After combining the (equal) first and last terms, we calculate

$$L_{ij}^{(2k_0)} = 2(\mathbf{p}_i - \mathbf{p}_j)(\mathbf{p}_i^{(2k_0)} - \mathbf{p}_j^{(2k_0)}) + \binom{2k_0}{k_0}(\mathbf{p}_i^{(k_0)} - \mathbf{p}_j^{(k_0)})(\mathbf{p}_i^{(k_0)} - \mathbf{p}_j^{(k_0)}).$$

Since for those $\{i, j\}$, $L_{ij}^{(2k_0)}$ is the first non-zero derivative of $(\mathbf{p}_i(t) - \mathbf{p}_j(t))^2$ at $t = 0$, it must be the correct sign or 0, and thus $[\mathbf{p}', \mathbf{p}'']$ as defined is a second-order motion of $(G, \mathbf{p})$. Thus if $(G, \mathbf{p})$ is second-order rigid, it is rigid. $\qquad\square$

We will see later that if a tensegrity is prestress stable in $\mathbb{E}^d$, then it is second-order rigid, but the converse is not always true.

6.5.3 Stress Analysis

We would like to see how the condition for being second-order rigid relates to stresses and stress coefficients. For a tensegrity $(G, \mathbf{p})$ with b members and only cables and struts, let $\mathbf{S}$ be the b-by-b diagonal matrix with a 1 corresponding to cables and bars, and -1 corresponding to struts. In terms of the rigidity matrix given in Chapter 4 the first-order condition becomes

$$\mathbf{SR}(\mathbf{p})\mathbf{p}' \leq \mathbf{0}$$

and the second-order condition becomes, for the rigidity matrix restricted to those rows with a 0 in the first-order part,

$$\mathbf{SR}(\mathbf{p}')\mathbf{p}' + \mathbf{SR}(\mathbf{p})\mathbf{p}'' \leq \mathbf{0}.$$

With this notation, ω is a proper stress for G if and only if $\omega\mathbf{S} \geq \mathbf{0}$. So we see that requiring ω to be a proper self-stress for $(G, \mathbf{p})$ it is equivalent to the condition

$$\omega\mathbf{SSR}(\mathbf{p}) = \omega\mathbf{R}(\mathbf{p}) = \mathbf{0}, \text{ for } \omega\mathbf{S} \geq \mathbf{0}.$$

Thus we see that if $[\mathbf{p}', \mathbf{p}'']$ is a second-order flex for $(G, \mathbf{p})$ in $\mathbb{E}^d$, then

$$0 \geq \omega R(\mathbf{p}')\mathbf{p}' = \sum_{ij} \omega_{ij}(\mathbf{p}'_i - \mathbf{p}'_j)^2 = (\mathbf{p}')^{\mathrm{T}}\mathbf{\Omega} \otimes \mathbf{I}^d \mathbf{p}', \qquad (6.12)$$

where $\omega = [\ldots, \omega_{ij}, \ldots]$ is the proper self-stress and $\mathbf{\Omega} \otimes \mathbf{I}^d$ its corresponding large stress matrix.

Now we can relate prestress stability and second-order rigidity.

Theorem 6.5.2. *If a tensegrity $(G, \mathbf{p})$ is prestress stable in $\mathbb{E}^d$, then it is second-order rigid in $\mathbb{E}^d$.*

Proof. If $\mathbf{p}'$ is a non-trivial first-order motion of $(G, \mathbf{p})$ in $\mathbb{E}^d$ and ω is a stabilizing proper self-stress, then $(\mathbf{p}')^{\mathrm{T}}\mathbf{\Omega} \otimes \mathbf{I}^d \mathbf{p}' > 0$, where $\mathbf{\Omega}$ is the corresponding stress matrix for ω by Theorem 6.4.1. So if $\mathbf{p}'$ were to extend to a second-order motion in $\mathbb{E}^d$ it would contradict (6.12). Thus $(G, \mathbf{p})$ is second-order rigid. $\qquad \square$

6.6 Calculating Prestressability and Second-Order Rigidity

We describe here a systematic way to set up the calculations to determine when a given tensegrity, including a bar framework, is prestress stable and second-order rigid. For prestress stability, if we are given the prestress, it is a routine calculation to determine if the tensegrity is prestress stable with the given self-stress; the challenge is to find this stabilizing self-stress, when the dimension of the stress space is large. Similarly, if a first-order motion is given, it is routine to calculate when it extends to a second-order motion; the challenge is to find which, if any, first-order motion does extend.

The method in this section leads to a system of equations that can be of higher order than linear, and this could turn out to be intractable. (This is opposed to the case for prestress stability that seems to be numerically tractable.) Nevertheless, these equations tend to be far more computationally feasible than, say, working with the original quadratic distance constraints. In some reasonable cases when the number of self-stresses or first-order motions is small, these methods are computationally quite reasonable.

We are given a tensegrity $(G, \mathbf{p})$ in $\mathbb{E}^d$, a basis $\omega(1), \ldots, \omega(s)$ of proper equilibrium self-stresses for $(G, \mathbf{p})$ and a basis $\mathbf{p}'(1), \ldots, \mathbf{p}'(r)$ of first-order motions for $(G, \mathbf{p})$ for a space complementary to the rigid-body motions.

Any first-order motion that is not a rigid-body motion can be written as some combination of $\mathbf{p}'(1), \ldots, \mathbf{p}'(r)$,

$$\mathbf{p}' = \hat{d}(1)\mathbf{p}'(1) + \hat{d}(2)\mathbf{p}'(2) + \ldots + \hat{d}(r)\mathbf{p}'(r),$$

which we write as

$$\mathbf{p}' = \mathbf{M}\hat{\mathbf{d}},$$

where $\mathbf{M}$ is an nd-by-r dimensional vector of rigid-body motions,

$$\mathbf{M} = [\mathbf{p}'(1), \mathbf{p}'(2), \ldots, \mathbf{p}'(r)],$$

and $\hat{\mathbf{d}}$ is an r-dimensional vector of components,

$$\hat{\mathbf{d}} = [\hat{d}(1); \hat{d}(2); \ldots; \hat{d}(r)].$$

For each $\omega(k)$, $k = 1, \ldots, s$ we have the corresponding large stress matrix $\mathbf{\Omega}(k) \otimes \mathbf{I}^d$, which we write as the r-by-r matrix $\hat{\mathbf{\Omega}}(k)$ when restricted to the first-order motions that are not rigid-body motions,

$$\hat{\mathbf{\Omega}}(k) = \mathbf{M}^\mathrm{T}\left(\mathbf{\Omega}(k) \otimes \mathbf{I}^d\right)\mathbf{M}.$$

We can calculate the coefficients of $\hat{\mathbf{\Omega}}(k)$ with respect to the basis $\mathbf{p}'(1), \ldots, \mathbf{p}'(r)$ using the definitions for the stress matrix in Section 5.6. The a, b entry $a = 1, \ldots, r$, $b = 1, \ldots, r$ of $\hat{\mathbf{\Omega}}(k)$ is

$$\hat{\mathbf{\Omega}}(k)_{(a,b)} = \mathbf{p}'(a)^\mathrm{T}\mathbf{\Omega}(k)\mathbf{p}'(b) = \sum_{ij} \omega_{ij}(k)(\mathbf{p}'_i(a) - \mathbf{p}'_j(a))(\mathbf{p}'_i(b) - \mathbf{p}'_j(b)). \quad (6.13)$$

The tensegrity $(G, \mathbf{p})$ is prestressable if and only if there is a linear combination of the matrices $\sum_{k=1}^{s} \alpha_k \hat{\mathbf{\Omega}}(k)$ which is positive definite, where the choice of weight α_k is restricted to ensure that $\sum_{k=1}^{s} \alpha_k \omega(K)$ is a proper stress.

To determine if the tensegrity $(G, \mathbf{p})$ is second-order rigid, calculate if there is a vector $\mathbf{p}'$ that serves as a common 0 for all the matrices $\mathbf{\Omega}(1), \ldots, \mathbf{\Omega}(s)$, in other words $(\mathbf{p}')^\mathrm{T}\mathbf{\Omega}(1)\mathbf{p}' = \cdots = (\mathbf{p}')^\mathrm{T}\mathbf{\Omega}(s)\mathbf{p}' = 0$. If there is no such $\mathbf{p}' \neq \mathbf{0}$, then $(G, \mathbf{p})$ is second-order rigid.

In Section 6.6.1 we apply this calculation to an example of bar framework in the plane that is second-order rigid, but not prestress stable.

6.6.1 Second-Order Rigidity Does Not Imply Prestress Stability

For a tensegrity, second-order rigid and prestressability are not always the same condition. Consider the bar framework shown in Figure 6.5.

For this framework, there are two independent self-stresses:

(i) $\omega(1)$ in the top bars, with $\omega_{12}(1) = 1/2$, $\omega_{23}(1) = 1$ and $\omega_{34}(1) = -1/2$ and all other bars unstressed;

(ii) $\omega(2)$ in the bottom horizontal bars with $\omega_{56}(1) = 1$, $\omega_{67}(1) = 1/2$ and $\omega_{78}(1) = -1$ and all other bars unstressed.

The two non-horizontal bars have 0 stress. There are two independent first-order motions, indicated in the figure. We apply the formula (6.13) to calculate the two restricted stress matrices:

$$\hat{\mathbf{\Omega}}(1) = \begin{bmatrix} 1 + \frac{1}{2} & 1 \\ 1 & 1 - \frac{1}{2} \end{bmatrix} = \frac{1}{2}\begin{bmatrix} 3 & 2 \\ 2 & -1 \end{bmatrix}, \quad \hat{\mathbf{\Omega}}(2) = \begin{bmatrix} 1 + \frac{1}{2} & \frac{1}{2} \\ \frac{1}{2} & -1 + \frac{1}{2} \end{bmatrix} = \frac{1}{2}\begin{bmatrix} 3 & 1 \\ 1 & -1 \end{bmatrix}.$$

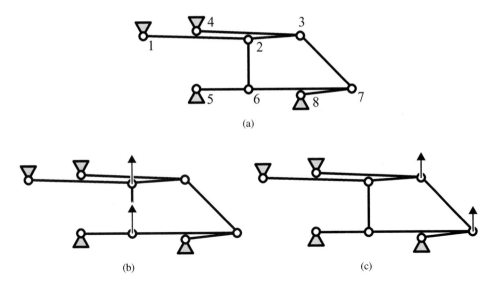

Figure 6.5 (a) A framework that is second-order rigid but not prestress stable, shown on page 50 of Kuznetsov (1991). All nodes lie on the corners of a square grid, and hence all bars should in fact be horizontal, but are shown slightly displaced so the connectivity is clear. (b) The first-order motion $\mathbf{p}'(1)$. (c) The first-order motion $\mathbf{p}'(2)$.

Note that each of $\hat{\boldsymbol{\Omega}}(1)$ and $\hat{\boldsymbol{\Omega}}(2)$ is semi-definite. For some combination

$$
\det(\alpha_1 \hat{\boldsymbol{\Omega}}(1) + \alpha_2 \hat{\boldsymbol{\Omega}}(2)) = \frac{1}{4} \det \begin{bmatrix} 3\alpha_1 + 3\alpha_2 & 2\alpha_1 + \alpha_2 \\ 2\alpha_1 + \alpha_2 & -\alpha_1 - \alpha_2 \end{bmatrix}
$$

$$
= \frac{1}{4}\left(-3(\alpha_1 + \alpha_2)^2 - (2\alpha_1 + \alpha_2)^2 \right) < 0.
$$

(6.14)

Thus $\alpha_1 \hat{\boldsymbol{\Omega}}(1) + \alpha_2 \hat{\boldsymbol{\Omega}}(2)$ is never singular for α_1 and α_2 not both 0, and is therefore indefinite: there is no self-stress for $(G, \mathbf{p})$ such that it is prestress stable for that stress.

On the other hand, it is easy to see that $(G, \mathbf{p})$ is second-order rigid. This is because there is no common zero set for the quadratic forms defined by $\hat{\boldsymbol{\Omega}}(1)$ and $\hat{\boldsymbol{\Omega}}(2)$. In other words, for any non-trivial first-order deformation $\mathbf{p}'$ of $(G, \mathbf{p})$, either $(\mathbf{p}')^{\mathrm{T}}\hat{\boldsymbol{\Omega}}(1)\mathbf{p}' \neq 0$ or $(\mathbf{p}')^{\mathrm{T}}\hat{\boldsymbol{\Omega}}(2)\mathbf{p}' \neq 0$. This is easy to check directly, but it also follows when all non-zero linear combinations of the two-by-two symmetric matrices are non-singular, as is the case from (6.14).

6.7 Second-Order Duality

Suppose we have a first-order motion $\mathbf{p}'$ for a tensegrity $(G, \mathbf{p})$ and we wish to know whether we can extend it to a second-order motion $[\mathbf{p}', \mathbf{p}'']$. One answer is just to solve the linear programming feasibility problem of Section 6.5. We have seen that self-stresses are relevant to the problem of whether $(G, \mathbf{p})$ is prestress stable, and it is natural to relate information about self-stresses to information about second-order rigidity.

More basically, there is the question of whether one would settle for a structure that was second-order rigid, but not prestressably rigid. This kind of structure, such as in Section 6.6.1 has the peculiar property that no particular prestress stiffens it. For the structure to have stiffness, it would have to know beforehand what the external loads will be and then adjust the prestress to compensate. This seems impracticable to build. On the other hand, if the structure is a finite mechanism, especially for a tensegrity with cables and struts that go slack at the second-order, the second-order motion can be very useful to know, or even to know that a second-order motion exists without an explicit construction. In this section we discuss how this second-order information can be obtained.

In order to make the duality statement for second-order rigidity in terms of the prestresses in the tensegrity, we need a somewhat different version of the Farkas Alternative Lemma 4.6.2. This can be found in Connelly and Whiteley (1996). This is a particularly sharp version that will be useful.

Lemma 6.7.1. *Let $\mathbf{A}_0$ and $\mathbf{A}_1$ be two matrices with n columns, and m_0 and m_1 rows, respectively, and $\mathbf{b}_0 \in \mathbb{E}^{m_0}, \mathbf{b}_1 \in \mathbb{E}^{m_1}$. Then there is a vector $\mathbf{x}$ in $\mathbb{E}^n$ such that*

$$\mathbf{A}_0 \mathbf{x} = \mathbf{b}_0$$
$$\mathbf{A}_1 \mathbf{x} < \mathbf{b}_1,$$

if and only if for all vectors $\mathbf{y}_0 \in \mathbb{E}^{m_0}, \mathbf{y}_1 \in \mathbb{E}^{m_1}$ such that when

$$\mathbf{y}_0^T \mathbf{A}_0 + \mathbf{y}_1^T \mathbf{A}_1 = \mathbf{0}$$
$$\mathbf{y}_1 \geq \mathbf{0},$$

then $\mathbf{y}_0^T \mathbf{b}_0 + \mathbf{y}_1^T \mathbf{b}_1 \leq 0$ with equality if and only if $\mathbf{y}_1 = \mathbf{0}$.

We next apply this lemma to the second-order situation. We say that $\mathbf{p}'$ or $\mathbf{p}''$ is *strict* on a member $\{i, j\}$ if the first-order or second-order inequality constraint, respectively, for $\{i, j\}$ is strict, i.e. not zero.

Theorem 6.7.2. *A first-order motion $\mathbf{p}'$ of a tensegrity $(G, \mathbf{p})$ in $\mathbb{E}^d$ extends to a second-order motion $[\mathbf{p}', \mathbf{p}'']$ if and only if for every proper self-stress $\boldsymbol{\omega}$ and associated stress matrix $\boldsymbol{\Omega}$, $(\mathbf{p}')^T \boldsymbol{\Omega} \otimes \mathbf{I}^d \mathbf{p}' \leq 0$. Furthermore, $\mathbf{p}''$ can be chosen to be strict on each cable and strut $\{i, j\}$ where $\mathbf{p}'$ is not strict if and only if for all proper self-stresses $\boldsymbol{\omega}$, $(\mathbf{p}')^T \boldsymbol{\Omega} \otimes \mathbf{I}^d \mathbf{p}' = 0$ implies $\omega_{ij} = 0$.*

Proof. We only consider those members of G where $\mathbf{p}'$ is not strict, since when $\mathbf{p}'$ is strict on a member $\{i, j\}$, it does not enter in to the calculation for $\mathbf{p}''$. Let $\mathbf{R}_0$ consists of those rows of the rigidity matrix corresponding to the bars of G and let $\mathbf{SR}_1$ consist of those rows of the rigidity matrix $\mathbf{SR}$ corresponding to the cables and struts of G, with the strut rows multiplied by -1, keeping in mind that the matrix $\mathbf{R}$ can be defined for any configuration. Similarly, let $\boldsymbol{\omega}_0$ and $\boldsymbol{\omega}_1$ correspond to the column vectors of stresses of bars and non-bars, respectively. Then we apply Lemma 6.7.1 to the following:

$$\mathbf{A}_0 = \mathbf{R}_0(\mathbf{p}) \qquad\qquad \mathbf{A}_1 = \mathbf{R}_1(\mathbf{p})$$
$$\mathbf{b}_0 = -\mathbf{R}_0(\mathbf{p}')\mathbf{p}' \qquad\qquad \mathbf{b}_1 = -\mathbf{SR}_1(\mathbf{p}')\mathbf{p}'$$
$$\mathbf{y}_0 = \boldsymbol{\omega}_0 \qquad\qquad \mathbf{y}_1 = \mathbf{S}\boldsymbol{\omega}_1.$$

We let the vector $\mathbf{p}''$ be $\mathbf{x}$ in the conclusion of Lemma 6.7.1, and this is what is desired. □

The idea is that each proper stress ω "blocks" a second-order motion on a member $\{i, j\}$, in that it prevents the second-order motion from being strict on that member if $\omega_{ij} \neq 0$ and $(\mathbf{p}')^{\mathrm{T}}\Omega \otimes \mathbf{I}^d \mathbf{p}' \geq 0$. In other words, when $\omega_{ij} \neq 0$, and $(\mathbf{p}_i - \mathbf{p}_j)(\mathbf{p}'_i - \mathbf{p}'_j) = 0$, then $(\mathbf{p}'_i - \mathbf{p}'_j)^2 + (\mathbf{p}_i - \mathbf{p}_j)(\mathbf{p}''_i - \mathbf{p}''_j) = 0$. But more to the point, if the second-order motion is not blocked in this way for any proper self-stress, then when $(\mathbf{p}_i - \mathbf{p}_j)(\mathbf{p}'_i - \mathbf{p}'_j) = 0$, $(\mathbf{p}'_i - \mathbf{p}'_j)^2 + (\mathbf{p}_i - \mathbf{p}_j)(\mathbf{p}''_i - \mathbf{p}''_j) \neq 0$.

A useful special case is the following.

Corollary 6.7.3. *Suppose that a tensegrity $(G, \mathbf{p})$ in $\mathbb{E}^d$ is such that for all proper stresses ω of $(G, \mathbf{p})$, the associated stress matrix Ω is negative semi-definite. Then every first-order motion $\mathbf{p}'$ of $(G, \mathbf{p})$ extends to a second-order motion $[\mathbf{p}', \mathbf{p}'']$ of $(G, \mathbf{p})$ in $\mathbb{E}^d$ that is strict (at either the first- or second-order) on each cable and strut.*

Note that this means that, assuming all proper stresses have Ω negative semi-definite, when $(G, \mathbf{p})$ has first-order motion in $\mathbb{E}^d$, then it extends to a second-order motion in $\mathbb{E}^d$ that is strict on each cable and strut. This is often enough to actually know that there is a finite motion such that all the cables and struts go slack, as in the following.

Corollary 6.7.4. *Suppose that a tensegrity $(G, \mathbf{p})$ in $\mathbb{E}^d$ is such that for all proper stresses ω of $(G, \mathbf{p})$, the associated stress matrix Ω is negative semi-definite and that $G0$ has only the $\mathbf{0}$ self-stress, where G_0 is the subgraph of bars of G. Then if $(G, \mathbf{p})$ is not first-order rigid, there is finite motion $\mathbf{p}(t)$, $t \geq 0$ of $(G, \mathbf{p})$ such that each cable and strut length changes for $t > 0$.*

Proof. By Corollary 6.7.3 there is a second-order motion $[\mathbf{p}', \mathbf{p}'']$ of $(G, \mathbf{p})$ in $\mathbb{E}^d$ such that $\mathbf{p}'$ is not a rigid-body motion. So $\mathbf{p}(t) = \mathbf{p} + t\mathbf{p}' + (t^2/2)\mathbf{p}''$ is a motion of the configuration that is strict at the first- or second-order, but we have to adjust for the bars of $(G, \mathbf{p})$. Since $G_0(\mathbf{p})$ has only the $\mathbf{0}$ self-stress, it means that the set in configuration space $M = \{\mathbf{q} \mid (\mathbf{q}_i - \mathbf{q}_j)^2 = (\mathbf{p}_i - \mathbf{p}_j)^2$, for all $\{i, j\}$ a bar$\}$ is a smooth manifold and $\mathbf{p}'$ is tangent to it. Let $T(M)$ be the tangent space to M in E^{nd}, where n is the number of nodes of G, and let $\exp: T(M) \to M$ be the standard exponential map. By adjusting $\mathbf{p}''$ so that it still remains in $T(M)$, we can arrange it so that $\exp(\mathbf{p}(t))$, for $t \geq 0$ is the motion desired. (See Connelly and Whiteley (1996) for details. The curvature of M has to be taken into account for this argument.) □

Specializing still further, we have the following, which solves a conjecture in Roth (1980).

Corollary 6.7.5. *Suppose $\mathbf{p}$ is a configuration that forms the vertices of a convex polygon P in the plane and $(G, \mathbf{p})$ is a tensegrity with the edges of P as bars of G,*

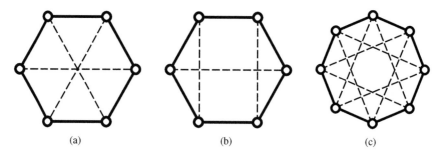

Figure 6.6 Three tensegrities that form regular polygons in the plane, with bars on the outside and cables inside as indicated. Examples (a) and (b) have too few members to be first-order rigid. Example (c) has enough members to be first-order rigid, but nevertheless there is still a first-order motion – see Exercise 6.9. So by Corollary 6.7.5 all these tensegrities are finite mechanisms. Note that the cables go slack at the second-order.

and cables as some of the interior diagonals. The tensegrity $(G, \mathbf{p})$ is rigid in $\mathbb{E}^2$ if and only if it is first-order rigid in $\mathbb{E}^2$.

Proof. Clearly if $(G, \mathbf{p})$ is first-order rigid in $\mathbb{E}^2$ it is rigid in $\mathbb{E}^2$ by our Theorem 3.8.1 of Chapter 3. The subgraph of the bars of $(G, \mathbf{p})$ have no non-zero self-stress, and by Theorem 5.14.4 of Chapter 5 any non-zero proper self-stress ω of $(G, \mathbf{p})$ has its stress matrix Ω negative semi-definite. So Corollary 6.7.3 applies, and thus any non-trivial first-order motion $\mathbf{p}'$ of $(G, \mathbf{p})$ extends to a strict finite motion where the cables all go slack for $0 < t$. ☐

Figure 6.6 shows flexible examples of the kind of tensegrity referred to in Corollary 6.7.5.

A useful consequence of the duality of Theorem 6.7.2 is the following:

Corollary 6.7.6. *If a bar framework $(G, \mathbf{p})$ in $\mathbb{E}^d$ is second-order rigid with either a one-dimensional space of equilibrium self-stresses or a 1-dimensional space of non-trivial infinitesimal flexes, then $(G, \mathbf{p})$ is prestress stable in $\mathbb{E}^d$.*

Another consequence of these ideas is the following:

Corollary 6.7.7. *Suppose that a bar framework $(G, \mathbf{p})$ in $\mathbb{E}^d$ is prestress stable in $\mathbb{E}^d \subset \mathbb{E}^D$ for all $D \geq d$. Then $(G, \mathbf{p})$ is super stable.*

See Exercise 6.9 for an application of Corollary 6.7.7.

6.8 Triangulated Spheres

Dehn's Theorem 3.9.4 and Alexandrov's Theorem 3.9.8 are concerned with triangulations of the surface of a convex polyhedral sphere, but there are restrictions on where the vertices of the triangulation can be placed. Namely, vertices are not permitted to be in the relative interior of any of the faces, since if any vertex is in the interior of any face, there is a non-trivial infinitesimal flex, and the bar framework would not be infinitesimally rigid in $\mathbb{E}^3$.

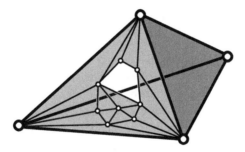

Figure 6.7 A tetrahedron with a quadrilateral hole removed from one triangular face. This is prestress stable in $\mathbb{E}^3$.

On the other hand, if one does triangulate a polytope with vertices in the interior of faces anyway, it should still be rigid. The following result in Connelly (1980) is one possible answer.

Theorem 6.8.1. *Any triangulation of the boundary of a convex polytope in $\mathbb{E}^3$ is second-order rigid.*

It is possible to do better, as mentioned in Connelly (1993). From the relative interior of each face of the polytope P remove the interior of a finite number of convex polygonal disks (the holes) where the boundaries of the holes are disjoint and in the interior of the faces of P. Assume that there is a convex polytope P_F sitting over each face F of P projecting down onto F orthogonally, such that each hole is a projection of one of the faces of P_F. Then any triangulation of P minus the holes is second-order rigid. Furthermore, this same triangulation is prestress stable. Figure 6.7 shows an example.

Notice that the holes can be quite large, occupying most of the area of each face of P. This points out that the rigidity of a polytope, regarded as a surface, is really concentrated in the neighbourhood of the edges of P.

Notice, also, that the signs of the stress that rigidifies the framework do not always have to be positive, but there always is a particular triangulation of each face with holes with the stress on the internal edges that is positive using the Maxwell–Cremona correspondence Theorem 3.5.2.

We also point out that a more recent result Connelly and Gortler (2017) is a strict improvement to Theorem 6.8.1 where second-order is replaced by prestress stability.

Theorem 6.8.2. *Any triangulation of the boundary of a convex polytope in $\mathbb{E}^3$ is prestress stable.*

This result also extends to the case when holes are appropriately placed in the faces of the polytope, and there is no restriction on how the resulting surface is triangulated. This idea of triangulating a surface with holes, and when there is a loss of tensile self-stress, is also discussed in Connelly et al. (2001) and Menshikov et al. (2002).

6.9 Exercises

1. Find an infinitesimal flex for each of the tensegrities shown in Figure 6.6.

2. In Section 8.8.3 a finite mechanism is described. There, a possible definition of third-order rigidity is given. Show that this cusp mechanism satisfies that definition of third-order rigidity, and yet it is flexible, i.e. a finite mechanism.

3. Consider the convex hexagon tensegrities as in Figure 6.8, where the vertices and external bars are the same, but the one internal cable is replaced by another. Show that not both tensegrities are rigid.

4. Suppose that $\mathbf{p} = ((\mathbf{p}_1, \mathbf{0}); \ldots; (\mathbf{p}_n, \mathbf{0}))$ is a configuration in $\mathbb{E}^d \times \mathbf{0}$ and $\mathbf{q} = ((\mathbf{0}, \mathbf{q}_1); \ldots; (\mathbf{0}), \mathbf{q}_n)$ is a configuration in $\mathbf{0} \times \mathbb{E}^d$ with the same number of corresponding vertices. Show that there is a smooth motion $\mathbf{p}(t)$ of configurations in $\mathbb{E}^d \times \mathbb{E}^d$ such that $\mathbf{p}(0) = \mathbf{p}$, $\mathbf{p}(1) = \mathbf{q}$, and for each $i, j = 1, 2, \ldots, n$, $|\mathbf{p}_i(t) - \mathbf{p}_j(t)|$ is a monotone function of t, for $0 \le t \le 1$. Hint: Use sine and cosine functions.

5. Suppose that $\mathbf{p} = ((\mathbf{p}_1, \mathbf{0}); \ldots; (\mathbf{p}_n, \mathbf{0}))$ and $\mathbf{q} = ((\mathbf{q}_1, \mathbf{0}); \ldots; (\mathbf{q}_n, \mathbf{0}))$ are two configurations in $\mathbb{E}^d \times \mathbf{0} \subset \mathbb{E}^d \times \mathbb{E}^d$. Show that $\mathbf{p}' = -\mathbf{p}$ and $\mathbf{p}'' = (\mathbf{0}, \mathbf{q}_1), \ldots, (\mathbf{0}, \mathbf{q}_n))$ is a second-order flex of a bar framework $(G, \mathbf{p})$ if and only if $(G, \mathbf{p})$ and $(G, \mathbf{q})$ have corresponding bar lengths the same. Furthermore, $\mathbf{p}$ is congruent to $\mathbf{q}$ if and only if $(\mathbf{p}', \mathbf{p}'')$ is a trivial second-order flex.

6. Define $(G, \mathbf{p})$ a bar framework in $\mathbb{E}^d$ to be *dimensionally rigid* if any equivalent framework $(G, \mathbf{q})$ (having corresponding bars the same length) in $\mathbb{E}^D$, for $D \ge d$, is such that the affine span of the vertices of $\mathbf{q}$ is at most d-dimensional. Suppose $(G, \mathbf{p})$ is rigid in $\mathbb{E}^d$. Then $(G, \mathbf{p})$ is dimensionally rigid if and only if it is universally rigid. (See Alfakih (2007) for more information.)

7. Show that if $(G, \mathbf{p})$, a bar framework in $\mathbb{E}^d$, is dimensionally rigid (see the previous problem), then any equivalent framework $(G, \mathbf{q})$ is an affine image of $(G, \mathbf{p})$. (See Alfakih (2007) and Connelly and Gortler (2015), Theorem 5.1.)

8. Consider the triangle with tabs, a planar bar framework, in Figure 6.9.

   a. Show that Figure 6.9 is second-order rigid and therefore rigid in $\mathbb{E}^3$ but not prestress stable in $\mathbb{E}^3$, while it is a finite mechanism in $\mathbb{E}^4$.

   b. Figure 6.10 shows a square with tabs in the plane. Show that it is a finite mechanism in $\mathbb{E}^3$ and indeed folds into the base square in the middle in several different ways.

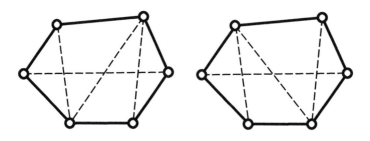

Figure 6.8

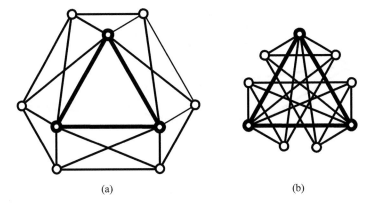

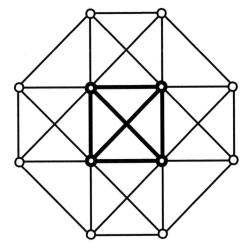

Figure 6.9 (a) A triangle with a rigid rectangle attached on each edge. (b) The same triangle with the rectangles, the tabs, folded over. Both are rigid in three-space, but not four-space.

Figure 6.10 This is similar to the framework in Figure 6.9, except that now the base is a rigid square.

9. Consider a chain of triangles in $\mathbb{E}^2$ such as those in Figure 6.11, where the line segment from one node to the last goes through the interior of all the triangles in the chain. It is clear that this *triangle-chain* bar framework is universally globally rigid, i.e. it is rigid in all higher dimensions.

   a. Show that a triangle-chain bar framework $(G, \mathbf{p})$ is second-order rigid in all $\mathbb{E}^D$, for $D \geq d$.
   b. Show that any triangle-chain framework $(G, \mathbf{p})$ is super stable, and determine the signs of the stresses for the framework in Figure 6.11.

10. It is possible to extend the result in Theorem 6.8.1 to the case when a hole touches the boundary of the face of the polytope, but only at one vertex. In $\mathbb{E}^3$ find a triangulation

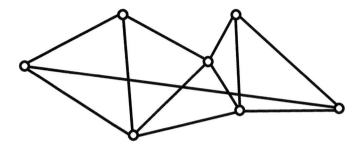

Figure 6.11

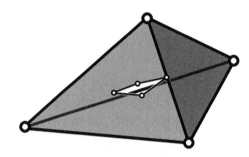

Figure 6.12

of the surface of the tetrahedron with a slit, as shown in Figure 6.12, that flexes. That is, it is a finite mechanism.

11. A finite graph G is called *d-realizable* if for every $D \geq d$ and every realization $(G, \mathbf{p})$ in $\mathbb{E}^D$, there is another configuration $\mathbf{q}$ in $\mathbb{E}^d$ such that $(G, \mathbf{p})$ is equivalent to $(G, \mathbf{q})$. In other words, for each bar $\{i, j\}$ of G, $|\mathbf{p}_i - \mathbf{p}_j| = |\mathbf{q}_i - \mathbf{q}_j|$. It is easy to see that the complete graph K_5 is not 3-realizable. Show that the edge graph of the regular octahedron is not 3-realizable. (For a characterization of 2-realizable and 3-realizable graphs, see Belk and Connelly (2007) and Belk (2007)).

7

Generic Frameworks

7.1 Introduction

Often a structure has the property (or one assumes that it has the property) that there is nothing "special" about the configuration of its nodes. For instance, there are no planes of reflection symmetry, no collinearities, and no four nodes that lie on a circle. For example, if we are interested in infinitesimal rigidity as in Chapters 2 and 3, we can simply assume that the rank of the rigidity matrix is maximal, namely the rank is $nd - d(d+1)/2$, for $n \geq d$ for n nodes in $\mathbb{R}^d$, which is equivalent to an appropriate determinant not being zero. (See also White and Whiteley (1983) for an algebraic criterion for configurations that are not generic, what they call the *pure-condition*.) It turns out that for bar frameworks, this explicit sense of being generic implies that it is only the abstract graph that determines the rigidity, and the (generic) configuration is not relevant. Further, in the plane (at least) there is an efficient algorithm to determine the rigidity combinatorially directly from the underlying graph.

However, for even a well-defined specific configuration $\mathbf{p}$, determining global rigidity, as defined in Subsection 2.4.5, is often difficult. On the other hand, if one assumes that the configuration $\mathbf{p}$ is generic in the general sense described below in Section 7.2, with a numerical calculation it is possible to determine whether $(G, \mathbf{p})$ globally rigid, as described in Section 7.8. Unfortunately, there does not seem to be a way to explicitly describe some equations that determine what it means for $\mathbf{p}$ to be generic in this case.

The present chapter will only give a fairly brief overview of this material: a much more extensive introduction is given in Graver (2001).

7.2 Definition of Generic

We will use a strong definition of generic, which is probably more constraining than necessary in any particular case, but will ensure that we don't miss any special conditions on the configuration. Suppose that $\mathbf{p} = [\mathbf{p}_1; \ldots; \mathbf{p}_n]$ is a configuration of nodes in $\mathbb{E}^d$. Consider all of the coordinates of $\mathbf{p}$ as a set of numbers, and consider any non-zero polynomial equation $f(x_1, x_2, \ldots, x_{dn})$ with integer coefficients and the numbers (the coordinates) substituted for the variables $x_1, x_2, \ldots, x_{dn}$. If $f(\mathbf{p}) \neq 0$ for every such f, we say that the configuration is *generic*.

An example that this definition calls non-generic is when three points in the plane lie on a line. If $\mathbf{p} = [\mathbf{p_1}; \mathbf{p_2}; \mathbf{p_3}] = [x_1; y_1; x_2; y_2; x_3; y_3]$, then the condition that those three points lie on a line is that

$$\det \begin{bmatrix} 1 & x_1 & y_1 \\ 1 & x_2 & y_2 \\ 1 & x_3 & y_3 \end{bmatrix} = x_1 y_2 + x_2 y_3 + x_3 y_1 - x_1 y_3 - x_2 y_1 - x_3 y_2 = 0.$$

Other examples that are non-generic by this definition include four points on a circle, six points on a conic section, a pair of points on a line with rational slope. The point is that the condition of being generic is more than adequate for ensuring that there is nothing "special" about the configuration. Almost anything that can be converted into some non-trivial algebraic condition will be ruled out when we say that the configuration is generic. We just eliminate them all to make sure. Clearly, the above definition is overkill, because, for example, if the configuration is a single node $[\sqrt{2}]$ in the line $\mathbb{E}^1$, that point is not generic, because it satisfies the polynomial equation $x_1^2 - 2 = 0$.

Although we have strong definition of generic, we would still like to be sure that "most" configurations are generic. A simple way to understand this is in terms of probability. Suppose that we put the standard uniform probability measure on the space of all configurations. This regards a configuration of n points in $\mathbb{E}^d$ as a single point in $\mathbb{E}^{nd}$, and if there are two sets of points $X \subset Y$ in $\mathbb{E}^{nd}$, then the probability that the configuration $\mathbf{p}$ chosen from Y is in X is just the ratio volume(X)/volume(Y), when volume(Y) is positive and finite. For example, Y could be a large box corresponding to those configurations in $\mathbb{E}^d$ whose coordinates all lie in some large interval. The following is a simple result that can be useful in justifying our definition of generic.

Theorem 7.2.1. *The probability that a randomly chosen configuration is generic is* 1.

The idea of the proof is that each polynomial defines a set of points in configuration space that has 0 volume. There are only a countable number of such polynomials and a countable union of sets with 0 volume have 0 volume. The probability of choosing a point in a set of 0 volume is 0. So the probability of not being in the set of 0 volume, the set of generic configurations, is 1.

Another advantage of looking at generic configurations is that they are *dense* in the space of all configurations. This means that for any configuration $\mathbf{p}$ in $\mathbb{E}^d$, whether generic or not, there is arbitrarily close to $\mathbf{p}$ another configuration $\mathbf{q}$ which is generic.

7.3 Infinitesimal Rigidity is a Generic Property

As we are concerned with first-order properties of bar frameworks, the following is an important and useful observation.

Theorem 7.3.1. *If $(G, \mathbf{p})$ is an infinitesimally rigid bar framework in $\mathbb{E}^d$ for the configuration $\mathbf{p} = [\mathbf{p_1}; \ldots; \mathbf{p_n}]$, then $(G, \mathbf{q})$ is also infinitesimally rigid at any generic configuration $\mathbf{q} = [\mathbf{q_1}; \ldots; \mathbf{q_n}]$.*

Proof. Any bar framework at a configuration $\mathbf{q}$ is infinitesimally rigid when the rank r of the rigidity matrix $\mathbf{R}(\mathbf{q})$ is maximal, $r = r_{\max}$, where $r_{\max} = dn - d(d+1)/2$ for $n \geq d$. The maximal rank is achieved when some $r_{\max}$-by-$r_{\max}$ submatrix has rank $r_{\max}$, which is equivalent to the determinant of that matrix being non-zero. This determinant is a polynomial in the coordinates of the configuration $\mathbf{q}$. The polynomial is not the zero polynomial, since there is at least one instance when that polynomial is not zero, namely when the coordinates of $\mathbf{p}$ are substituted. Since generic configurations $\mathbf{q}$ do not satisfy any such polynomial, all such frameworks $(G, \mathbf{q})$ must be infinitesimally rigid. $\square$

As an example of an application of this theorem consider the framework obtained from a convex polyhedral surface with all triangular faces. We showed in Chapter 2 that such a bar framework is infinitesimally rigid in $\mathbb{E}^3$. By Theorem 7.3.1, any generic configuration with bars corresponding to the same graph is infinitesimally rigid in $\mathbb{E}^3$, even if the framework does not come from a convex polyhedral surface. The only drawback to this observation is that, for a particular example of such a framework, you are never quite sure, without further work, that they are not in some sort of special position that would allow the framework to be infinitesimally flexible.

Another point is that rigidity of a generic configuration of a framework is a property of the abstract graph G. In a sense, any special consideration for the geometry of any configuration has been deliberately extracted. All that is needed is for $(G, \mathbf{p})$ to be infinitesimally rigid for one configuration, and Theorem 7.3.1 guarantees that any generic configuration is infinitesimally rigid. With this in mind, we say that a graph G is *generically rigid in E^d* if $(G, \mathbf{p})$ is infinitesimally rigid for all generic configurations $\mathbf{p}$, or equivalently for any one configuration $\mathbf{p}$, generic or not.

The property of being generic also has very strong consequences with regard to the property of being either rigid or a finite mechanism.

Theorem 7.3.2. *Let G be any bar graph. Then either $(G, \mathbf{p})$ is infinitesimally rigid for all generic configurations $\mathbf{p}$ in $\mathbb{E}^d$ or $(G, \mathbf{p})$ is a finite mechanism for all generic configurations $\mathbf{p}$ in $\mathbb{E}^d$.*

Proof. By Theorem 7.3.1 if $(G, \mathbf{p})$ is infinitesimally rigid for any generic configuration $\mathbf{p}$ in $\mathbb{E}^d$, it is infinitesimally rigid for all such generic configurations, and $(G, \mathbf{p})$ is not a finite mechanism for any of these configurations from our discussion in Chapter 2.

Suppose that $(G, \mathbf{p})$ is not infinitesimally rigid for a generic configuration $\mathbf{p}$. Then is not infinitesimally rigid for all generic configurations. Then the rank of the rigidity matrix at $\mathbf{p}$ is less than $dn - d(d+1)/2$ when G is not a simplex. So by the inverse function theorem (Spivak, 1965), since the dimension of the image of the rigidity map f is less than $dn - d(d+1)/2$ for all configurations, the dimension of the inverse image $f^{-1}(f(\mathbf{p})) = \{\mathbf{q} \in \mathbb{E}^{nd} \mid f(\mathbf{q}) = f(\mathbf{p})\}$ is more than $d(d+1)/2$, and thus it must contain more than congruent copies of $\mathbf{p}$ near $\mathbf{p}$. Thus $(G, \mathbf{p})$ is a finite mechanism. $\square$

We see that we can equivalently define a graph G to be generically rigid in $\mathbb{E}^d$ if $(G, \mathbf{p})$ is simply rigid (i.e. not a finite mechanism) for some generic configuration $\mathbf{p}$ in $\mathbb{E}^d$. Sometimes

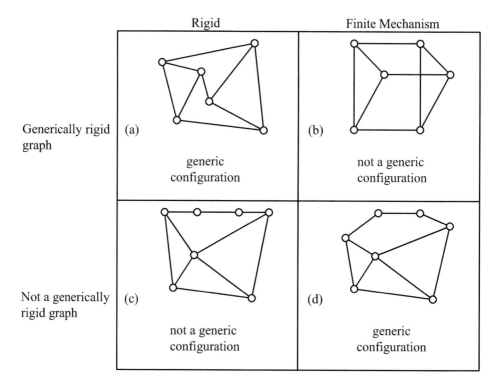

Figure 7.1 Two configurations for each of two frameworks with different underlying graphs. One of the frameworks is generically rigid in the plane, and one is not. In each case, one configuration is rigid, and one is a finite mechanism.

this terminology is shortened to saying that G is *d-rigid*. But we must bear in mind that when rigidity is referred to a graph, it is implicit that it means the rigidity or infinitesimal rigidity at a generic configuration **p**.

7.3.1 Examples

Figure 7.1 shows a generic and non-generic configuration of each of two frameworks with different graphs. Example (a) is infinitesimally rigid in the plane, but Example (b) is in a special configuration, and is not infinitesimally rigid (indeed, it is actually a finite mechanism), even though it has the same underlying graph as Example (a). Thus, the abstract graph is generically rigid in the plane (equivalently, 2-rigid).

On the other hand the graph in Examples (c) and (d) is not generically rigid in the plane. In the generic configuration in Example (d), the framework is a finite mechanism. However, in the special configuration of Example (a) the framework is rigid in the plane (but, of course, not infinitesimally rigid).

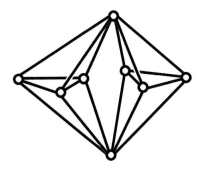

Figure 7.2 The "double banana" bar framework that has the property that the conclusion of Theorem 7.4.1 holds but nevertheless is not generically rigid in $\mathbb{E}^3$. It has eight nodes and $18 = 3 \cdot 8 - 6$ bars, and every subgraph with $n_0 \geq 3$ nodes has $b_0 \leq 3n_0 - 6$, yet clearly the framework is not generically rigid, since the two bananas can rotate relative to each other about their common ends.

7.4 Necessary Conditions for Being Generically Rigid

Since the property of a graph being generically rigid in $\mathbb{E}^d$ is an entirely combinatorial property of the graph G, we can hope that there is a reasonable entirely combinatorial characterization. In the plane, there is such a characterization and it also leads to a computationally efficient method of determining generic rigidity. Unfortunately, in higher dimensions, in particular dimension 3, there is no known computationally efficient purely combinatorial algorithm to determine generic rigidity.

A starting point for the discussion of the combinatorics of generic rigidity is to consider the case when a graph G has at least d nodes (so that their affine span is at least $(d-1)$-dimensional) and has just enough bars to be infinitesimally rigid in $\mathbb{E}^d$, namely $b = dn - d(d+1)/2$. For any configuration $\mathbf{p}$, if $(G, \mathbf{p})$ has a non-zero self-stress, then by our discussion in Chapter 3, $(G, \mathbf{p})$ cannot be infinitesimally rigid in $\mathbb{E}^d$. Further, if there is a subgraph G_0 of G with $n_0 \geq d$ nodes, and $b_0 > dn_0 - d(d+1)/2$ bars, then that subgraph alone will have a non-zero self-stress. Thus we get the following:

Theorem 7.4.1. *If a bar graph G with $n \geq d$ nodes and $b = dn - d(d+1)/2$ bars is generically rigid in $\mathbb{E}^d$, then for any subgraph G_0 of G with $n_0 \geq d$ nodes, and b_0 bars, $b_0 \leq dn_0 - d(d+1)/2$.*

This says that in the plane, if G is 2-rigid, then every subgraph of G with $n_0 \geq 2$ nodes, and b_0 bars must have $b_0 \leq 2n_0 - 3$. If G is 3-rigid, then every subgraph with $n_0 \geq 3$ nodes and b_0 bars must have $b_0 \leq 3n_0 - 6$.

But is the converse of Theorem 7.4.1 true? For $\mathbb{E}^3$ consider the infamous "double banana" in Figure 7.2. It satisfies the conclusion of Theorem 7.4.1 but is clearly not generically rigid.

7.5 Generic Rigidity in the Plane

The converse of Theorem 7.4.1 in the plane is true. In order to get to this, we need some simple results. The *union of two graphs* is the graph on the union of the vertices of each graph, and two vertices in the union have an edge (a bar) between them if and only if there is an edge between those vertices in either one of the graphs. The following is immediate.

> **Lemma 7.5.1.** *If two graphs G_1 and G_2 are 2-rigid and have exactly two vertices in common, then their union $G_1 \cup G_2$ is 2-rigid.*

Note that some edges in the union may come from both G_1 and G_2. Next we need a simple counting result. Recall that the *degree of a vertex* in a graph is the number of edges that contain it.

> **Lemma 7.5.2.** *If a graph G (having no loops or multiple edges) with $n \geq 2$ nodes has $e = 2n - 3$ (or fewer) edges, then the average degree of a vertex $\bar{n}$ is strictly less than 4, and hence there must be a vertex of degree 3 or less.*

Proof. The sum of the degrees of the vertices counts each vertex twice, so the average degree is

$$\bar{n} = 2e/n \leq (4n - 6)/n = 4 - 6/n < 4,$$

as desired.   □

The next result is an important theorem in Pollaczek-Geiringer (1927) and Laman (1970). It is the converse to Theorem 7.4.1 in the plane. (For years, Laman (1970) had been cited as the first paper stating and proving Theorem 7.5.3 below, but indeed, Pollaczek-Geiringer (1927) had essentially the same result much earlier. See also her results in dimension 3 Pollaczek-Geiringer (1932).) At first sight it seems not to be useful as an algorithm to determine 2-rigidity, since it appears to be necessary to check an exponential number of graphs, but we will see later that this is not the case.

> **Theorem 7.5.3** (Pollaczek-Geiringer, Laman). *If a bar graph G with n nodes and $b = 2n - 3$ bars is such that for all subgraphs G_0 of G, $b_0 \leq 2n_0 - 3$, where G_0 has $n_0 \geq 2$ nodes, and b_0 bars, then G is generically rigid in $\mathbb{E}^2$.*

Proof. We proceed by considering how we can remove nodes and bars in reverse-Henneberg moves to end up with a single bar with $n = 2$ that is itself 2-rigid. We make an induction hypothesis that we can always remove a node from the graph without changing the count on any subgraph, and hence without changing rigidity. We call the condition $b_0 \leq 2n_0 - 3$, for each G_0 a subgraph of G as above, the *bar count*.

We know from Lemma 7.5.2 that there must be at least one node, which we label p_0, with degree 3 or less, which we wish to show that we can remove. The node p_0 cannot have degree 1, else by removing the node and its incident edge we would obtain a subgraph with $b_0 > 2n_0 - 3$, contradicting the hypothesis in the theorem.

If p_0 has degree 2, we can remove it and the two incident edges to get a subgraph G_0 with unchanged bar count for all subgraphs. Then, if G_0 is 2-rigid, G will also be 2-rigid

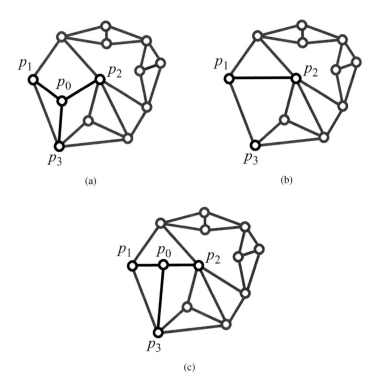

Figure 7.3 The node p_0 in (a) has degree 3. In (b), this node and its three incident edges have been removed and replaced with a single edge. In this example, after removing the three edges in (a), this is the is only way to insert the new edge in (b) to leave a graph that remains 2-rigid. To reverse the operation, a Type II Henneberg operation on the new edge in (b) gives (c); if this new vertex is perturbed to the position in (a) then the configuration is generic.

by applying a Henneberg Type I operation. Recall the Henneberg operations defined in Section 3.2.3.

Now consider if p_0 has degree 3. Label the three adjacent vertices as p_1, p_2, and p_3. We would like to remove p_0 and its three incident edges and include a new edge between p_i and p_{i+1} for some $i = 1, 2, 3$ (mod 3) to give a new graph G_0 which has an unchanged bar count for all subgraphs. If G_0 is 2-rigid, G will also be 2-rigid by applying a Henneberg Type II operation taking G_0 to G after moving the new vertex to a generic configuration, as shown in Figure 7.3.

We now assume the converse, that it is not possible to form G_0 from G by removing a degree 3 p_0 and adding a new edge between p_i and p_{i+1} for some $i = 1, 2, 3$ (mod 3), and will arrive at a contradiction.

Let G_i be the graph obtained by removing p_0, its three incident edges, and inserting the edge $\{p_{i+1}, p_{i-1}\}$ (indices modulo 3). Some of these graphs could have a multiple edge. But in any case our assumption implies by induction, for each $i = 1, 2, 3$, that there is a subgraph H_i of G_i with b_i bars and n_i nodes including p_{i+1} and p_{i-1}, where

$2n_i - 3 + 1 = b_i$, i.e. one bar more than required for rigidity. Each H_i must include the added edge $\{p_{i+1}, p_{i-1}\}$, otherwise it would contradict the assumption of the hypothesis of the Theorem. So if we remove that added edge we get a subgraph K_i of H_i, which has exactly $2n_i - 3$ edges, and K_i is a subgraph of G which includes the nodes p_{i+1} and p_{i-1}. (Note that K_i could include the edge $\{p_{i+1}, p_{i-1}\}$, if it is doubled in G_i.)

By the induction hypothesis, each K_i is 2-rigid. If any K_i and K_j intersect in more than one of the original p_i, by Lemma 7.5.1, $K_i \cup K_j$ is also 2-rigid. But when p_0 and its three incident edges are added to $K_i \cup K_j$, the node count increases by one, and the edge count increases by three, so we have one too many edges for a subgraph of G, and this graph contradicts the hypothesis of the theorem. So K_i and K_j intersect in just one node from p_1, p_2, p_3.

Lastly look at the graph $K_1 \cup K_2 \cup K_3$. It has $n_1 + n_2 + n_3 - 3$ nodes and $2n_1 - 3 + 2n_2 - 3 + 2n_3 - 3 = 2(n_1 + n_2 + n_3 - 3) - 3$ edges. Again when p_0 and its three incident edges are added to $K_1 \cup K_2 \cup K_3$ we obtain a subgraph of G with one more node and three more edges, and hence this subgraph also contradicts the hypothesis of the theorem as it has one too many edges as a subgraph of G. This is the final contradiction from our converse assumption, and hence at least one of the pairs of points of p_1, p_2, p_3 can be used to perform a Henneberg Type II operation. For an example, see Figure 7.4. $\qquad\square$

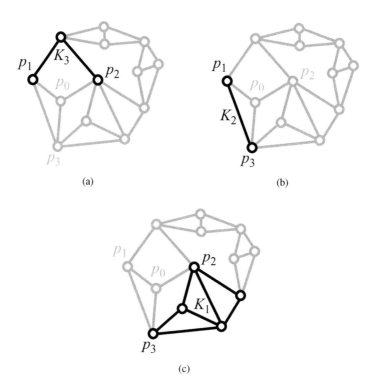

(a)

(b)

(c)

Figure 7.4 The subgraphs K_i shown bold on the original graph of Figure 7.3. K_1 and K_2 are 2-rigid, while K_3 is not, and so the additional edge in the reverse-Henneberg Type II operation is placed between the nodes p_1 and p_2, which are members of K_3.

Note that the proof above does provide an algorithm for finding a sequence of Henneberg operations to build the 2-rigid graph, but it is not clear that this algorithm will take a polynomial number of steps in terms of n the number of nodes in G. On the other hand, once this sequence of operations has been determined, it can be checked that G is 2-rigid in a linear number of steps, no more than the number of edges of G starting from a single edge.

7.6 Pebble Game

The "pebble game," introduced by Jacobs and Hendrickson (1997) provides an efficient algorithm for determining 2-rigidity. For a graph G with n vertices and $2n - 3$ edges, it will either find a subgraph G_0 with $n_0 \geq 2$ nodes and more than $2n_0 - 3$ edges (we say that G_0 is *over-counted*) or determine that there is no such subgraph, thus demonstrating that G is 2-rigid by Laman's Theorem 7.5.3. The following is a description of the *pebble game* algorithm.

In the pebble game, each edge e of G has to be tested to determine whether it is part of a subgraph G_0 with n_0 nodes, and more than $2n_0 - 3$ edges. For each edge e of G, this is done by adding three extra copies of e, one at a time. Call the graph with the three additional edges (four edges in all between its incident nodes) G_e.

For each node of G_e imagine a small container that initially contains two pebbles. The object of the game is to place all the pebbles on edges, where each pebble from each node is placed on an adjacent edge, and each edge has exactly one pebble on it. If this is possible, it turns out that there can be no over-counted subgraph of G that contains e. Thus, if this algorithm is applied to each edge of G, Laman's Theorem 7.5.3 implies that G is 2-rigid. An example of the application of the pebble game is shown in Figure 7.5.

We attempt to cover all the $2n$ edges of G_e, but saving at least one of the four copies of e to be covered last. When a pebble from node i is placed on an adjacent edge $\{i, j\}$, we put it next to the incident node i that it came from, and orient the edge from i to j. So during the algorithm, all the edges that are covered by a pebble are oriented. Figure 7.5(c) shows the game part-way through the process.

The process starts easily, but as more edges get covered, there may be a need to backtrack to cover some of the uncovered edges. For an uncovered edge, if there are no pebbles immediately available for its incident nodes, search in the direction of the oriented edges for a pebble that has not been placed yet. This is a greedy search along the directed edges. A depth first search, for example, will do. This means that once an edge has been searched for pebbles at its nodes, it is kept in a list and never again asked if its nodes have a free pebble. At each stage the algorithm looks for unvisited edges coming from the last node visited. If it runs out of new edges oriented away from its last node, it backs up and looks for another new unexplored edge. The algorithm never goes down a path it has previously visited.

If it turns out that the algorithm finds a node with a free pebble, it exchanges pebbles back along the oriented path to free up a pebble to cover the edge in question. For example, in Figure 7.5(d) the straight diagonal edge, uncovered in Figure 7.5(c), is covered by the

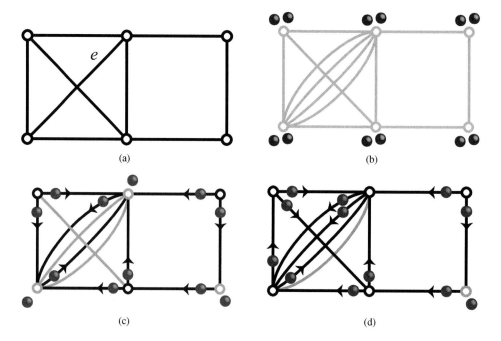

Figure 7.5 An example of the pebble game, to determine if the edge e for graph G with $2n - 3$ edges, shown in (a), is part of an over-counted subgraph. (b) shows G_e, in which edge e has been replaced with four edges between its incident edges. Two "pebbles" have been given to each node, and remain tied to that node. Each node that has spare pebbles remaining is shown in grey, as is each uncovered edge. (c) shows a partially completed stage of the pebble game. Some of the edges have been covered and appropriately directed, with an arrow pointing away from the covering pebble. In (d) the game has terminated, and one of the duplicated edges remains uncovered, while one pebble is left unused: the edge e is part of an over-counted subgraph of G – the leftmost square with its diagonals. By Laman's Theorem 7.4.1, G is not 2-rigid.

pebble that had previously covered the left vertical edge, and the free pebble on the lower left in Figure 7.5(c) is instead used to cover that edge.

If it turns out that one of the e edges cannot be covered this way, it means that there is no directed path to a free pebble. The algorithm terminates declaring that e is contained in an over-counted subgraph. The edges that have been explored by the previous depth-first search do not reach all the vertices of G_e, and these edges form a subgraph G_0 of G_e with n_0 nodes and at least $2n_0 + 1$ edges. The corresponding subgraph of G then has at least $2n_0 - 3 + 1$ edges, and thus G is not 2-rigid, and indeed, the edge e is part of an over-counted subgraph.

If the edge e is contained in an over-counted subgraph G_0 of G, it is not possible for its $2n_0$ pebbles to cover all the edges of G_0, and so the algorithm will terminate leaving some subgraph containing an e uncovered, declaring that G is not 2-rigid. On the contrary, if the algorithm covers all of the $2n$ edges of G, e is not in any over-counted subgraph, and the algorithm will terminate with this result. This completes the description of the algorithm.   □

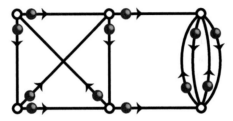

Figure 7.6 An example where the pebble game terminates with all edges covered, and yet the graph is not 2-rigid, since an edge chosen to be duplicated is not in an over-counted subgraph.

Note that it is possible that the algorithm can terminate with all the edges of G_e covered even when G is not 2-rigid. This happens if the duplicated edge e is not part of an over-counted subgraph. The other edges of G have not yet been tested to determine if any of them were part of an over-counted subgraph. Figure 7.6 shows such an example.

For the pebble game, we wish to estimate how many elementary steps it will take until the algorithm determines that a given graph G is either 2-rigid or not. We assume that G is given as a database, where once a vertex is accessed, its adjacent vertices and edges can be found in a constant number of steps, say one step. In the pebble game, when an edge has been chosen, and quadrupled, the process of placing the pebbles and orienting the covered edges involves no more than a constant times the number of nodes n in G, even with the search involved in finding an available pebble. Since the game is played with each edge being quadrupled in turn and the total number of edges is $2n - 3$, the total number of steps is then at most cn^2, for some constant c.

In a more general setting, it is often desired to deal with graphs that have a number of edges other than $2n - 3$. One can ask for maximal 2-rigid subgraphs, subgraphs that have a non-zero self-stress in a generic configuration, and subgraphs that are not generically rigid. These can all be determined using variations of the pebble game in at worst cnb steps, where n is number of nodes of G, and b is the number of bars of G. For example, start with a graph that is *generically independent*, which means that it has only the **0** self-stress in a generic configuration. By Laman's Theorem 7.5.3 this is equivalent to there being no subgraph with $b' > 2n' - 3$ bars, where n' is the number of nodes. Then edges are added back one at a time, checking each time for a pebble covering, using the algorithm described previously. If the pebble game ends with a successful pebble covering, add the bar. If not, the algorithm has determined that this bar is redundant. For example, for the framework in Figure 7.5(a), one could start with the one of the triangles on the left, and keep adding bars. When one of the diagonal bars on the left is determined to be redundant, there are not enough bars left for the entire framework to be rigid, but the four nodes on the left form a rigid subgraph. This is explained in Jacobs and Hendrickson (1997). Some similar algorithms predating the pebble game are Lovász and Yemini (1982); Crapo (1979); and Asimow and Roth (1978, 1979).

7.7 Vertex Splitting

It is useful to be able to build rigid frameworks from simpler ones: the Henneberg Type I and II operations for bar frameworks of Section 3.2.3 are examples of this. Another useful operation is *vertex splitting*, originally described in Whiteley (1988b). An example is shown in Figure 7.7. Vertex splitting takes a graph that is generically rigid in $\mathbb{E}^d$ and adds one new vertex with appropriate rearranging and the addition of new edges to give a new generically rigid bar framework, as described next.

Consider a node of degree $n-1$ labelled 1 in a graph G for a framework in $\mathbb{E}^d$. The node is connected by edges to vertices labelled $2, \ldots, n$. We will split the vertex labelled 1. Partition the nodes connected to 1 into three sets: $[2, \ldots, d]$; $[d+1, \ldots, d+k]$; and $[d+k+1, \ldots, n]$ for some value of k; either or both of the latter two sets are allowed to be empty. Remove the k edges connecting node 1 to the nodes in the set $[d+1, \ldots, d+k]$, shown in grey for the example in Figure 7.7(a) (where $d = 3$, $k = 3$ and $n = 8$). Now add a new vertex labelled 0 and connect that vertex with $k+d$ new edges: $\{0, 1\}$; edges between 0 and the set of nodes $[2, \ldots, d]$; and edges between 0 and the set of nodes $[d+1, \ldots, d+k]$. The new edges are shown in grey for the example in Figure 7.7(b). Call this new graph the *d-dimensional vertex split* of G.

Another way of considering this is to remove the vertex 1 and its adjacent edges, write its neighbours as the union of two sets A and B with exactly $d-1$ vertices in $A \cap B$, and join 0 to A and 1 to B. Either $A - B$ or $B - A$ may be empty. Note that one new vertex and a total of d additional edges are added.

The value of vertex-splitting follows from the following theorem, from Whiteley (1988b).

Theorem 7.7.1. *If a graph G is generically rigid in $\mathbb{E}^d$, then so is G_1 obtained from G by the d-dimensional vertex splitting of any vertex.*

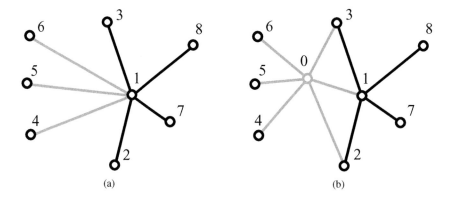

(a)   (b)

Figure 7.7 An example of vertex splitting for a framework in $\mathbb{E}^3$. The vertex labelled 1 in (a) is replaced by two vertices labelled 0 and 1 in (b). The edges shown in grey in (a) are deleted during the operation, and the edges and node shown in grey in (b) are added. In total one node and $d = 3$ edges are added to the framework.

This theorem was applied by Whiteley (1988b) to give an entirely combinatorial proof that a generic configuration of a triangulated sphere is rigid. Then Fogelsanger (1988) showed that any triangulated oriented surface is generically rigid in $\mathbb{E}^3$. But he showed much more. A *minimal d-cycle* is a simplicial complex that is a cycle in the sense of algebraic topology, and no proper subset is a cycle. For $d = 2$, being a *cycle* just means that it consists of oriented triangles as in Section 3.5.1 where, for each edge $e = \{i, j\}$, the number of triangles with e oriented (i, j) is the same as the number of triangles with e oriented (j, i). We call the *graph of a minimal cycle* as the vertices and edges of such a minimal cycle, sometimes called the one-skeleton. Fogelsanger's result is the following:

Theorem 7.7.2. *For $d \geq 2$ the graph of a minimal $(d - 1)$-cycle is generically rigid in $\mathbb{E}^d$.*

7.8 Generic Global Rigidity

Recall the definition in Section 2.4.5 of global rigidity of a framework $(G, \mathbf{p})$ in $\mathbb{E}^d$. If a framework $(G, \mathbf{p})$ has a self-stress with a stress matrix Ω that satisfies the conditions of Theorem 5.14.1, then $(G, \mathbf{p})$ is not only globally rigid in $\mathbb{E}^d$ but in $\mathbb{E}^N$, for all $N \geq d$. A crucial condition of Theorem 5.14.1 is that Ω be positive semi-definite, as well as being of maximal rank $n - d - 1$, where n is the number of nodes of G. It can happen that for a given graph G, there are configurations with a positive semi-definite stress matrix, and other configurations with an indefinite stress matrix. But, as we will see, for generic global rigidity in a fixed dimension, that does not matter, only the rank matters.

7.8.1 Generic Global Rigidity in All Dimensions

If the configuration $\mathbf{p}$ is generic in $\mathbb{E}^d$, we can still say more about global rigidity, as in the following result of Connelly (2005).

Theorem 7.8.1. *Let $(G, \mathbf{p})$ be a bar framework in $\mathbb{E}^d$, where $\mathbf{p}$ is generic, and ω is a self-stress for $(G, \mathbf{p})$ whose stress matrix Ω has rank $n - d - 1 \geq 1$, where n is the number of nodes of G. Then $(G, \mathbf{p})$ is globally rigid in $\mathbb{E}^d$.*

The idea here is to look at the rigidity map $f : \mathbb{E}^{nd} \to \mathbb{E}^b$, defined in Equation (3.3) as the vector of squared bar lengths, where b is the number of members (bars) of G. There is another configuration $\mathbf{q}$ such that $f(\mathbf{p}) = f(\mathbf{q})$ if and only if $(G, \mathbf{p})$ and $(G, \mathbf{q})$ are equivalent as bar frameworks. It is possible to show that if the configuration $\mathbf{p}$ is generic, then there is, at least, a small neighbourhood $U_{\mathbf{p}} \subset \mathbb{E}^{dn}$ of $\mathbf{p}$ and a small neighbourhood $U_{\mathbf{q}} \subset \mathbb{E}^{dn}$ of $\mathbf{q}$ and a differentiable homeomorphism $h : U_{\mathbf{p}} \to U_{\mathbf{q}}$ (i.e. h is one-to-one and onto, and its inverse function is continuous), such that $f = fh$, for f restricted to $U_{\mathbf{p}}$ and $U_{\mathbf{q}}$, respectively. Taking the derivative of this equation at the configuration $\mathbf{p}$, we get $df_{\mathbf{p}} = df_{\mathbf{q}} \, dh_{\mathbf{p}}$. Since the derivative of the rigidity map is the rigidity matrix, up to a constant, this gives $\mathbf{R}(\mathbf{p}) = \mathbf{R}(\mathbf{q}) dh_{\mathbf{p}}$, where $\mathbf{R}(\mathbf{p})$ and $\mathbf{R}(\mathbf{q})$ are the rigidity maps for $(G, \mathbf{p})$ and $(G, \mathbf{q})$ respectively, and this implies that the cokernels (left nullspaces) of $\mathbf{R}(\mathbf{p})$ and $\mathbf{R}(\mathbf{q})$ are the same. This means that a self-stress ω for $(G, \mathbf{p})$ is also a self-stress for $(G, \mathbf{q})$, and vice versa. (In other words, $\omega \mathbf{R}(\mathbf{p}) = \mathbf{0}$ if and only if $\omega \mathbf{R}(\mathbf{q}) = \mathbf{0}$.) If the rank of the corresponding stress matrix Ω is $n - d - 1$, then $\mathbf{p}$ is a universal configuration for ω, and so $\mathbf{q}$ must be an affine image of $\mathbf{p}$. Further, if the configuration $\mathbf{p}$ is generic, then the member directions do not lie on a conic at infinity, and so $\mathbf{q}$ must be congruent to $\mathbf{p}$.

One of the problems with the proof, sketched above, is that it is not really easy to determine what the conditions are that will ensure that the configurations $\mathbf{p}$ and $\mathbf{q}$ will have neighbourhoods that will map onto one another, and not just have neighbourhoods that intersect on a lower dimensional set. This is where, in Connelly (2005), there is an appeal to a "Tarski-Seidenberg"-type elimination theory. This guarantees that there is a finite set of polynomial equations that are to be avoided that will ensure generic global rigidity, but it does not supply a polynomial-time algorithm that will calculate those polynomials. Indeed, it is reasonable to expect that there is no such polynomial-time algorithm to detect global rigidity, or even a way to calculate polynomial equations whose complement in the space of configurations are globally rigid. This is further discussed below.

It is, however, possible to determine when a graph G has the property that all generic configurations $\mathbf{p}$ are such that the conditions of Theorem 7.8.1 are satisfied and thus $(G, \mathbf{p})$ is globally rigid, even if you may not know of any given configuration that is globally rigid. The following corollary provides a practical way of determining generic global rigidity.

> **Corollary 7.8.2.** *Let $(G, \mathbf{p})$ be an infinitesimally rigid bar framework in $\mathbb{E}^d$, where ω is a self-stress for $(G, \mathbf{p})$ whose stress matrix Ω has rank $n - d - 1$. Then for any generic configuration $\mathbf{q}$, $(G, \mathbf{q})$ is globally rigid in $\mathbb{E}^d$.*

Proof. The rigidity matrix $\mathbf{R}(\mathbf{q})$ is a continuous polynomial function of the configuration $\mathbf{q}$, and it has maximal rank, since that is the case when $\mathbf{q} = \mathbf{p}$. Furthermore the nullspace of $\mathbf{R}(\mathbf{q})$, namely the space of self-stresses ω and the corresponding stress matrices Ω for $(G, \mathbf{q})$ are rational functions of $\mathbf{q}$, or, more concretely, there is a basis for these spaces that are rational functions of $\mathbf{q}$. So the rank will be maximal at a generic configuration if it is maximal at one configuration $\mathbf{p}$. $\qquad\qquad\square$

When there is a framework $(G, \mathbf{p})$ that satisfies the hypothesis of Corollary 7.8.2, we say that it is a *certificate for generic global rigidity*. Note that a certificate for generic global rigidity $(G, \mathbf{p})$ may not itself be globally rigid, as in Figure 7.8.

The question remains as to the converse of Theorem 7.8.1. This is answered very nicely by the following result of Gortler et al. (2010).

> **Theorem 7.8.3.** *Let $(G, \mathbf{p})$ be a globally rigid bar framework in $\mathbb{E}^d$, where $\mathbf{p}$ is generic with n nodes. Then either $(G, \mathbf{p})$ is a simplex or there is a non-zero self-stress ω for $(G, \mathbf{p})$ such that the corresponding stress matrix Ω has rank $n - d - 1$.*

Very roughly, the idea of the proof is to show that, if the rigidity map, modulo rigid congruences, and minus some low-dimensional subsets and restricting the image appropriately, is not one-to-one, then it has a well-defined topological degree that is not ± 1. So at a generic configuration there must be at least one other configuration that maps to it.

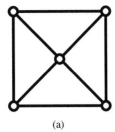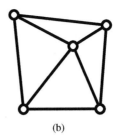

(a)   (b)

Figure 7.8 The framework in (a) is not globally rigid in $\mathbb{E}^2$ (consider folding one corner node across a diagonal), but it has a self-stress whose stress matrix $\mathbf{\Omega}$ has rank $n - d - 1 = 5 - 2 - 1 = 2$. Thus, by Corollary 7.8.2, the framework is a certificate for generic global rigidity, and a generic configuration of the framework, such as that in (b), will be globally rigid in $\mathbb{E}^2$.

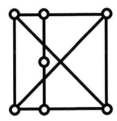

Figure 7.9 This framework is universally globally rigid in $\mathbb{E}^2$, but it has a single state of self-stress whose stress matrix $\mathbf{\Omega}$ has rank $2 < n - d - 1 = 6 - 2 - 1 = 3$, i.e. it does not have maximal rank. It was generated from 3.2(c) with an additional pair of co-linear members connecting an original node to the opposite diagonal member. The new members are unstressed by the self-stress.

7.8.2 Generic Universal Global Rigidity

As mentioned above, if the stress matrix $\mathbf{\Omega}$ of a bar framework $(G, \mathbf{p})$ is positive semi-definite as well as having maximal rank $n - d - 1$, and the affine motions are well-behaved, then the bar framework can be thought of as a tensegrity, and will be globally rigid. However, there are many examples where a framework has no self-stress whose stress matrix has rank $n - d - 1$, and yet the framework is still universally globally rigid – one such example is shown in Figure 7.9. But in the generic case the following result of Gortler and Thurston (2014) makes matters clearer.

Theorem 7.8.4. *Let $(G, \mathbf{p})$ be a universally globally rigid bar framework in $\mathbb{E}^d$ with n nodes, where $\mathbf{p}$ is generic in $\mathbb{E}^d$. Then either $(G, \mathbf{p})$ is a simplex, or there is a non-zero self-stress ω for $(G, \mathbf{p})$ such that the corresponding stress matrix $\mathbf{\Omega}$ has rank $n - d - 1$ and is positive semi-definite.*

The idea of the proof is to look at the convexity properties of the image of the rigidity map whose target is $\mathbb{E}^N \supset \mathbb{E}^d$, for N large. (Note that in Gortler and Thurston (2014) universal global rigidity is shortened to *universal rigidity*.)

From Theorems 7.8.1 and 7.8.3 one way to determine if a framework in generic position is globally rigid is to check if there is a stress with maximal rank, but there is no assurance that if the configuration **p** is not generic, whether it is globally rigid at **p**. Finding the configurations that are not singular (i.e. not special in the sense of global rigidity) can be difficult even if you know that a generic configuration is globally rigid for a given graph G. On the other hand, to determine if a framework $(G, \mathbf{p})$ is universally globally rigid, in some cases, can often be determined with certainty. Namely when the conditions of Theorem 5.14.1 hold, in particular if the PSD property of the stress matrix holds for $(G, \mathbf{p})$, then universal global rigidity holds for $(G, \mathbf{p})$. Of course, for many configurations **p**, global rigidity holds, but not universal rigidity. The following results show how to find some configurations which are indeed universally globally rigid, when it is known that generically $(G, \mathbf{p})$ is globally rigid.

In Alfakih (2017) the following result is shown.

Theorem 7.8.5. *Let G be a graph on n nodes that is $(d + 1)$-vertex connected, where $1 \leq d \leq n-1$. Then there is a d-dimensional bar framework $(G, \mathbf{p})$ in general position in $\mathbb{E}^d$ such that $(G, \mathbf{p})$ is universally globally rigid.*

From the abstract "The proof is constructive, and is based on a theorem by Lovász et al. concerning orthogonal representations and connectivity of graphs Lovász et al. (1989, 2000)." Note that this result is independent of whether the graph G is generically globally rigid or not. Indeed, it holds even if G is not generically rigid. It just has the connectivity conditions. When we have generic global rigidity for G, then the following holds from Connelly et al. (2020).

Theorem 7.8.6. *If the graph G is generically globally rigid in $\mathbb{E}^d$, then there exists a framework $(G, \mathbf{p})$ in R^d that is infinitesimally rigid in $\mathbb{E}^d$ and super stable. Moreover, every framework in a small enough neighbourhood of $(G, \mathbf{p})$ will be infinitesimally rigid in $\mathbb{E}^d$ and super stable, and thus must include some generic framework.*

7.8.3 The Henneberg Constructions

In Section 3.2.3 it is shown that the Henneberg Type I and Type II constructions preserve infinitesimal rigidity. Thus, it is clear that these constructions preserve generic rigidity as well. In the Type II case the additional node is not initially placed by the Henneberg operation in a generic position; however, the framework is infinitesimally rigid, so that if this node is perturbed to a generic position, the rank of the rigidity matrix is preserved, since it is already maximal.

Considering global rigidity, it is clear that the Henneberg Type I construction does not preserve global rigidity, since the vertex that is attached can be reflected about the hyperplane determined by its neighbours. On the other hand, generically a Henneberg Type II construction does preserve generic global rigidity, as shown by the following.

Theorem 7.8.7. *Suppose* $(G, \mathbf{p})$ *is a framework in* $\mathbb{E}^d$ *with* n *nodes, a self-stress* ω, *and an associated stress matrix* Ω *of nullity* N. *If a member is subdivided, then the new framework has a corresponding self-stress, whose stress matrix also has nullity* N.

Proof. When the member is subdivided, the self-stress can be maintained if each of the two new members formed by subdivision carry the same internal force as the original member, so that the force coefficients are inversely proportional to their lengths, thus defining the "corresponding" self-stress in the theorem. The subdivided framework is not infinitesimally rigid as there is no restraint placed on the new node in a direction perpendicular to the original member, thus allowing an infinitesimal flex. Now recall from Theorem 5.8.1 that $(G, \mathbf{p})$ must have a universal configuration $\tilde{\mathbf{p}}$ in $\mathbb{E}^k$, where $k = N - 1$. The subdivided framework must have the same universal configuration, where equilibrium requires the extra node to lie on the line determined by the nodes of the original member. If the universal configuration of the subdivided framework was higher dimensional, then we could simply replace the subdivided member with the original member without affecting the dimension spanned by the framework, contradicting the universality of the configuration $\tilde{\mathbf{p}}$ in $\mathbb{E}^k$. Thus the subdivided framework must have a universal configuration in $\mathbb{E}^k$, and its stress matrix must have nullity $N = k + 1$. $\qquad \square$

Applying Corollary 7.8.2 we get the following.

Corollary 7.8.8. *Suppose* $(G, \mathbf{p})$ *is an infinitesimally rigid framework in* $\mathbb{E}^d$ *with* n *nodes and a self-stress* ω *with an associated stress matrix* Ω *of maximal rank* $n - d - 1$. *Let* G_{II} *be the graph obtained by a Henneberg Type II operation. Then for any configuration* $\mathbf{q}$ *sufficiently close to* $\mathbf{p}$ *together with the subdivided node,* $G_{II}(\mathbf{q})$ *is infinitesimally rigid and the corresponding stress matrix has rank* $(n + 1) - d - 1$. *Thus* G_{II} *is generically globally rigid in* $\mathbb{E}^d$.

In terms of just generic global rigidity, Theorem 7.8.3 gives:

Corollary 7.8.9. *If a graph* G *is generically globally rigid in* $\mathbb{E}^d$, *so is the graph obtained by a Henneberg Type II operation.*

It is interesting to note that in Jordán and Szabadka (2009), Corollary 7.8.8 is proved without recourse to stress matrices, Theorem 7.8.3, or Corollary 7.8.2.

7.8.4 Hendrickson's results

In Hendrickson (1995) some fundamental observations about generic global rigidity are made.

Theorem 7.8.10. *Suppose that a bar framework* $(G, \mathbf{p})$ *is globally rigid in* $\mathbb{E}^d$, *that the framework is other than a simplex, and that the configuration* $\mathbf{p}$ *is generic. Then the following conditions must hold.*

    (i) *The graph G is vertex* $(d + 1)$ *connected. (It takes the removal of at least* $d + 1$ *vertices to disconnect the graph.)*

    (ii) $(G, \mathbf{p})$ *is redundantly rigid. (*$(G, \mathbf{p})$ *is rigid, and remains so after the removal of any bar.)*

Property *i* is clear, since if G is not $(d + 1)$-connected, then part of the configuration can be reflected about a hyperplane containing d or fewer nodes of $(G, \mathbf{p})$.

Property *ii* is more subtle. The idea is that if the framework flexes after the removal of a bar $\{i, j\}$, then keep track of the distance $|\mathbf{p}_i - \mathbf{p}_j|$ during the motion. The motion must increase and decrease $|\mathbf{p}_i - \mathbf{p}_j|$, and when it comes back to the same distance, it has to be in a different configuration. Connecting $\mathbf{p}_i$ and $\mathbf{p}_j$ again gives an equivalent non-congruent configuration.

In fact Hendrickson originally conjectured that Properties *i* and *ii* were sufficient as well as necessary for any $d \geq 1$. The case $d = 1$ is easy. The case $d = 2$ is correct (as we will see), but the conjecture is false for all $d \geq 3$. In dimension 3, the only known counterexample, as described by Connelly (1991), was the complete bipartite graph $K_{5,5}$. (This can be easily shown using Bolker and Roth (1980), where there is essentially a complete description of the stress matrices for bipartite graphs in any dimension.) Then in Jordán et al. (2014) other quite simple examples were found. Further counterexamples were found by Frank and Jiang (2011), and indeed for each $d \geq 5$, they have infinitely many counterexamples to Hendrickson's conjecture.

7.8.5 Combinatorial Global Rigidity

In dimension 2, there is a purely combinatorial algorithm to determine generic rigidity, as explained in Section 7.6. There is a similar combinatorial algorithm for generic global rigidity in dimension 2. This is a proof of Hendrickson's conjecture in dimension 2 and uses Corollary 7.8.8. The first step was conjectured by Connelly and proved in Berg and Jordán (2003) when the number of members was $2n - 2$ and n is the number of nodes. The general case was proved in Jackson and Jordán (2005).

    Theorem 7.8.11. *Let G be a vertex* 3-*connected graph that is generically redundantly rigid in* $\mathbb{E}^2$. *Then G can be obtained by a sequence of Henneberg Type II operations and edge insertions on a triangle.*

One can detect generic global rigidity by Laman's Theorem (Theorem 7.4.1), the pebble game, and vertex connectivity by standard graph theoretical methods. The trick is to combine the methods. As a consequence, Hendrickson's Conjecture in dimension 2 follows.

    Corollary 7.8.12. *In the plane, let G be a graph that satisfies Conditions i and ii of Theorem 7.8.10. Then G is generically globally rigid.*

For further related results along these lines see Jackson and Jordán (2010); Jackson et al. (2006, 2007); and Tanigawa (2015).

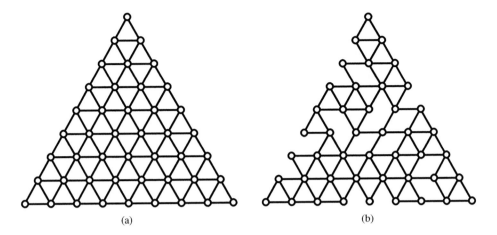

(a)   (b)

Figure 7.10 An example of rigidity percolation. Start with a triangular grid (a), and remove each edge with probability $(1 - p)$ to give the graph (b); this is assumed to be in a generic position, and so rigidity can be calculated using the pebble game.

7.9 Applications

In many small examples of graphs, it is fairly easy to determine their 2-rigidity by finding a Henneberg sequence of Type I or Type II operations. But as the graph gets larger and more complicated it becomes more difficult. One example is the following "rigidity percolation" problem described by Jacobs and Thorpe (1996). Take a large, but finite, triangular grid G as shown in Figure 7.10(a). Choose a probability $0 < p < 1$: each edge in G is retained with probability p, otherwise (with probability $(1 - p)$) it is removed. The resulting graph G_p is a model for a 2-dimensional glass. An example is shown in Figure 7.10(b). The rigidity is calculated using the pebble game. One of the interesting results is that there is a critical probability p_0, where for $p > p_0$, G_p is almost surely 2-rigid, and for $p < p_0$, G_p is almost surely floppy. If $p < 2/3$, from Lemma 7.5.2 it is expected that G_p is floppy. One might guess that $p_0 = 2/3$, but it seems that the critical probability is actually slightly larger.

7.10 Exercises

1. Another model for rigid structures is the following in Tay and Whiteley (1984, 1985) and Tay (1984). A *bar-and-body* framework is a bar framework where the nodes are partitioned into sets called *bodies*, where each body has enough bars to ensure that, by itself, it is (infinitesimally) rigid. Each pair of bodies is itself connected by some collection of bars from nodes of one body to nodes of other bodies. The essential concept is the notion of being generic. A bar-and-body framework is *generic* if the configuration of all the nodes of all the bodies are generic as before, and each node for each body is incident to at most only one bar connecting different bodies. See also the results of Katoh and Tanigawa (2011) about the "Molecular Conjecture" which are bodies connected by co-dimension two hinges, which solved an outstanding

conjecture and allowed the combinatorial characterization of their generic rigidity with a polynomial algorithm.

Show that a body-and-bar framework is generically rigid in $\mathbb{E}^d$ if and only if the associated multigraph, obtained by identifying the nodes of each body to a vertex, is the union of $d(d+1)/2$ edge disjoint spanning trees. Conclude that there is a polynomial-time algorithm to decide whether a bar-and-body framework is generically rigid in $\mathbb{E}^d$.

2. Show that vertex splitting preserves generic rigidity.
3. Show that any triangulated 2-dimensional sphere can be obtained by a sequence of vertex splits starting with the complete graph K_4.
4. Show that a Henneberg Type II operation preserves generic global rigidity. (It is a conjecture of W. Whiteley that vertex splitting preserves generic global rigidity, for $k \geq 1$ and $n \geq d+k+1$, i.e. neither of the sets of nodes $A - B$ or $B - A$ are empty.)
5. Show that the complete bipartite graph $K_{5,5}$ is vertex 4-connected and generically redundantly rigid in $\mathbb{E}^3$, but it is not globally rigid in $\mathbb{E}^3$. A result of Bolker and Roth (1980) is that the stress matrix for $K_{5,5}$ in $\mathbb{E}^3$ has every diagonal entry zero.

8

Finite Mechanisms

8.1 Introduction

If a bar framework is not rigid, it is called a finite mechanism. From the discussion in Chapter 3, this means that there is an analytic path in configuration space that describes the motion of each coordinate of each node in the tensegrity in such a way that the distance constraints are satisfied. The reason it is called a *finite* mechanism is to distinguish it from an *infinitesimal* mechanism. An infinitesimal mechanism can mean that the framework $(G, \mathbf{p})$ *only* has an infinitesimal motion $\mathbf{p}'$ as defined in (2.3), but that it does not have a continuous/analytic flex as defined in Section 2.3. Most of the discussion in other chapters of the present book is concerned with finding ways to tell when a structure is rigid, and how to find its response under load; but in this chapter, we will be concerned with finding techniques to tell when a structure becomes a finite mechanism. In some cases we will find a way of parametrizing the actual motion, or at least describing it in some useful way – although it is usually easier just to determine that there is a motion, rather than finding it explicitly.

There are several techniques for determining when a structure is a finite mechanism. One way is to use the rigidity map for rigid structures and vary one (or more) of the edge lengths in the rigid structure. This is explained in Section 8.2. Another method is to create mechanisms that might normally be rigid, but have some members that maintain their distance constraints because they match other symmetric partners. Some examples of this are explained in Section 8.3, and this technique is developed more fully and systematically in Part II. There are other techniques to create mechanisms that rely on algebraic techniques that do not seem to have any obvious symmetry properties, but depend on an understanding of the algebraic structure. Some examples of this are explained in Section 8.4.

In some cases, it can be difficult to decide if an object is rigid, or is a finite mechanism. For instance, a conjecture dating back to Euler (Gluck, 1975) is that any embedded triangulated surface in three-space is rigid, that is it, is not a finite mechanism; but the conjecture is false, and in Section 8.6 we present a counterexample. Another example: for some years, it was an open question as to whether it was always possible to continuously open an embedded chain of edges in the plane without overlap and fixing the edge lengths. This came to be known as the Carpenter's Rule Problem. In Section 8.7 we outline how to show that such an opening motion is always possible. Any non-overlapping configuration can be opened to straight line configuration.

Much of this chapter is closely related to the subject of kinematics, which has a long and distinguished history relating to design of machines, robotics, and protein folding, which are a few of the many applications. One of the topics of kinematics is the design of finite mechanisms that do specific tasks, and one of those tasks can be to have one node of the mechanism trace out a predetermined part of a curve that is given by polynomial equations. For a very extensive and interesting collection of linkages and references to a wide variety of kinematic models, see the Cornell University's Kinematic Models for Design Digital Library at `http://kmoddl.library.cornell.edu/`.

8.2 Finite Mechanisms Using the Rigidity Map

In the present section we explore what happens if we start with an isostatic framework and delete members.

In Section 3.2 the rigidity map $f : \mathbb{E}^{nd} \rightarrow \mathbb{E}^b$ is defined, where $\mathbb{E}^{nd}$ is the space of configurations of $\mathbf{p} = [\mathbf{p}_1; \dots ; \mathbf{p}_n]$, each $\mathbf{p}_i$ is in $\mathbb{E}^d$, and b is the number of bars of a bar graph G. For simplicity, assume that the configuration $\mathbf{p}$ is *full dimensional*, i.e. not all of the nodes $\mathbf{p}_i$ lie in a $(d-1)$-dimensional hyperplane in $\mathbb{E}^d$.

We say that a bar framework $(G, \mathbf{p})$ in $\mathbb{E}^d$ is *isostatic* if it is kinematically and statically determinate. This means that the framework is infinitesimally rigid, has no state of self-stress, and has number of bars $b = nd - d(d+1)/2$. If $(G, \mathbf{p})$ is isostatic and the configuration is full dimensional, the derivative of the rigidity map, the rigidity matrix $\mathbf{R}(\mathbf{p})$, has rank $nd - d(d+1)/2$, since the rigid-body motions form a $d(d+1)/2$-dimensional subspace. Thus the image of the derivative of the rigidity map $df_{\mathbf{p}} = \mathbf{R}(\mathbf{p})$ is the same as the dimension of the target space $\mathbb{E}^b$. By the Inverse Function Theorem 3.8.3, the image of the rigidity map f contains a neighbourhood $f(\mathbf{p})$.

Define as $X_{\mathbf{p}}$ the space of configurations near the configuration $\mathbf{p}$ satisfying the member constraints. We say that a framework (or tensegrity) $(G, \mathbf{p})$ has k *degrees of freedom* if the dimension of $X_{\mathbf{p}}$, modulo rigid-body motions, is k-dimensional. In other words, if we define an affine linear subspace $L_{\mathbf{p}}$ that is the orthogonal complement to the rigid-body images in the space of configurations at $\mathbf{p}$, then $L_{\mathbf{p}} \cap X_{\mathbf{p}}$, is k-dimensional near $\mathbf{p}$. This leads us to the following.

> **Theorem 8.2.1.** *Let $(G, \mathbf{p})$ be an isostatic bar framework with b bars in $\mathbb{E}^d$, and let G_k denote the subgraph of G obtained by deleting k bars, where k is a number chosen from $\{1, \dots, b\}$. Then $G_k(\mathbf{p})$ is a finite mechanism with k degrees of freedom.*

Proof. When the configuration is not of full dimensional, the only isostatic frameworks are those that are simplices (in dimension 3, these are a single node, a single bar, or a triangle), and the conclusion clearly holds for these cases. So we can consider full dimensional cases. From the discussion above, near the configuration $\mathbf{p}$ the image of the rigidity map f contains all other configurations sufficiently close to $f(\mathbf{p})$. So this means that if we vary the k lengths corresponding to the k bars that were deleted by a sufficiently small amount, there will be configurations corresponding to those perturbed lengths. This provides the k degrees of freedom motion as desired.   □

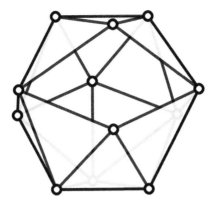

Figure 8.1 A regular icosahedron with an edge (and two associated faces) missing; the framework shown is then a finite mechanism with one degree of freedom.

Corollary 8.2.2. *Let P be a convex polytope in $\mathbb{E}^3$ with all its faces triangles, and let $G_k(\mathbf{p})$ be the framework with nodes at the vertices and bars along the edges, but with k bars removed. Then $G_k(\mathbf{p})$ is a finite mechanism with k degrees of freedom.*

Proof. By Theorem 3.9.4 the framework obtained by using all the edges of P as bars is isostatic. Then the result follows from Theorem 8.2.1.   □

Figure 8.1 shows an example of a bar framework as described in Corollary 8.2.2.

We can use the same method to consider the degrees of freedom of framework constructed from convex polytopes that do not have all faces triangular. We start with an isostatic framework generated by constructing shallow cones over all of the non-triangular faces. We then delete all the edges of the cone, but do not count the three degrees of freedom associated with each isolated cone point. Thus, in total each face with k vertices contributes $k - 3$ degrees of freedom. For example, a framework constructed with bars along the edges of a cube has six degrees of freedom, one from each of the six quadrilateral faces.

8.3 Finite Mechanisms Using Symmetry

In Part II we will see a systematic method of using symmetry to analyse the possible motions of a tensegrity or bar framework. Here we instead describe some very special examples which will be useful later in the chapter.

8.3.1 Bricard Line-Symmetric Octahedron

Following the work of Bricard (1897), we describe some examples of finite mechanisms where the graph is combinatorially the graph of the edges of a regular octahedron, but the geometric realization has some self-intersections.

First we describe an elementary geometric result, illustrated by Figure 8.2.

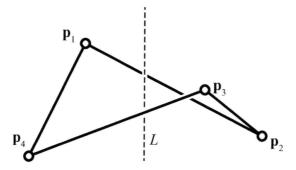

Figure 8.2 A quadrilateral in three-space with opposite sides of the same length and line of symmetry L guaranteed by Lemma 8.3.1.

Lemma 8.3.1. *Let $\mathbf{p} = [\mathbf{p}_1, \mathbf{p}_2, \mathbf{p}_3, \mathbf{p}_4]$ be a quadrilateral of distinct points in $\mathbb{E}^3$ with opposite edges the same length. That is $|\mathbf{p}_1 - \mathbf{p}_2| = |\mathbf{p}_3 - \mathbf{p}_4|$ and $|\mathbf{p}_2 - \mathbf{p}_3| = |\mathbf{p}_4 - \mathbf{p}_1|$. Then $\mathbf{p}$ is symmetric about a line L. In other words, rotation by $180°$ about L takes $\mathbf{p}_1$ to $\mathbf{p}_3$ and $\mathbf{p}_2$ to $\mathbf{p}_4$.*

Proof. Consider the permutation of the points where $\mathbf{p}_1$ and $\mathbf{p}_3$ are interchanged, and $\mathbf{p}_2$ and $\mathbf{p}_4$ are interchanged. The lengths l_{13} and l_{24} are preserved by this permutation, and the conditions on the other four edge lengths imply that all the pairwise distances are preserved. Thus this permutation is a congruence of a bar framework consisting of all six bars between the four points, and this therefore extends to a congruence of all of $\mathbb{E}^3$. Since all congruences are affine linear functions of Euclidean space, the geometric centre $(\mathbf{p}_1 + \mathbf{p}_2 + \mathbf{p}_3 + \mathbf{p}_4)/4$ is fixed, and if we choose a coordinate system so that $\mathbf{p}_1 + \mathbf{p}_2 + \mathbf{p}_3 + \mathbf{p}_4 = \mathbf{0}$, the congruence is given by an orthogonal matrix $\mathbf{A}$ of $\mathbb{E}^3$, where $\mathbf{A}^2 = \mathbf{I}$. Since $\mathbf{A}$ is orthogonal $\mathbf{A}^T\mathbf{A} = \mathbf{I}$, and thus $\mathbf{A} = \mathbf{A}^T$ is a symmetric matrix. Thus $\mathbf{A}$ has only ± 1 as eigenvalues.

Since the points are distinct, $\mathbf{A}$ is not the identity, and not all eigenvalues of $\mathbf{A}$ are $+1$. Thus there may be up to three choices for the congruence, with A having two, one, or three negative eigenvalues:

(i) If there are two -1 eigenvalues, A is a rotation by $180°$ about L, which is defined by the eigenvector associated with the $+1$ eigenvector.

(ii) If there is one -1 eigenvalue, A is a reflection in the plane defined by the eigenvectors with eigenvalue $+1$. Thus the four points lie in a plane (possibly not unique), and the congruence may be replaced by a rotation by $180°$ about L, which is perpendicular to the plane.

(iii) If there are three -1 eigenvalues, A is reflection through the origin. Thus the four points lie in a plane (possibly not unique), and the congruence may be replaced by a rotation by $180°$ about L, which is perpendicular to the plane.

Thus A can always be chosen to have two -1 eigenvalues.   □

Note that when the four vertices are full dimensional, that is they do not lie in a plane, then the line of symmetry is unique and varies continuously with small changes of configuration

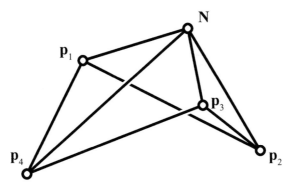

Figure 8.3 A cone over a quadrilateral, which is a finite mechanism in three-space.

around **p**. This remains true even when the configuration lies in a plane, as long as it does not lie on a line.

We now describe a class of finite mechanisms that were first investigated by Bricard (1897). First, in $\mathbb{E}^3$ consider joining a node **N** to the four nodes of a quadrilateral, which we call an equator, as shown in Figure 8.3. Considered as a framework in a generic position, this has $j = 5$ nodes and $b = 8 < 3j - 6$ bars, and is hence a finite mechanism.

Suppose that the equator of Figure 8.3 is a quadrilateral with opposite edges of the same length, and so by Lemma 8.3.1 it has a line L of symmetry. Now generate an additional node **S** by rotating **N** about L by $180°$, and connect **S** to every node on the equator as in Figure 8.4. The graph of this framework is the same as the graph of a regular octahedron where the north and south pole nodes, **N** and **S**, are joined to each of the four vertices on the equator. If the nodes were in a generic configuration, we would expect this framework to be rigid; but because it is symmetric, the bars connected to the additional node **S** allow the same motion of the equator as the bars connected to the original node **N**, and hence this is a finite mechanism.

8.3.2 $K(4, 4)$ in the Plane

Figure 8.5 shows an example of a finite mechanism in the plane, due to Bottema (1960) (see also Wunderlich, 1976). The framework is left unchanged by four symmetry operations, the identity, reflection across the x- and y-axes, and 2-fold rotation: the effect of these symmetry operations on a point is described by the matrices $\mathbf{E}$, $\mathbf{R}_x$, $\mathbf{R}_y$, and $\mathbf{C}_2$ respectively, where

$$\mathbf{E} = \begin{bmatrix} 1 & 0 \\ 0 & 1 \end{bmatrix}, \quad \mathbf{R}_x = \begin{bmatrix} 1 & 0 \\ 0 & -1 \end{bmatrix}, \quad \mathbf{R}_y = \begin{bmatrix} -1 & 0 \\ 0 & 1 \end{bmatrix}, \quad \mathbf{C}_2 = \begin{bmatrix} -1 & 0 \\ 0 & -1 \end{bmatrix}.$$

The framework has eight nodes and sixteen bars, with an underlying graph of $K(4,4)$, the complete bipartite graph on two sets (the partitions) of four nodes each. Each node in one partition is connected to all the nodes in the other partition. Each partition forms one orbit of nodes, where all the nodes in one orbit are symmetric copies of each other. The are four orbits of bars. If the configuration were generic, there would be three more bars than the

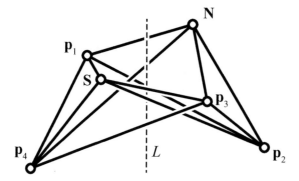

Figure 8.4 The mechanism of Figure 8.3 with the south pole **S** symmetrically attached. The framework has the graph of a regular octahedron, but the 2-fold symmetry about the line *L* forces the geometric realization to have intersecting faces. If the node **S** were moved down in the figure, a non-intersecting realization would be reached, but this would destroy the symmetry, and the framework would not be a finite mechanism.

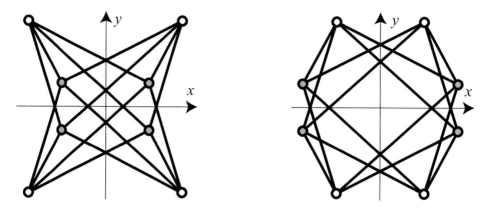

Figure 8.5 A framework with an underlying graph of $K(4,4)$ that has reflection symmetry in the *x*- and *y*-axes; one of the sets of four nodes is shown shaded. The framework is a finite mechanism with one degree of freedom: two different configurations are shown.

$2 \times 8 - 3$ required to make the system isostatic, and hence four more than would be expected in the generic case for a finite mechanism.

We show here that one member of each orbit of bars is redundant. Choose a node 1 from one partition, and a node 2 from the other. The squared lengths of the bars connected to node 1 are given by $(\mathbf{p}_1 - \mathbf{p}_2)^2 = (\mathbf{p}_1 - \mathbf{E}\mathbf{p}_2)^2$, $(\mathbf{p}_1 - \mathbf{R}_x\mathbf{p}_2)^2$, $(\mathbf{p}_1 - \mathbf{R}_y\mathbf{p}_2)^2$, and $(\mathbf{p}_1 - \mathbf{C}_2\mathbf{p}_2)^2$. However, the constraints provided by these four bars are not independent, as they are related by the equation

$$(\mathbf{p}_1 - \mathbf{E}\mathbf{p}_2)^2 + (\mathbf{p}_1 - \mathbf{C}_2\mathbf{p}_2) - (\mathbf{p}_1 - \mathbf{R}_x\mathbf{p}_2)^2 - (\mathbf{p}_1 - \mathbf{R}_y\mathbf{p}_2)^2 = 0.$$

Thus any one of the orbits of bars (containing four bars) can be removed without changing the constraints on the nodes, and the framework is a finite mechanism.

8.4 Algebraic Methods for Creating Finite Mechanisms

In this section, we present two examples of finite mechanisms, where there does not seem to be any symmetry to show that framework is a finite mechanism.

8.4.1 Bipartite Graph

The first example is a complete bipartite graph, partitioned so that one partition of the nodes is on the x-axis and the other partition is on the y-axis as shown Figure 8.6. Suppose that the starting configuration of points is $\mathbf{p} = [\mathbf{p}_1, \ldots, \mathbf{p}_n, \mathbf{p}_{n+1}, \ldots, \mathbf{p}_{n+m}]$, where $\mathbf{p}_i = [a_i, 0]$ for $i = 1, \ldots, n$ and $\mathbf{p}_j = [0, b_j]$ for $j = n + 1, \ldots, n + m$, and all the $a_i \neq 0$, all the $b_j \neq 0$. The following is a description of the motion of the configuration, for $0 \leq t \leq \min\{b_{n+1}^2, \ldots b_{n+m}^2\}$:

$$\mathbf{p}_i(t) = \left[\frac{a_i}{|a_i|} \sqrt{a_i^2 + t} \; ; \; 0 \right] \text{ for } i = 1, \ldots, n \text{ and}$$

$$\mathbf{p}_j(t) = \left[0 \; ; \; \frac{b_j}{|b_j|} \sqrt{b_j^2 - t} \right] \text{ for } j = n + 1, \ldots, n + m.$$

It is easy to check that this motion is a finite mechanism.

8.4.2 Suspension

The second example is obtained by taking a closed polygon in $\mathbb{E}^3$, the *equator*, together with two other nodes $\mathbf{N}$, the *north pole*, and $\mathbf{S}$, the *south pole*, both of which are joined to each node on the equator. This whole framework is called a *suspension* (and sometimes it is also called a *bipyramid*). In Connelly (1974) a complete description of when such suspensions

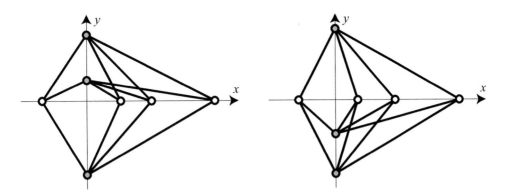

Figure 8.6 A framework with underlying graph $K(3,4)$, where the three vertices (shown shaded) lie along the y-axis, and four vertices along the x-axis. This is a finite mechanism, for which two configurations are shown. Notice that one node has crossed the x-axis in the motion between the two configurations – only one node from each partition will do this during the motion (see Exercise 8.9).

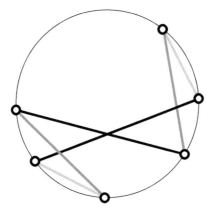

Figure 8.7 This shows the equator of a suspension, where the bars are paired such that paired bars have the same length, but go in opposite directions around the circle. (During a walk along the cycle of bars, for each pair of bars, the walk will go clockwise around the circle for one bar, counter clockwise for the other.) When the north and south poles are attached one gets a suspension that is a finite mechanism.

are finite mechanisms is given. (See also Connelly, 1978.) Here we present one special case, where the equator is planar.

Let $\mathbf{p} = [\mathbf{p}_1; \ldots; \mathbf{p}_n]$ be the nodes on the equator in order, so $\{1,2\}, \{2,3\}, \ldots,$ $\{n-1,n\}, \{n,1\}$ are the bars of the equator. Place the nodes on a circle in such a way that the equatorial bars come in pairs of the same length, and no bar is a diameter of the circle. But one must also arrange it so that as one proceeds around the equator in the order of the indices; the direction of each bar projected along the circle is the opposite of the other. Figure 8.7 shows an example with $n = 6$. When $\mathbf{N}$ and $\mathbf{S}$ are placed on the line perpendicular to the plane of the equator, the resulting suspension is a finite mechanism. This is easy to see because angles on the circle subtended by the bars add to 0 since they cancel in pairs. So as the circle is expanded and contracted the bars stay the same length when they are constrained to lie on the circle, and the $\mathbf{N}$ and $\mathbf{S}$ nodes simply move up and down on the perpendicular axis.

Note that although the pairing of the equatorial bars is needed to construct this mechanism, there is no overall symmetry of this framework.

8.5 Crinkles

8.5.1 The Bricard Octahedral Mechanisms

In Bricard (1897), three types of mechanisms are described, whose graph is the graph of an octahedron with each vertex of degree four. They are all self-intersecting since the generalized volume that they bound is zero. The first, Type 1, is symmetric about a line and was already described in Section 8.3.1.

The second Bricard octahedron, Type 2, is symmetric about a plane π_0, where the north $\mathbf{N}$ and south $\mathbf{S}$ poles are in π_0, and the equator is in a plane π_1 perpendicular to π_0 so the

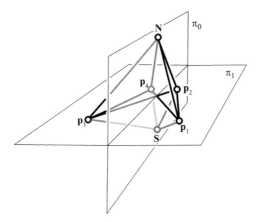

Figure 8.8 A Bricard Type 2 mechanism, shown in a perspective view. Four vertices lie in a plane π_1, which in general is not a plane of reflection, whereas the plane π_0 is a plane of reflection.

opposite vertices of the equator are reflected through π_0. Figure 8.8 shows this arrangement. The equator, the *bow-tie*, is a finite mechanism in the plane π_1 that flexes, maintaining reflective symmetry with respect to the plane π_0. As the line $\mathbf{p}_1\mathbf{p}_2$ moves, it is possible to find the position of node $\mathbf{N}$ by rotating the triangle $\mathbf{N}, \mathbf{p}_1, \mathbf{p}_2$ around the $\mathbf{p}_1, \mathbf{p}_2$ axis until the $\mathbf{N}$ vertex intersects the π_0 plane. The π_0 plane acts as a mirror and so a similar arrangement for $\mathbf{p}_3\mathbf{p}_4$ will be identical, and the four lengths connecting $\mathbf{N}$ the four vertices of the equator $\mathbf{p}_1, \mathbf{p}_2, \mathbf{p}_3, \mathbf{p}_4$ are constant. A similar construction works for the south pole $\mathbf{S}$.

Note that if the $\mathbf{N}$ and $\mathbf{S}$ vertices are chosen to be symmetric to the intersection line of $\pi_0 \cap \pi_1$, the framework will be of both Type 1 and Type 2. For a Type 2 Bricard octahedron, a pair of opposite edges will intersect during the motion, while that may not happen with a Type 1 Bricard octahedron.

Bricard's Type 3 octahedron has the property that twice during the motion it flexes so that all six vertices line in a plane, but it generally does not have any plane or line symmetry.

8.5.2 Crinkles

We describe how to create embedded triangulated surfaces with a simple quadrilateral boundary that are finite mechanisms, but so that they have a pair of nonadjacent vertices that remain fixed during the motion. We call these *crinkles*. Start with one of the line symmetric or planar Bricard octahedra, such that when one of the edges on the equator and the two adjacent triangles are removed, it becomes embedded. For example, when the equator is a self-intersecting quadrilateral (with opposite edges of equal length), the *bow-tie*, or with line symmetry sufficiently close to a bow-tie, one can remove one of the long edges, and the deleted surface will be embedded. Figure 8.9(a) shows the planar or near-planar realization of the equatorial quadrilateral. Then Figure 8.9(b) shows the completed crinkle.

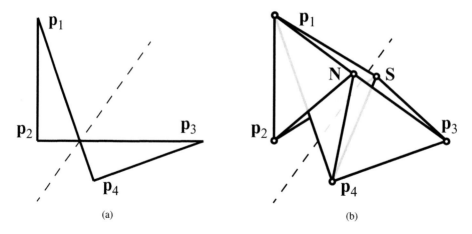

Figure 8.9 A construction for a crinkle. (a) A bow-tie is drawn in the plane of the paper π_1, as shown in Figure 8.8, symmetric about the dashed line shown. (b) The line $\mathbf{p}_2\mathbf{p}_3$ is deleted (the implied edge). North and south poles are added above and below the plane, symmetric in the line of symmetry and connected to the other nodes to form the six faces shown, giving the completed crinkle.

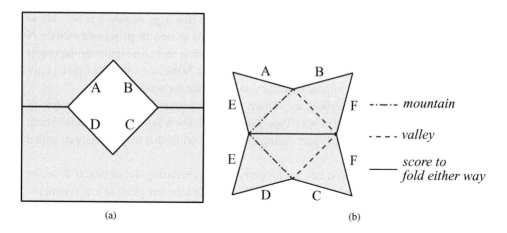

Figure 8.10 How to make an example crinkle. A sheet with a fold line and a square hole is shown in (a). The net for the crinkle is shown in (b), and can be folded with two triangular flaps folded forward, two backward, and inserted in the sheet so that edges are joined A-A, B-B, etc. The resultant object is shown in Figure 8.11.

Crinkles are useful, because they can replace a portion of a hinged surface without affecting the kinematics of any mechanisms. Figure 8.10(a) shows a portion of a surface, where two rigid planar polygons are joined along a line that serves as a hinge allowing the two polygons to flex relative to each other. Figure 8.10(b) shows an unfolded net of one example of a crinkle, such that when the designated pairs of edges are attached to each other (one up and one down), the other pairs of edges can be attached and inserted to the

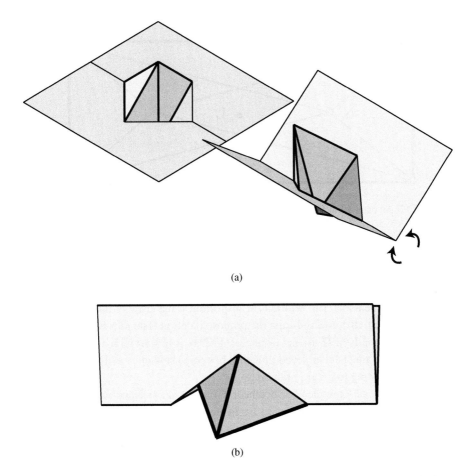

(a)

(b)

Figure 8.11 The crinkle manufactured from the net in Figure 8.10, shown in isometric view (a), both when the surface is flat and flexed, and shown in side view in (b) when the surface is flexed. Note that both the crinkle, its configuration, and the choice of viewpoint shown in (b) has been chosen to match the construction of the crinkle given in Figure 8.9. A key feature shown in (b) (and Figure 8.9(b)) is the "notch" in the implied edge ($\mathbf{p}_2\mathbf{p}_3$) that allows other material to intrude on this edge without clashing; this will be used for the construction in Section 8.6.

corresponding edges of 8.10(a) yielding the final crinkle. Figure 8.11 shows the crinkle embedded in the surface.

8.6 A Triangulated Surface that is a Finite Mechanism

The previous examples of triangulated surfaces without boundary in this chapter have all been self-intersecting. There was a long-standing conjecture that all embedded surfaces in $\mathbb{E}^3$ were rigid. Here we show how to construct counterexamples along with a proof that they are, indeed, finite mechanisms. Several examples have been proposed as such

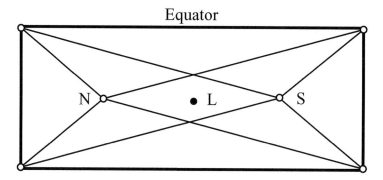

Figure 8.12 An octahedral finite mechanism, shown in a planar configuration, with a line of symmetry which is perpendicular to the plane of the rectangle forming the equator. The north and south poles are chosen symmetrically about the line of symmetry inside the rectangle in the plane.

counterexamples, and upon building them out of paper or other materials, it can seem that they are finite mechanisms. But with closer inspection it has usually turned out that they have some very small distortions during the proposed motion. The idea here, following the construction in Connelly (1979b) and Connelly (1979a), is to start with a framework that is known to be a finite mechanism, even though it has some self-intersections, and then alter it by cutting and pasting to get rid of them.

Figure 8.12 shows an example of an octahedral framework, a suspension over a quadrilateral which is a finite mechanism, that starts out in the plane, but during the motion it becomes 3-dimensional. The equator starts out as a rectangle in a plane, and the **N** and **S** nodes are placed also in that plane as shown. By Lemma 8.3.1, there is a line of symmetry, which, in this case, is a line perpendicular to the plane through the centre of the rectangle. If every triangle forms a face, the resultant surface is topologically a sphere, but geometrically, every point in the northern hemisphere, except the equator, intersects a corresponding point in the southern hemisphere. In addition, two pairs of bars intersect. Things get a little better as the framework moves out of the plane, but it turns out that, although one of the intersections of a pair of triangles disappears, it always happens that another pair of triangles link. This self-intersection prevents this surface from being an embedding. However, we can *almost* create an embedding in the following way, shown in Figure 8.13: for every point in the northern hemisphere replace each triangular face by the upper three triangles of a tetrahedron. This puts four triangular based "tents" on the triangle bases in the northern hemisphere. Then do the same, but for downward shaped tents for the southern hemisphere as in Figure 8.13.

We have now rid ourselves almost completely of the offending self-intersections. If we look into the surface near the intersection of one pair of the bars, it is at the bottom of a valley and looks something like the surfaces in Figure 8.14(a), two wedges touching at one point in each of their edges.

We now turn to the cut-and-paste part of the construction. Near one of the intersection points, remove a small quadrilateral in, say, the top surface, creating a hole as shown in Figure 8.14(b). Now insert a crinkle, such as the one shown in Figure 8.11(b) so that

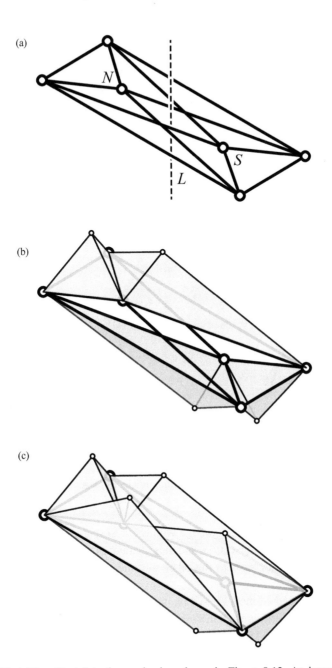

Figure 8.13 Adding "tents" to the mechanism shown in Figure 8.12. An isometric view without tents is shown in (a); two tents are added both top and bottom in (b); all eight tents have been added in (c). The resultant surface is flexible, but self-intersects at two points. As the surface flexes, one intersection moves apart, but the clash at the other intersection becomes worse, with increasing portions of the surface intersecting as the magnitude of the flex increases.

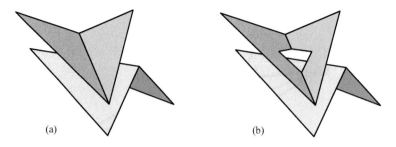

Figure 8.14 (a) A detail showing the point of intersection between the north and south tents shown in Figure 8.13. (b) The intersection can be removed by cutting away some of the surface; the resultant hole can be filled in with a crinkle.

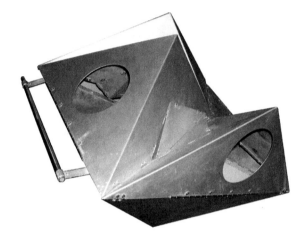

Figure 8.15 A physical model of a flexible surface based on the recipe given in this chapter. (Photograph courtesy of Institut des Hautes Études Scientifiques.)

the "notch" fits over the underlying wedge with room to spare. This allows space for the surface to flex without self-intersection, and thus guarantees a triangulated flexible surface. Of course, adding a crinkle near to the other self-intersection will increase the possible range of motion.

The first well-engineered model based on the tents construction was built at the Institut des Hautes Études Scientifiques (IHÉS) in 1977, and it still present in their library. The model is shown in Figure 8.15.

Figure 8.16 shows how to build another polyhedral surface that is a finite mechanism, due to Klaus Steffen. The model attaches two crinkles, made from Bricard Type 1 mechanisms, to each other, and to a rigid tetrahedron. The model is very efficient, in that is uses only 9 nodes and 14 triangles, and this is the flexible triangulated embedded surface with the smallest known number of nodes or triangles.

There is an interesting property of these surfaces that are finite mechanisms in $\mathbb{E}^3$. The "bellows" conjecture says that the volume bounded by such surfaces is constant during

A symmetric flexible Connelly sphere with only nine vertices

by Klaus Steffen (I.H.E.S.)

1.) Make 14 rigid triangles and attach them to each other in a flexible fashion as indicated in fig.1,2 (two copies!); a good choice of parameters is e.g. $a := 6$, $b := 5$, $c := 2.5$, $d := 5.5$, $e := 8.5$.

2.) Connect (in a flexible way!) the two edges marked ① in fig.1 by rotating the corresponding triangles upward and the two edges marked ② by rotating the corresponding triangles downward (in either copy!).

3.) Attach the two aggregates of 6 triangles to each other as indicated by ③,④ in fig.3.

4.) Connect the two remaining single triangles (fig.2) along edge e thereby making a "roof" which is attached to the configuration of 12 triangles from step 3.) as indicated by ⑤,⑥,⑦,⑧ in fig.3.

5.) If you did not mess up everything the resulting sphere looks like fig.4 and flexes by about 30° as indicated by the arrows. (It is a good idea to cut a "window" in the "roof" to make the inside visible.)

fig. 1
(2 copies)

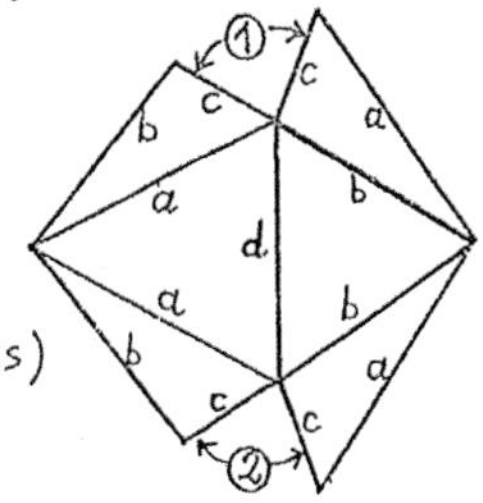

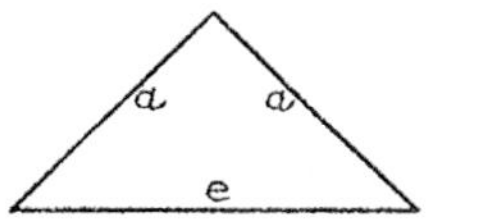

fig. 2 (2 copies)

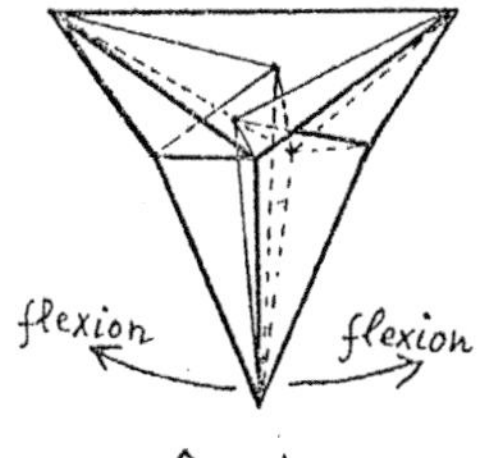

flexion flexion

fig. 4

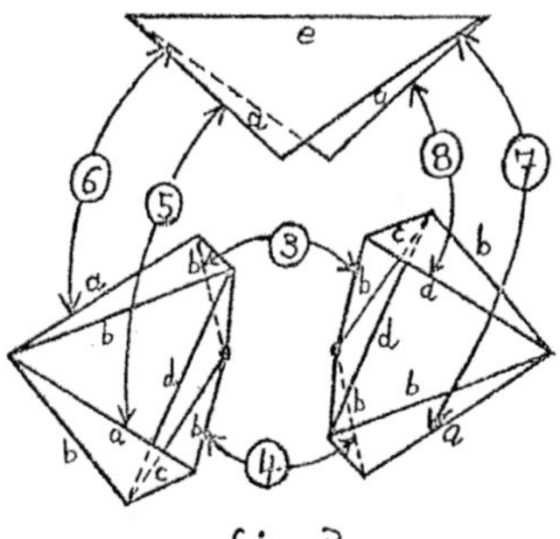

fig. 3

Figure 8.16 A description of the Steffen flexible polyhedron, written by Klaus Steffen at IHÉS in 1977, and reproduced here with his permission.

the motion. This was first proved in Sabitov (1998). See also another proof in Connelly et al. (1997). So it is not possible to build a perfect bellows in the sense that it blows air in and out through a hole in its side. Actual bellows must rely on small distortions in the material of the surface to allow the air to flow in and out or use a structure such as Figure 8.1, which is a surface with boundary. More recently, in Gaifullin (2014), it is shown for arbitrarily high dimensions, that a flexible surface bounds a constant volume as it flexes.

8.7 Carpenter's Rule Problem

Suppose that we have a closed polygon with no self-intersections in the plane, such as that shown in Figure 8.17(b). Steve Shanuel, Sue Whitesides, and Joe Mitchell (somewhat independently) asked whether any such closed polygon can be continuously opened to a convex shape (see Kirby (1995), problem 5.18 for an early statement of the problem). A closely related question was whether an embedded open chain can be continuously opened to a straight segment, keeping the edge lengths fixed – this being where the carpenter's rule name came from, as the open polygon resembles a carpenter's rule (or ruler) laid flat. However, if the end points of an open chain are connected to create an embedded closed chain, the open

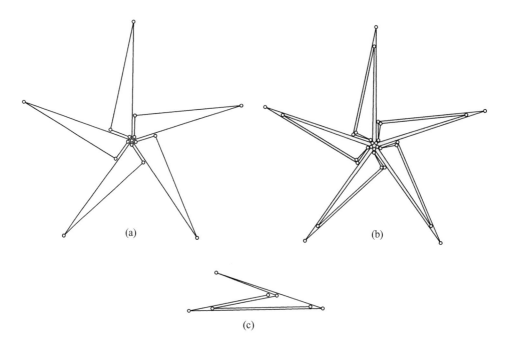

Figure 8.17 Framework (a), from Biedl et al. (2002), shows that a tree can lock in the plane. It was thought that a small neighborhood of such a tree (shown in (b), with the interior coloured grey) would also lock, but the closed chain can open to a convex polygon. Notice that if the inside portions are squeezed close enough together, no one of the arms can open without bumping into its neighbors. They all have to move together. The chain (c) is from Kirby (1995) and was also thought to be locked, but it can open to a straight chain.

chain can be opened to be part of a convex closed chain. Then the open part can be easily moved to a straight line configuration, and so we concentrate on the closed chain case.

The following key observation was conjectured by Günter Rote and proved in Connelly et al. (2003).

Theorem 8.7.1. *Let $(G, \mathbf{p})$ be the tensegrity on the plane formed by placing struts between every pair of non-adjacent nodes of an embedded non-convex bar polygon, where no two adjacent bars are in a straight line. Then there is an infinitesimal motion of $(G, \mathbf{p})$ that is strict on every strut.*

It is easy to construct a triangulation of the convex hull of the polygonal chain or closed curve such that all the edges of the triangulation are struts, which want to increase in length, and this will ensure the expansive nature of the final motion. If there were a proper self-stress on the planar tensegrity, it would block any infinitesimal expansive motion of the tensegrity that is desired, and conversely, if there is no such non-zero proper self-stress, then there is a strict infinitesimal motion.

The key idea here is to use the Maxwell–Cremona correspondence Theorem 3.5.2 that associates to each self-stress for a framework in the plane, a surface in three-space that projects onto it. If there were a proper non-zero self-stress, there would be a lift to a surface, where the sign of the force coefficients determines the convexity at the corresponding edges of the polyhedral surface. By looking at the maximum point(s) of that surface, it is possible to derive a contradiction, unless the surface is flat, and the self-stress is zero.

For example, if the maximum occurs at single vertex, $\mathbf{p}_i$, the level curve for the level just below the maximum is a closed polygon, where, as you proceed in a counterclockwise way around the curve, all the exterior angles except possibly at, at most two, vertices are positive, i.e. the level curve turns left. This is not possible. The only way that a vertex on a polyhedral surface can be at a (local) maximum point is when there are at least three convex edges adjacent to the maximum point, while the others are concave, as shown in Figure 8.18.

Corollary 8.7.2. *With the same hypothesis as given in Theorem 8.7.1, there is a finite motion of the constructed tensegrity that is increasing on every strut.*

Proof. This is similar to the proof of Theorem 6.7.4 of Chapter 6, except that here it is for first-order motions. Restricted to the bars of the polygon, there is only the $\mathbf{0}$ self-stress, so we can arrange the motion to fix the length of those bars, and increase strictly on all the struts. $\square$

We say that an infinitesimal motion of two sets is *expanding* if it remains an infinitesimal motion when struts are placed between all the pairs of points, one in one set and one in the other set.

Lemma 8.7.3. *If an infinitesimal motion of two bars is expanding when restricted to their end points, then it is expanding between all pairs of points in the two sets. It is strictly expanding between a pair of points, one an interior point and not both from the same interval, if the inequality for the infinitesimal motion on any pair of end points on the corresponding strut is strict.*

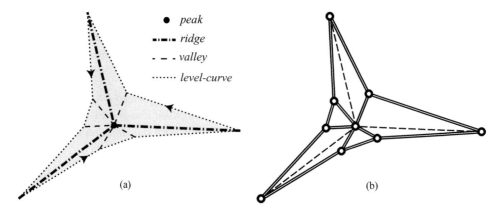

Figure 8.18 (a) An oriented level curve (dotted) just below the maximum point of a surface. (b) The corresponding tensegrity, showing struts and cables. There must be at least three points at which the level curve turns left (three ridges to the peak). The Maxwell–Cremona construction for the carpenter's rule allows at most two ridges, and hence there can be no maximum point at a single vertex.

Proof. It is enough to consider the case where one point is in the interior of one interval and the other points are end points of the other interval. Say $[\mathbf{p}_1, \mathbf{p}_2]$ is the bar and $\mathbf{p}_3$ is the other point. Assuming the first-order motion $\mathbf{p}'$ is increasing for those three points, we wish to show that the infinitesimal motion is increasing for $t\mathbf{p}_1 + (1-t)\mathbf{p}_2$ and $\mathbf{p}_3$, for $0 < t < 1$. We can assume $\mathbf{p}'_1 = \mathbf{p}'_2 = \mathbf{0}$ since it is a bar. But it is clear that

$$(\mathbf{p}_3 - \mathbf{p}_1)(\mathbf{p}'_3 - \mathbf{p}'_1) = (\mathbf{p}_3 - \mathbf{p}_1)\mathbf{p}'_3 \geq 0 \quad \text{and} \quad (\mathbf{p}_3 - \mathbf{p}_2)(\mathbf{p}'_3 - \mathbf{p}'_2) = (\mathbf{p}_3 - \mathbf{p}_2)\mathbf{p}'_3 \geq 0$$

implies

$$t(\mathbf{p}_3 - \mathbf{p}_1)\mathbf{p}'_3 + (1 - t)(\mathbf{p}_3 - \mathbf{p}_2)\mathbf{p}'_3 = (\mathbf{p}_3 - (t\mathbf{p}_1 + (1 - t)\mathbf{p}_2))\mathbf{p}'_3 \geq 0. \qquad \square$$

Thus it is clear that under the corresponding finite motion, there will be no self-intersections that will be created, if the motion is expanding on the nodes of the polygon. This leads us to the solution of the Carpenter's Rule Problem.

> **Theorem 8.7.4.** *For any closed bar polygon embedded in the plane, there is an expanding finite motion that opens it to a convex configuration.*

The idea of the proof is that it is possible to define a canonical expanding infinitesimal motion for any configuration that is not convex (thus giving a *vector field* in the local configuration space). Then the finite motion is obtained by integrating this vector field, which amounts to solving a set of differential equations. During the motion some pairs of adjacent bars may become collinear. When this happens, just fuse those edges into one longer edge and continue the process. This is explained carefully in Connelly et al. (2003). For examples of the animation of the open motion, see Demaine (1993).

One interesting consequence of the expanding nature of the opening motion is that the area enclosed by the closed polygon increases during the motion until the polygon becomes

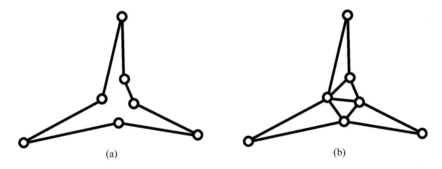

(a)   (b)

Figure 8.19 (a) A polygon that is a pseudo-triangle with $k = 7$ vertices: there are three vertices with a positive external angle and four others that have a negative external angle. (b) The pseudo-triangle can be converted to a triangulation by the addition of $k - 3$ bars.

convex. The idea is to subdivide the region bounded by the polygon into non-obtuse triangles and extend the expanding motion to the edges of that triangulation. Then it possible to see that the area of the non-acute triangles is increasing. This is also explained in Connelly et al. (2003).

8.7.1 Pseudo-Triangulations

There have been some extensions and modifications of the ideas above. One of the most interesting is the idea of inserting additional bars into the convex hull of the polygon in such a way that there is only one degree of freedom in the expansive motion. For a polygon $\mathbf{p} = [\mathbf{p}_1; \ldots; \mathbf{p}_n]$ embedded in the plane, where the vertices are counted cyclically in order in a counter clockwise direction, the *external angle* at a vertex $\mathbf{p}_i$ is the angle from the vector $\mathbf{p}_i - \mathbf{p}_{i-1}$ to $\mathbf{p}_{i+1} - \mathbf{p}_i$. A *pseudo-triangle* in the plane is a polygon, where the external angle is positive at exactly three vertices. Figure 8.19(a) shows an example. If a pseudo-triangle has k nodes, we can convert the pseudo-triangulation to an actual triangulation without adding any new nodes by adding bars. The number of additional bars that are added is $k - 3$ – this can be shown by induction on the number of vertices in each region. We define a node with a negative external angle to be an *edge node* for that pseudo-triangle. There are $k - 3$ edge nodes for each pseudo-triangle, equalling the number of bars added to make the actual triangulation – see Figure 8.20(b).

A *pointed pseudo-triangulation* of a convex polygon in the plane is a bar framework that include the edges of the convex polygon, has no crossing bars that include the edges of the convex polygon, where all the regions inside the convex polygon are pseudo-triangles, and where for each interior node $\mathbf{p}_i$, the adjacent bars all lie on one side of a line through $\mathbf{p}_i$. Figure 8.20(a) shows an example of a pseudo-triangulation.

The following two results can be found in Streinu (2005).

Lemma 8.7.5. *Any closed (embedded) polygon in the plane, with no adjacent edges collinear, can be extended to be part of a pointed pseudo-triangulation without adding any new nodes.*

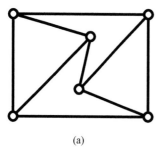
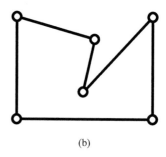

(a)   (b)

Figure 8.20 (a) A pointed pseudo-triangulation of a rectangle. Note that this pseudo-triangulation includes the non-convex closed polygon shown in (b). An alternative carpenter's rule proof to the one given in Section 8.7 would have as a first step the generation of the pseudo-triangulation in (a) from the closed polygon in (b).

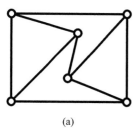
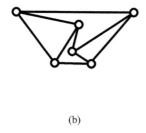
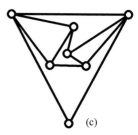

(a)   (b)   (c)

Figure 8.21 Starting with the polygon in (a) we first deform it by a projective transformation to get the polygon (b), and then attach the vertex of degree two as shown in (c) to place the polygon in a triangle. We thus obtain a pseudo-triangulation with one more pseudo-triangle than the original polygon, and a triangular boundary.

The idea is to add a maximal number of edges maintaining the pointed property. The crucial point of this approach is the following, which again uses the Maxwell–Cremona lifting argument outlined in Figure 8.18.

Theorem 8.7.6. *Any pointed pseudo-triangulation of a convex planar polygon is an isostatic bar framework, and if one of the boundary bars is converted to a strut, there is a finite expansive motion of the configuration as the length of the strut increases.*

Proof. We assume that the polygon is a triangle for simplicity. The general case can always be converted to this case by attaching a single vertex of degree two to two vertices on the boundary, and using a projective transformation of the original polygon so that the polygon is inside the triangle formed with just one more pseudo-triangle. Figure 8.21 shows this conversion.

We first show that if there are n nodes in the pseudo-triangulation and b bars, then $b = 2n - 3$. To do this we complete the pseudo-triangulation to an actual triangulation without adding any new nodes. Let b^* be the total number of bars in the resultant triangulation of the triangle. As each of the $n-3$ internal nodes is pointed, it must be an edge-node for precisely one pseudo-triangle, and will correspond to one additional bar in the triangulation.

Thus the number of added bars equals the number of internal nodes, $n - 3 = c = b^* - b$. An Euler characteristic argument gives that the number of bars in a triangulation of the triangle is $b^* = 3n - 6$. Thus $n - 3 = 3n - 6 - b$, which implies that $b = 2n - 3$.

As $b = 2n - 3$ the pseudo-triangulation will be isostatic if and only if it has only the zero self-stress. But the Maxwell–Cremona lift of each pseudo-triangle must be planar and cannot have a maximum at a node with negative external angle. Thus the maximum cannot occur at any internal node. We can assume that the triangle boundary is horizontal, and thus the lift must be flat, and the corresponding self-stress must be zero.

To show the expanding nature of the infinitesimal motion when one of the external bars is a strut, we extend the Maxwell–Cremona argument to the case when the additional members used to create the triangulation are also struts, similar to the argument of Theorem 8.7.1. In other words, there is no proper self-stress in this tensegrity. So all the triangles of the triangulation are expanding under the resulting infinitesimal motion. By extending the triangulation to include a subdivision between all pairs of nodes of the pseudo-triangulation, we can be sure that the motion is expanding. The motion here is strictly expanding between any two nodes whose line segment does not pass through only triangles of the pseudo-triangulation. □

So combining Lemma 8.7.5 and Theorem 8.7.6 we can find a one degree of freedom expansion of any closed non-convex polygon. This only stops when the framework ceases to be a pseudo-triangulation. Then we can re-pseudo-triangulate.

This provides an extension of the ideas of Theorem 8.7.4 to alter the expanding motion, allowing one degree of freedom. The expanding nature of the motion is assured by the pointed nature of the bar framework and the existence of at least one strut on the boundary.

A *Laman graph* is a graph G with $n \geq 2$ nodes, $b = 2n - 3$ bars, and such that every subgraph with n^* nodes and b^* bars has $b^* \leq 2n^* - 3$. This is precisely the condition that a corresponding bar framework $(G, \mathbf{p})$ be infinitesimally rigid in the plane, as shown in Chapter 7. Another interesting property of pseudo-triangulations related to Laman graphs is the following from Haas et al. (2005).

Theorem 8.7.7. *Any planar Laman graph can be realized in the plane as a pseudo-triangulation.*

8.7.2 Energy Methods

One very natural method to convexify a closed chain for the carpenter's rule problem is to use energy functions defined on the configuration space and then proceed following a gradient flow to a local minimum. For example, let G be the graph that is a closed chain of bars with struts between every pair of nodes that are not joined by a bar. Then define $E(\mathbf{p}) = \sum_{ij} f(|\mathbf{p}_i - \mathbf{p}_j|)$ on the space of configurations with the given bar lengths, where $f(x)$ is real-valued function such that it is strictly monotone decreasing and goes to infinity or at least gets very large as x goes to 0. The problem with this approach, as with others of this sort, is that there seems to be no guarantee that the final configuration will be convex. However, that is not the case. It is pointed out in Cantarella et al. (2004) that such a method always works. The point is that the energy function always has the option of increasing the distances between all the nodes not connected by bars by Theorem 8.7.1, and so there never can be a non-convex convex configuration that is at a local minimum. The motion may decrease some pairs of distances during the intermediate motion, but ultimately it will arrive at a convex configuration. It is also pointed out in Cantarella et al. (2004) that implementations of this algorithm seem to be somewhat faster, more accurate, and more efficient than those using the algorithm in Connelly et al. (2003).

8.8 Algebraic Sets and Semi-Algebraic Sets

An *algebraic set* is simply the set of points that are satisfied by a finite set of polynomial equations,

$$\{(x_1, \ldots, x_N) \mid p_1(x_1, \ldots, x_N) = 0, \ldots, p_M(x_1, \ldots, x_N) = 0, \ p_i \text{ polynomials}\}.$$

On the other hand a *semi-algebraic set* is the finite union of sets defined by polynomial inequalities,

$$\bigcup_j \{(x_1, \ldots, x_N) \mid p_{1j}(x_1, \ldots, x_N) \leq 0, \ldots, p_{Mj}(x_1, \ldots, x_N) \leq 0, \ p_{ij} \text{ polynomials}\},$$

where some of the inequalities could be strict.

It is also important to realize that the projection of an algebraic set is not necessarily an algebraic set, but it is always a semi-algebraic set. For example, a hyperbola $\{(x, y) \mid xy = 1\}$ is an algebraic set in the plane. But its projection onto the x-axis is the set of non-zero points, which is not an algebraic set. Finite unions, finite intersections, boundaries, and complements of semi-algebraic sets are again semi-algebraic sets. See Bochnak et al. (1998) for a discussion of such matters.

Even one-dimensional algebraic sets can be topologically different from semi-algebraic sets. For example, an algebraic set cannot be topologically a closed line segment, although clearly a closed line segment can be defined by inequalities and so is a semi-algebraic set. See, for example, a more general result along these lines in Sullivan (1971).

The relevance of algebraic sets here is that the configuration space defined by a set of fixed bar lengths (giving second-order polynomial equations) is an algebraic set. One consequence of this is that if a finite mechanism is such that the motion seems to stop at some point, and then must retrace the same path used to get to that configuration, then it is not exactly a true finite mechanism.

Although an algebraic set cannot have an end point, it is possible for an algebraic set to have a cusp – for example, $\{(x, y) \mid x^2 = y^3\}$ has cusp at $(0, 0)$. A mechanism where a given configuration corresponds to a cusp will "stop" and proceed back in the direction that it came, but along a different path. In Section 8.8.3 we will see an example of such a mechanism, which also serves as a simple example showing that the natural definition of third-order rigidity is not what one would want.

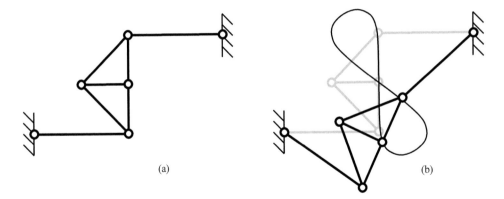

Figure 8.22 The Watt mechanism, where the end nodes are fixed in space. (a) The mechanism in its central configuration. To first order, the motion of the central node is vertical about this point, but to higher order, the node moves to the right as it moves down. (b) The mechanism in a displaced configuration, with the full path of the central node also shown. The central node has moved slightly to the right.

Even more generally, it is true that a wide variety of algebraic sets can be realized as the configuration space of a mechanism. One old problem was to define a finite mechanism called a *linkage* such that one of the nodes of the linkage would trace out a given curve. Indeed, in the 19 century, it was an open problem for a time to find a linkage in the plane that would trace out a portion of a straight line. It was eventually solved by Peaucellier, see Coxeter and Greitzer (1967), and independently by Lipkin. See Davis (1983) for an interesting account of this story. We discuss this in Section 8.8.2, but first explore an early, simple, approximate straight line mechanism.

8.8.1 Watt Mechanism

In the 18th century, James Watt invented a mechanism that allowed a point to trace out an approximation to a straight line, for guiding a piston within a steam engine. A framework reproducing the kinematics of the Watt mechanism is shown in Figure 8.22.

8.8.2 Peaucellier mechanism

In order to understand the geometry of the Peaucellier straight-line linkage to be described later, we first describe a function α called *inversion*, from the plane minus the origin to itself, that has the property that it takes circles through the origin to straight lines. It also takes circles not through the origin to other circles not through the origin, and lines through the origin to themselves. We will only be concerned with circles through the origin. For $\mathbf{p}_1 \in \mathbb{E}^2 - \mathbf{0}$ we define

$$\mathbf{p}_2 = \alpha(\mathbf{p}_1) = \frac{l^2 \mathbf{p}_1}{|\mathbf{p}_1|^2}.$$

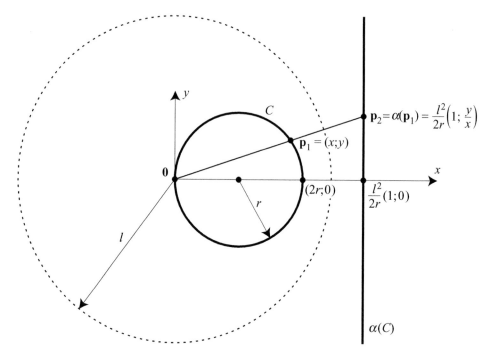

Figure 8.23 A circle C of radius r through $\mathbf{0}$ is inverted through a circle of radius l into a straight line $\alpha(C)$.

In other words, $\mathbf{p}_2 = \alpha(\mathbf{p}_1)$ is the point on the ray from $\mathbf{0}$ to $\mathbf{p}_1$ that is at a distance $l^2/|\mathbf{p}_1|$ from the origin. The points inside the circle of radius l and outside the circle of radius l are interchanged, while the points on the circle of radius l are fixed by α.

> **Lemma 8.8.1.** *For any circle C through the origin, inversion interchanges the points on $C - \mathbf{0}$ and a line perpendicular to the line through the centre of C and $\mathbf{0}$.*

Proof. Choose a coordinate system so that the centre of the circle is at $(r; 0)$, and hence the equation for C is $(x - r)^2 + y^2 = r^2$, for $r > 0$. Thus $x^2 + y^2 = 2xr$, and if $\mathbf{p}_1 = (x; y)$, then

$$\alpha(\mathbf{p}_1) = \frac{l^2 \mathbf{p}_1}{(x^2 + y^2)} = \frac{l^2}{2r}\left(1; \frac{y}{x}\right),$$

which describes the points on a straight line. See Figure 8.23.   □

We now describe the Peaucellier linkage, first described in 1864, for drawing a portion of a straight line. The idea is to define a linkage such that, as one node moves in the plane, another is positioned at its inverted point.

> **Lemma 8.8.2.** *In Figure 8.24(a) the lengths a and b are the lengths of the corresponding bars, and $a^2 - b^2 = l^2$. Then the points $\mathbf{p}_1$ and $\mathbf{p}_2$ are inverted into each other through a circle of radius l.*

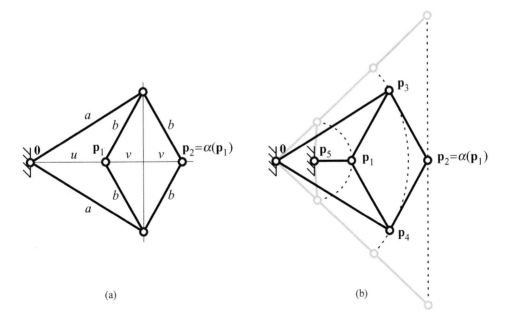

(a)   (b)

Figure 8.24 (a) A linkage with six bars that inverts $\mathbf{p}_1$ and $\mathbf{p}_2$ about the point $\mathbf{0}$ through a circle of radius l. The bar lengths a and b are fixed, with $a^2 - b^2 = l^2$, and the lengths u and v can vary. (b) The Peaucellier linkage, where the points $\mathbf{0}$ and $\mathbf{q}$ are fixed, and the possible range of points for $\mathbf{p}_1, \mathbf{p}_2, \mathbf{p}_3, \mathbf{p}_4$ are shown. The configurations that correspond to the limiting positions of the points are shown in grey.

Proof. With u and v as indicated, we calculate the square of the altitude of the isosceles triangle with side lengths b in two ways. It is $a^2 - (u + v)^2 = b^2 - v^2$. Thus $l^2 = a^2 - b^2 = u^2 - 2uv = u(u + 2v)$, and $|\alpha(\mathbf{p}_1)| = u + 2v = l^2/u = l^2/|\mathbf{p}_1|$. □

If we force the node $\mathbf{p}_1$ to lie in a circle through $\mathbf{0}$ by attaching a bar to $\mathbf{p}_1$ and fixing $\mathbf{0}$ and the point $\mathbf{p}_5$, as in Figure 8.24, the node $\mathbf{p}_2$ will trace a part of line that is the inversion of a circle, namely a straight line by Lemma 8.8.1 and Lemma 8.8.2.

The paths traced by $\mathbf{p}_1$, $\mathbf{p}_2$, $\mathbf{p}_3$ and $\mathbf{p}_4$ are a good illustration of the point made about algebraic and semi-algebraic sets at the start of the present section. Each of the paths are semi-algebraic sets that have end points. But each of the paths does not describe the entire range of configurations $\mathbf{p}$. The configuration $\mathbf{p}$ is an 8-dimensional vector giving the position of all four non-fixed nodes, and the set of possible configurations is given by a 1-dimensional algebraic set that cannot be, topologically, a closed line segment. Each of the individual points $\mathbf{p}_1 \ldots \mathbf{p}_4$ is a projection of $\mathbf{p}$, and the sets of possible $\mathbf{p}_i$ for $i = 1 \ldots 4$ are semi-analytical sets that are topologically equivalent to a closed-line segment.

For the Peaucellier linkage, all four non-fixed nodes reach their limit points simultaneously, when all bars but the one connected to $\mathbf{p}_5$ are aligned. The configuration path is continuous either side of this configuration, where comparing the two sides, the mechanism looks geometrically identical, but with points $\mathbf{p}_3$ and $\mathbf{p}_4$ having swapped places. In fact, the

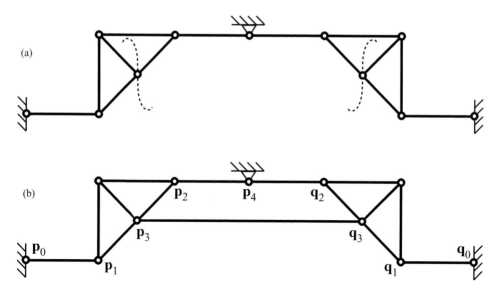

Figure 8.25 Generating a cusp mechanism. (a) Two Watt mechanisms, shown back to back. This configuration has two mechanisms where the central node of each mechanism will locally trace the paths shown, to first order moving vertically, but moving apart as they move up, and together as they move down. (b) Adding an additional bar between the nodes p_3 and q_3 adds a state of self-stress, but it does not prevent the infinitesimal mechanisms. However, it does imply that nodes p_3 and q_3 cannot move up along the final path. They must go down, and either one can move down the more. Each case represents a branch of the cusp.

configuration where all nodes reach their limit points is a bifurcation in the configuration space. Because p_3 and p_4 are aligned, another set of configurations is possible, where p_2 rotates around this point, but we will not explore that further here.

8.8.3 Cusp Mechanism

Although a set of possible configurations cannot have an end point, it is possible for an algebraic set to have a cusp. For example, $\{(x, y) \mid x^2 = y^3\}$ has cusp at $(0, 0)$. A mechanism where a given configuration corresponds to a cusp will "stop" and proceed back in the direction that it came, but along a different path. Here we will show an example of such a mechanism, taken from Connelly and Servatius (1994), which also serves as a simple example showing that the natural definition of third-order rigidity is not what one would want.

We start with two Watt mechanisms similar to the one shown in Figure 8.22, arranged back to back as shown in Figure 8.25(a). An additional bar is then added to couple the two mechanisms, as shown in Figure 8.25(b). This configuration represents a cusp point in the configuration space, because the infinitesimal motion of the central vertices both have to be in the same direction, since the path of the central vertex in the Watt mechanism is at an inflection point.

An interesting property of this finite mechanism is that for any analytic parametrization of the path, the initial velocity has to be **0**, since the configuration has to reverse direction at the cusp point. Indeed, if one tries to extend the definition of second-order rigidity to the natural analogous definition of third-order rigidity there is a problem. The following are the equations that one gets after differentiating the distance (squared) equations three times for a bar framework for each bar $\{i, j\}$:

$$(\mathbf{p}_i - \mathbf{p}_j)(\mathbf{p}'_i - \mathbf{p}'_j) = 0 \tag{8.1}$$

$$(\mathbf{p}_i - \mathbf{p}_j)(\mathbf{p}''_i - \mathbf{p}''_j) + (\mathbf{p}'_i - \mathbf{p}'_j)^2 = 0 \tag{8.2}$$

$$(\mathbf{p}_i - \mathbf{p}_j)(\mathbf{p}'''_i - \mathbf{p}'''_j) + 3(\mathbf{p}'_i - \mathbf{p}'_j)(\mathbf{p}''_i - \mathbf{p}''_j) = 0, \tag{8.3}$$

where $\mathbf{p}', \mathbf{p}'', \mathbf{p}'''$ are the k-order motions of $\mathbf{p}$ for $k = 1, 2, 3$ respectively. The "natural" definition of third-order rigidity would say that for every third-order motion $[\mathbf{p}', \mathbf{p}'', \mathbf{p}''']$ that satisfies (8.1), (8.2), (8.3), $\mathbf{p}'$ is trivial. But we show in Appendix A.2 that the finite mechanism of Figure 8.25(b) satisfies this definition, yet it is not rigid. So this "natural" definition of third-order rigidity is not what one would want.

8.9 Exercises

1. Show that there is a chain of three non-overlapping triangles in the plane, hinged at their vertices, that cannot be opened up continuously to another non-overlapping chain, unless some pair is forced to overlap. See Connelly et al. (2010) for a condition, *slender adornments*, on chains of shapes that allow them to be opened continuously without overlap.

    Figure 8.26 shows a chain of hinged polygons in the plane due to Dudeney (1902) that take an equilateral polygon to a square of the same area. See the book Demaine and O'Rourke (2007) for more related information.
2. Find a finite mechanism in $\mathbb{E}^4$ whose graph is the same as the edges of a *cross-polytope*, i.e. the complete graph on eight nodes with four pairs of disjoint edges deleted. Make sure that all the missing edge lengths vary during the motion.
3. Corollary 8.2.2 implies that the regular icosahedron with one edge deleted is a finite mechanism, as shown in Figure 8.1. Each internal edge of the remaining surface is adjacent to two other triangular facets which form a dihedral angle between the planes that they define. As the length of the removed edge decreases, determine which dihedral angles increase and which decrease. (It is helpful to place cables or struts between certain pairs of vertices and check to see if the resulting tensegrity has a proper stress or not. If there is no such proper stress the distance between those vertices must decrease or increase in accord with the cable or strut placement. It is also helpful to use the symmetry of the icosahedron.)
4. Show that for a generic triangulated polyhedral spherical surface with one edge missing and no separating triangle, not only is the system a finite mechanism, but all the internal dihedral angles change during the motion. An idea is to use vertex splitting as

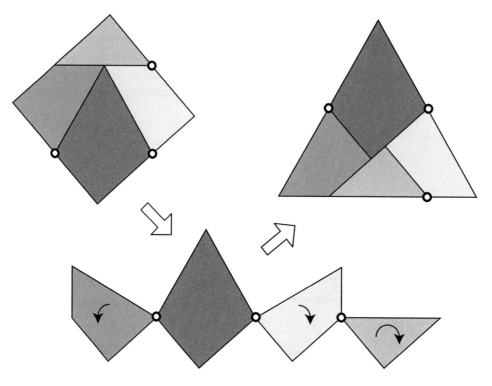

Figure 8.26

in Section 7.7 and the idea of adding extra members as in Problem 8.9. See Whiteley (1988a) for details.

5. Show that, for K(4,3) shown in Figure 8.6, with generic positions of the nodes along the axes, only one node will cross the x-axis during the finite mechanism, and one node will cross the y-axis. What other behaviour is possible if the generic condition is relaxed?

6. Consider the framework consisting of the vertices and edges of the regular dodecahedron with 30 extra bars inserted as follows. In a rotationally consistent way, at each pentagonal facet F, insert a bar from each vertex $\mathbf{p}_i$ of F to another vertex of a pentagonal facet adjacent to F (but not adjacent to $\mathbf{p}_i$) as in Figure 8.27. Each pair of adjacent facets describes the same additional bar, so there are 30 in all. Show that this bar framework is a finite mechanism. (Hint: Use some but not all of the symmetry of the dodecahedron. There are at least two ways to do this. See Connelly et al. (1994), Stachel (1992), and Guest (1999) for the history of this problem suggested by Branko Grünbaum.) This problem is a bit of a preview of Part II.

7. A bar framework $(G, \mathbf{p})$ in $\mathbb{E}^2$ is said to have a *parallel redrawing* if there is another configuration $\mathbf{q}$ such that for each member $\{i, j\}$ of G, $\mathbf{p}_i - \mathbf{p}_j$ is parallel to $\mathbf{q}_i - \mathbf{q}_j$.

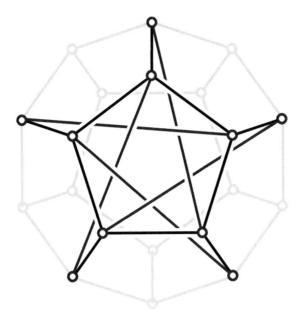

Figure 8.27 This is a portion of a dodecahedron with additional bars inserted. The entire mechanism has 60 bars and 20 vertices, more than enough to be rigid, but the mechanism is not.

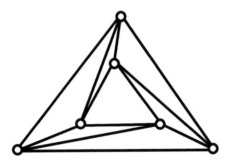

Figure 8.28

Show that $(G, \mathbf{p})$, with all non-zero length members, is infinitesimally rigid in $\mathbb{E}^2$ if and only if $\mathbf{q}_i = c\mathbf{p}_i + \mathbf{b}$, for all i, where c is a scalar constant, and $\mathbf{b}$ is a fixed vector in $\mathbb{E}^2$, are the only parallel redrawings. (See Crapo and Whiteley (1994a) for more information.)

8. Show that the tensegrities of Figure 5.10, Cauchy polygons, and those of Figures 5.12 (a) and (b), are projections of convex polyhedra such that the sign of the resulting stress in $\mathbb{E}^2$ from the Maxwell–Cremona Theorem 3.5.2 is proper for the tensegrities.

9. Show that the triangulation of a triangle in Figure 8.28 does not have a stress that is positive on all of the internal members. Thus by the Maxwell–Cremona Theorem 3.5.2

there is no lift to a convex polytope that projects to it. (See Connelly and Henderson (1980) for more information, and an example where there is an extension of the triangulation to the interior of the tetrahedron, which does not allow a strictly convex realization of the solid 3-dimensional body.)

Part II

Symmetric Structures

9

Groups and Representation Theory

9.1 Introduction

In Part II, we are interested in structures that have more than just trivial symmetry; many symmetric frameworks have interesting properties that can be elegantly explained using that symmetry. (Here "interesting" properties are those that differ from the properties of corresponding generic frameworks.) In particular, in Chapter 10 we will examine the first-order analysis of symmetric structures to find states of self-stress and infinitesimal mechanisms, and will develop a symmetry version of Maxwell counting; and in Chapter 11 we will explore the properties of symmetric tensegrity structures, leading to a simple algorithm to generate symmetric tensegrities.

We are fortunate that there is a tool with a long tradition that ideally suits our purpose in "symmetrizing" the theory presented in Part I: the representation theory of finite groups, although we only need the basic concepts of this vast theory for the groups in which we are interested. We will spend the rest of this present chapter sketching out the relevant parts of group representation theory for our purpose, but we note that there are many books that will present this theory in greater depth, such as the texts by Bishop (1973) and James and Liebeck (2001).

A difficulty in the presentation here is that the language used in mathematical representation theory, as presented in, for example, James and Liebeck (2001), differs in a number of important aspects from that used in applied group theory, as presented in, for example, Bishop (1973). Much of the previous literature in this area (e.g. Kangwai et al. (1999), Kangwai and Guest (2000), Fowler and Guest (2000)) is written using the language of applied group theory; here, however, we will use the language of mathematical group theory to make this work in the wider context that we need here, while maintaining a suitable "translation" between those familiar with the language of applied group theory and the mathematical theory.

9.2 What is Symmetry?

We will restrict our use of the word "symmetry" to only concern geometric symmetry. Thus, we define a *symmetry operation* to be an isometry of Euclidian space that leaves a framework

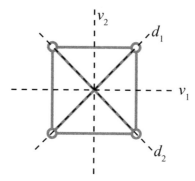

Figure 9.1 The eight symmetry operations for the symmetric structure originally introduced in Figure 3.2(c): the identity E; anticlockwise rotations C_4, $C_4^2 = C_2$, C_4^3; reflections σ_{v1}, σ_{v2}, σ_{d1}, σ_{d2}. The figure labels the vertices 1–4, and the lines of reflection as v_1, v_2, d_1, d_2.

geometrically unaltered. Note that the isometry may not preserve orientation, so a left hand might be converted to a right hand. In $d = 2$, the possible symmetry operations are: rotations by an angle $2\pi/n$, C_n; reflections across a line, σ; and the zen-like identity, E, that leaves everything untouched. The identity will be as important to our theory as the number 1 is for multiplication, or the number 0 for addition. Note that $E \equiv C_1$.

In $d = 3$, we also have the inversion i and improper rotations S_n. The inversion is a symmetry operation where every point is reflected through an origin, whereas an improper rotation is equivalent to a rotation about an axis by $2\pi/n$, followed by a reflection in a plane normal to the axis. Note that $i = S_2$.

We will term a structure to be *symmetric* when more than just the identity is a symmetry operation for the structure; otherwise we will say the structure only has trivial symmetry. As an example, Figure 9.1 shows the symmetry operations for the planar symmetric framework introduced in Figure 3.2(c).

9.3 What is a Group?

We will first answer the question "What is a group?" in a suitably general way, before giving a number of concrete examples of group and types of groups – all of which will be relevant to the purposes of this book.

Consider a set $\mathcal{G}$ with an operation defined so that if g_1 and g_2 are two elements of $\mathcal{G}$, $g_1 g_2$ is another element in $\mathcal{G}$ defined by applying the group operation to g_1 and g_2. The set $\mathcal{G}$, with its group operation, forms a *group* if it satisfies the following properties:

(i) Closed: If g_1, g_2 are any two elements of $\mathcal{G}$, $g_1 g_2$ is in $\mathcal{G}$.
(ii) Identity: There is a unique element, called the identity element, E in $\mathcal{G}$, such that for all g in $\mathcal{G}$, $Eg = gE = g$.

(iii) Inverses: For every element g in $\mathcal{G}$, there is a unique element g^{-1} in $\mathcal{G}$, called the *inverse* of g, such that $g^{-1}g = gg^{-1} = E$.

(iv) Associativity: For every three elements g_1, g_2, g_3 in $\mathcal{G}$, $g_1(g_2g_3) = (g_1g_2)g_3$.

9.3.1 Terminology and Comments

The *order of a group* $\mathcal{G}$ is the number of elements in $\mathcal{G}$ – we will only be dealing with finite groups, where the order is finite.

The group operation is often called multiplication. So g_1g_2 is the *product* of g_1 and g_2. Sometimes it is convenient to write out a multiplication table for a group with small order, sometimes called a "Cayley table" – we shall see some examples shortly.

We write g^n for g multiplied by itself n times, $n = 1, 2 \ldots$, and $g^{-n} = (g^{-1})^n$. We say the *order of an element* is the smallest positive value of n for which $g^n = E$.

Note that for, a group, g_1g_2 may or may not be equal to g_2g_1, but if $g_1g_2 = g_2g_1$, we say that they *commute*. If $\mathcal{G}$ has the property that every pair of its elements commutes, we say $\mathcal{G}$ is a *commutative* group. Often such a commutative group is also called *abelian* (in honour of the Norwegian mathematician Abel). For example, the group of rotations of a square in the plane is abelian. Sometimes, when the group is abelian, the group operation is written with a $+$ sign (or a $-$ minus sign for the inverse operation), but we will tend to stick to the multiplicative notation.

There are several little facts about groups that are handy from time to time. One is that for any two group elements g_1 and g_2 in $\mathcal{G}$, $(g_1g_2)^{-1} = g_2^{-1}g_1^{-1}$, $(g_1^{-1})^{-1} = g_1$.

A subset of $\mathcal{H}$ of a group $\mathcal{G}$ is called a *subgroup* if the multiplication of any two elements in $\mathcal{H}$ is another element in $\mathcal{H}$ and is given by their product in $\mathcal{G}$, and with this multiplication $\mathcal{H}$ satisfies the four properties of being a group (the last property of associativity being automatic since $\mathcal{H}$ is a subset of $\mathcal{G}$).

Suppose that $\mathcal{G}$ and $\mathbb{H}$ are two groups. Consider the set $\mathcal{G} \times \mathbb{H} = \{(g, h) \mid g \in \mathcal{G}, h \in \mathbb{H}\}$. We define a group multiplication on $\mathcal{G} \times \mathbb{H}$ by $(g_1, h_1) \cdot (g_2, h_2) = (g_1g_2, h_1h_2)$. It is easy to check that this is a proper definition of a group.

In order not to confuse the notation of a group with the symbol of a graph G, we will use calligraphic letters both for the group $\mathcal{G}$, and in particular examples of groups written below.

9.3.2 Symmetry Groups: Point Groups in 3D and 2D

The set of all symmetry operations that leave an object unchanged are called point groups – because they will always leave one point unshifted. We will use the Schoenflies notation, commonly in use in chemistry, to describe these groups (there is also a Hermann–Mauguin notation commonly used in crystallography, which we won't describe). Each point group is denoted by a letter (which we write in a calligraphic script) followed by a subscript that

Table 9.1. The multiplication table for the point group C_{4v}: the column is the first operation, the row the second

	E	C_4	$C_4^2 = C_2$	C_4^3	σ_{v1}	σ_{v2}	σ_{d1}	σ_{d2}
E	E	C_4	C_2	C_4^3	σ_{v1}	σ_{v2}	σ_{d1}	σ_{d2}
C_4	C_4	C_2	C_4^3	E	σ_{d1}	σ_{d2}	σ_{v2}	σ_{v1}
$C_4^2 = C_2$	C_2	C_4^3	E	C_4	σ_{v2}	σ_{v1}	σ_{d2}	σ_{d1}
C_4^3	C_4^3	E	C_4	C_2	σ_{d2}	σ_{d1}	σ_{v1}	σ_{v2}
σ_{v1}	σ_{v1}	σ_{d2}	σ_{v2}	σ_{d1}	E	C_2	C_4^3	C_4
σ_{v2}	σ_{v2}	σ_{d1}	σ_{v1}	σ_{d2}	C_2	E	C_4	C_4^3
σ_{d1}	σ_{d1}	σ_{v1}	σ_{d2}	σ_{v2}	C_4	C_4^3	E	C_2
σ_{d2}	σ_{d2}	σ_{v2}	σ_{d1}	σ_{v1}	C_4^3	C_4	C_2	E

Table 9.2. The elements of two isomorphic permutation groups, derived from the nodes and members of the framework shown in Figure 9.1, using the numbering shown in Figure 3.2, with the corresponding symmetry operations for the framework. The set of operations in each column forms a group.

Symmetry operation	Nodal permutation	Element permutation
E	(1)(2)(3)(4)	(1)(2)(3)(4)(5)(6)
C_4	(1234)	(1234)(56)
C_2	(13)(24)	(13)(24)
C_4^3	(1432)	(1432)(56)
σ_{v1}	(14)(23)	(13)(56)
σ_{v2}	(12)(34)	(24)(56)
σ_{d1}	(24)	(14)(23)
σ_{d2}	(13)	(12)(34)

may be a number, or a letter, or a number and a letter. Many of the possible point groups are listed in Table 9.5.

As an example, the eight symmetry operations shown in Table 9.2 form a point group C_{4v} – the point group for the structure shown in Figure 9.1. The multiplication table for this group is shown in Table 9.1.

9.3.3 Permutation Groups

The second common way of defining a group is as a collection of permutations of the elements of a given set. (A permutation of a finite set is a one-to-one function of the elements of the set to itself.) The group multiplication is given by the composition of these permutation functions. For example, the set of all permutations of a set with four elements forms a finite group of order 24 denoted by $\mathcal{S}_4$ and is called the *symmetric group* on four symbols (although note that this standard use of the word "symmetric" is at odds with our geometric definition of symmetry in Section 9.2).

For permutation groups, there are two very convenient notations for the group elements. For the first notation: If $g_i : \{1, 2, \ldots, m\} \rightarrow \{1, 2, \ldots, m\}$ is a permutation of m elements, then we can denote it simply as $g_i = [g_i(1), g_i(2), \ldots, g_i(m)]$, where we take advantage of the order of the numbers. For example, $[2, 1, 4, 3]$ is a permutation operator on four elements that has order 2. The advantage of this notation is that each element has a unique description, and it is easy to multiply permutations, which is just the composition of functions.

For the second notation, we write permutation operations as the product of disjoint cycles. Since our groups have finite order, for every permutation operation g_i, every element of the set, for instance x, is part of one cycle of elements $x \rightarrow g_i(x) \rightarrow g_i(g_i(x)) \rightarrow \cdots$ $g_i(\ldots g_i(x)\ldots)$, where the image under g_i of the last element of the set is the first. These cycles partition the set $\{1, 2, \ldots, m\}$ into disjoint cycles, and we can describe any group element as the product of these cyclic permutations. Note that the starting point for this description is arbitrary; elements that are fixed by the permutation are usually left out of the description, although cycles of length 1 can be included if desired; and the order of the cycles is arbitrary. For example, the permutation $[2, 1, 4, 3]$ is written as $(12)(34)$, and the permutation $[2, 3, 1, 4]$ is written as (123), or more fully as $(123)(4)$. The advantage of this notation is that the order of the element can be seen quickly and easily, and other properties of the permutation can be seen easily. For example, (12) can be quickly distinguished from $(12)(34)$, even though they both have the same order. It is also easy to check whether a permutation is even or odd in disjoint cycle notation. Cycles of odd length are even permutations, and cycles of even length are odd permutations. So a permutation is *even* if and only if it has an even number of odd cycles (of even length), and any number of even cycles (of odd length). Otherwise a permutation is called *odd*.

In both of these notations, when we multiply permutations, we read from the right to the left as in functional notation. For example, consider $(123)(234)$. The first operation (on the right) shifts 4 to 2, and the second operation (on the left) shifts 2 to 3, so the composite operation shifts 4 to 3. Repeating this for each element shows $(123)(234) = (12)(34)$. By contrast, $(234)(123) = (13)(24)$.

Two examples of permutation groups can be naturally derived from the framework shown in Figure 9.1, one for the nodes, and one for the members, and these are shown in Table 9.2.

There are four particular sets of permutation groups that we will be interested in. The group that consists of all possible permutation groups on n symbols is called the symmetric group $\mathcal{S}_n$. It is also straightforward to show that all even permutations on n symbols form a group (clearly a subgroup of $\mathcal{S}_n$), known as the alternating group $\mathcal{A}_n$. A further subgroup of $\mathcal{S}_n$ is the *cyclic group*. Consider permutations of the numbers $\{1, 2, \ldots, n\}$, i.e. $1 \rightarrow k + 1$, $2 \rightarrow k + 2, \ldots, n \rightarrow k + n$, where each number is taken modulo n and $0 \leq k \leq n - 1$. We write the cyclic group as $\mathcal{C}_n$ to fit with our general notation scheme (but note that the group is often written as $\mathbb{Z}_n$). Finally, another subgroup of $\mathcal{S}_n$ is the *dihedral group*, which consists of all permutations of the n vertices of a regular polygon corresponding to isometries of the polyhedron. We write the dihedral permutation group as $\mathcal{D}_n$.

Table 9.3. Transformation matrices T(g) for each of the symmetry operations $g \in C_{4v}$.

$$\mathbf{T}(E) = \begin{bmatrix} 1 & 0 \\ 0 & 1 \end{bmatrix} \qquad \mathbf{T}(C_4) = \begin{bmatrix} 0 & -1 \\ 1 & 0 \end{bmatrix} \qquad \mathbf{T}(C_4^2) = \begin{bmatrix} -1 & 0 \\ 0 & -1 \end{bmatrix} \qquad \mathbf{T}(C_4^3) = \begin{bmatrix} 0 & 1 \\ -1 & 0 \end{bmatrix}$$

$$\mathbf{T}(\sigma_{v1}) = \begin{bmatrix} 1 & 0 \\ 0 & -1 \end{bmatrix} \qquad \mathbf{T}(\sigma_{v2}) = \begin{bmatrix} -1 & 0 \\ 0 & 1 \end{bmatrix} \qquad \mathbf{T}(\sigma_{d1}) = \begin{bmatrix} 0 & 1 \\ 1 & 0 \end{bmatrix} \qquad \mathbf{T}(\sigma_{d2}) = \begin{bmatrix} 0 & -1 \\ -1 & 0 \end{bmatrix}$$

Table 9.4. Permutation matrices $P_X(g)$ that describe the permutation of the nodes for each of the symmetry operations $g \in C_{4v}$ for the example framework. All other entries in the matrices are zero.

$$\mathbf{P}_X(E) = \begin{bmatrix} 1 & & & \\ & 1 & & \\ & & 1 & \\ & & & 1 \end{bmatrix} \qquad \mathbf{P}_X(C_4) = \begin{bmatrix} & & & 1 \\ 1 & & & \\ & 1 & & \\ & & 1 & \end{bmatrix}$$

$$\mathbf{P}_X(C_4^2) = \begin{bmatrix} & & 1 & \\ & & & 1 \\ 1 & & & \\ & 1 & & \end{bmatrix} \qquad \mathbf{P}_X(C_4^3) = \begin{bmatrix} & 1 & & \\ & & 1 & \\ & & & 1 \\ 1 & & & \end{bmatrix}$$

$$\mathbf{P}_X(\sigma_{v1}) = \begin{bmatrix} & & & 1 \\ & & 1 & \\ & 1 & & \\ 1 & & & \end{bmatrix} \qquad \mathbf{P}_X(\sigma_{v2}) = \begin{bmatrix} & & 1 & \\ 1 & & & \\ & & & 1 \\ & 1 & & \end{bmatrix}$$

$$\mathbf{P}_X(\sigma_{d1}) = \begin{bmatrix} 1 & & & \\ & & & 1 \\ & & 1 & \\ & 1 & & \end{bmatrix} \qquad \mathbf{P}_X(\sigma_{d2}) = \begin{bmatrix} & & 1 & \\ & 1 & & \\ 1 & & & \\ & & & 1 \end{bmatrix}$$

9.3.4 Matrix Groups

A third common way of defining a group is where each element is defined as an invertible *n*-by-*n* matrix. The set of all such matrices for some *n* defines an infinite group, the general linear group, denoted $GL(n, \mathbb{R})$ if the entries are real, or $GL(n, \mathbb{C})$ if the entries are complex. We will be interested in finite subgroups of $GL(n, \mathbb{R})$, and sometimes of $GL(n, \mathbb{C})$.

One natural way of defining a matrix group is to consider a set of transformation matrices $\mathbf{T}(g)$ that correspond to the symmetry operations *g* in some symmetry group $\mathcal{G}$. The $d \times d$ transformation matrices describe the effect of the symmetry operations in $\mathbb{E}^d$ on a point $\mathbf{p}_i$ defined in terms of *d* basis vectors

$$\mathbf{p}_i \xrightarrow{g} \mathbf{T}(g)\mathbf{p}_i.$$

As an example, Table 9.3 shows transformation matrices $\mathbf{T}(g)$ for the eight symmetry operations shown in Figure 9.1 when a horizontal and vertical orthogonal basis is used. It is straightforward to confirm that these eight matrices $\mathbf{T}(g)$ form a group.

Permutation matrices form another natural way of defining a matrix group. For any permutation operation g acting on a set, if we consider the members of the set written as column vectors, then we can write a permutation matrix $\mathbf{P}(g)$ that appropriately permutes the set. It is straightforward to write down $\mathbf{P}(g)$: consider that g permutes a set containing n members; then the matrix $\mathbf{P}(g)$ will be found by taking an $n \times n$ identity matrix and permuting the columns according to g. For any permutation matrix, there will be a single 1, with all other entries 0, in each column, and in each row.

As an example, Table 9.4 shows permutation matrices $\mathbf{P}_X(g)$ that describe the permutation of nodes for the eight symmetry operations shown in Figure 9.1. It is again straightforward to confirm that these eight matrices $\mathbf{T}(g)$ form a group.

9.4 Homomorphisms and Isomorphisms of Groups

From the various descriptions of groups above, it is clear that some groups are essentially the same as others, and some are not. In general, we are interested in how groups are transformed from one to another. A very fundamental concept is the following. Consider two groups $\mathcal{G}$ and $\mathcal{H}$ and a function ρ that assigns an element $\rho(g)$ in $\mathcal{H}$ for every element g in $\mathcal{G}$. We write this as $\rho : \mathcal{G} \to \mathcal{H}$. We say that ρ is a *homomorphism* if for each pair of elements g_1 and g_2 in $\mathcal{G}$ the following holds:

$$\rho(g_1 g_2) = \rho(g_1)\rho(g_2).$$

Incidentally, it is easy to show that this definition of a group homomorphism is such that $\rho(E) = E$ (where E denotes the identity element in each respective group) and $\rho(g^{-1}) = \rho(g)^{-1}$ for each element g in $\mathcal{G}$. The "morphism" in the name comes from the change from $\mathcal{G}$ to $\mathcal{H}$, and "homo" means the likeness or similarity between $\mathcal{G}$ and $\mathcal{H}$.

The *kernel* of a group homomorphism $\rho : \mathcal{G} \to \mathcal{H}$ is the set of all elements $g \in \mathcal{G}$ such that $\rho(g) = 1$, the identity of $\mathcal{H}$. This is related to the notion of the kernel of matrix A, which are those vectors $\mathbf{v}$ such that $A\mathbf{v} = 0$, which is not necessarily in the context of a group.

If, further, a homomorphism $\rho : \mathcal{G} \to \mathcal{H}$ is one-to-one as a function and for every h in $\mathcal{H}$ there is a g in $\mathcal{G}$ such that $\rho(g) = h$, then we say that it is an *isomorphism* and that $\mathcal{G}$ and $\mathcal{H}$ are *isomorphic*. In terms of the formal group properties, $\mathcal{G}$ and $\mathcal{H}$ are the "same", even though the symmetry elements themselves might be very different. Many of the examples that we have seen in the present chapter are isomorphic groups: the symmetry group C_{4v}; the permutation groups defined in Table 9.2; and the matrix groups defined in Table 9.3 and 9.4.

We give, in Table 9.5, some examples of group isomorphisms between permutation groups and symmetry groups that will be important later, particularly in Chapter 11.

Table 9.5. Examples of isomorphic permutation groups and symmetry groups.

Permutation group	Symmetry group	Order	Description
$\mathcal{S}_1$	$\mathcal{C}_1$	1	The trivial group, consisting of just the identity.
$\mathcal{S}_2$	$\mathcal{C}_2; \mathcal{C}_s; \mathcal{C}_i$	2	The group consisting of just two elements, the identity and one other element. So $\mathcal{G} = \{E, g_2\}$, with $g_2^2 = E$. As a symmetry group, the element g_2 might be one of three symmetry operations: for $\mathcal{C}_2$ it is a half turn, $g_2 = C_2$; for $\mathcal{C}_s$ it is a reflection in a plane, $g_2 = \sigma$; and for $\mathcal{C}_i$ it is the inversion, $g_2 = i$. There are many isomorphic matrix groups: for instance, if $\mathbf{I}$ is the n-by-n identity matrix ($n \geq 1$), then the set of matrices $C_i = \{\mathbf{I}, -\mathbf{I}\}$ forms a group isomorphic to S_2.
$\mathcal{C}_n$	$\mathcal{C}_n$	n	The cyclic group.
$\mathcal{D}_n$	$\mathcal{D}_n; \mathcal{C}_{nv}$	$2n$	The dihedral group is the group of symmetries of a regular polygon in the plane. The symmetries that reverse the cyclic orientation of the vertices can be realized in two ways. If that symmetry is given by reflection in a plane perpendicular to the plane of a polygon, the geometric group is denoted as $\mathcal{D}_n$, while if it is given by a rotation about a line in the plane of the polygon, it is denoted as $\mathcal{C}_{nv}$.
$\mathcal{A}_4$	$\mathcal{T}$	12	The symmetry group $\mathcal{T}$ contains the rotations, but not reflections, of the regular tetrahedron; which is isomorphic to the group of even permutations on four objects $\mathcal{A}_4$. The isomorphism can clearly be seen if the permutations are considered as acting on the vertices of the tetrahedron: every even permutation corresponds either to a 3-fold rotation of the tetrahedron about an axis through a vertex, or a 2-fold rotation of the tetrahedron about an axis acting through the centre of two opposite edges.
$\mathcal{S}_4$	$\mathcal{T}_d; \mathcal{O}$	24	The permutation group $\mathcal{S}_4$ is isomorphic to two different symmetry groups. The first is $\mathcal{T}_d$, the symmetries of the tetrahedron; considering the permutation operations as acting on the vertices, every permutation operation is in one-to-one correspondence with either a rotation or a reflection of the tetrahedron. The second isomorphic symmetry group is $\mathcal{O}$, the rotations, but not reflections, of a cube. The permutations can then be considered as acting on the four space diagonals of the cube.
$\mathcal{A}_4 \times \mathcal{S}_2$	$\mathcal{T}_h$	24	The symmetry group $\mathcal{T}_h$ combines all rotation symmetries of a regular tetrahedron together with either *inversion* or *identity*. The inversion is reflection through the origin, taking each vector to its negative.

Table 9.5. (*Cont.*)

Permutation group	Symmetry group	Order	Description
$S_4 \times S_2$	$\mathcal{O}_h$	48	The symmetry group $\mathcal{O}_h$ contains all symmetries of the cube (equivalently: the regular octahedron). Note that this is not isomorphic to the permutations of the vertices of the cube (S_8) – for instance, there is no Euclidean isometry of order eight on the vertices of a cube.
A_5	$\mathcal{I}$	60	The symmetries of the regular dodecahedron (equivalently: regular icosahedron) that are rotations, i.e. those that are orientation preserving.
$A_5 \times S_2$	$\mathcal{I}_h$	120	All symmetries of the regular dodecahedron (equivalently: the regular icosahedron).

9.5 Representations

A *representation* of a group $\mathcal{G}$ is a group homomorphism $\rho : \mathcal{G} \to GL(n, \mathbb{R})$, and, similarly, a *complex representation* is a group homomorphism $\rho : \mathcal{G} \to GL(n, \mathbb{C})$. The *dimension* of the representation is n, the size of the matrices. In other words a representation ρ assigns to each group element g a non-singular n-by-n matrix $\rho(g)$, such that for any two elements g_1, g_2 in $\mathcal{G}$, $\rho(g_1 g_2) = \rho(g_1)\rho(g_2)$.

Our primary interest may be in one particular representation or other, but we will see that it is very useful to include in our considerations other representations as well. We will primarily be interested in real representations, but complex representations arise in a natural way, and simplify our understanding considerably; some key theorems only apply to complex, and not real representations. In fact, complex representations are so natural, we even regard a real representation as being in $GL(n, \mathbb{C})$, and when we speak of a representation, the default meaning will be that we regard it as a complex representation.

9.5.1 Equivalent Representations

Given one n-dimensional representation ρ, there are very easy ways of creating other representations that are almost the same. We say that two n-dimensional representations ρ_1 and ρ_2 are *equivalent* if there is a matrix $\mathbf{P}$ in $GL(n, \mathbb{C})$ such that $\mathbf{P}\rho_1\mathbf{P}^{-1} = \rho_2$. So for each element g in $\mathcal{G}$, $\mathbf{P}\rho_1(g)\mathbf{P}^{-1} = \rho_2(g)$. We can regard $\mathbf{P}$ as simply giving a change in coordinate system. For example, the particular representation given in Table 9.3 was derived using a particular set of basis vectors in Figure 9.1; a different choice of basis vectors would give a different, but equivalent, representation.

Note that a subgroup of $GL(n, \mathbb{C})$ or $GL(n, \mathbb{R})$ is not a representation by itself. It is important to provide a correspondence, a labelling, in order to complete the definition. Later, we need to fix one group, for example a permutation group, such as A_4, S_4, or A_5, and consider several representations of that one fixed "base group." Then it becomes clear that we must be careful to consider the labelling as part of the definition of a representation, since we will need to compare different representations of the same group.

9.5.2 Direct Sum of Representations

Given two representations ρ_1 and ρ_2 of the group $\mathcal{G}$, where ρ_1 is n_1-dimensional and ρ_2 is n_2-dimensional, we create another (n_1+n_2)-dimensional representation, denoted by $\rho_1 \oplus \rho_2$. For every element g in $\mathcal{G}$, $(\rho_1 \oplus \rho_2)(g)$ is the following (n_1+n_2)-by-(n_1+n_2) matrix written in block form:

$$\begin{bmatrix} \rho_1(g) & 0 \\ 0 & \rho_2(g) \end{bmatrix}.$$

So a more complicated representation may be decomposed in this way to simpler lower-dimensional ones. If a representation ρ is equivalent to the direct sum of lower-dimensional representations, we say that ρ is *reducible*. Otherwise ρ is *irreducible*. The surprising fact about finite groups is that it always has only a finite number of irreducible representations, up to equivalence.

9.5.3 Traces and Characters

There are some surprising tools that allow one to distinguish between the various representations of a group, up to equivalence. First we recall some basic definitions involving matrices and their properties. Let $\mathbf{A} = (a_{ij})$ be an n-by-n matrix. The *trace* of $\mathbf{A}$ is $tr(\mathbf{A}) = \sum_i a_{ii}$, the sum of the diagonal entries. The following are some basic properties, easily proved about the trace. Let $\mathbf{A}$ and $\mathbf{B}$ two n-by-n matrices. Then

$$tr(\mathbf{A} + \mathbf{B}) = tr(\mathbf{A}) + tr(\mathbf{B}) \quad \text{and} \quad tr(\mathbf{AB}) = tr(\mathbf{BA}).$$

If ρ is a representation of the group $\mathcal{G}$, then the *character* of ρ, denoted as χ, is the function that assigns $\chi(g) = tr(\rho(g))$ to each element g of the group $\mathcal{G}$. Note that a character is a function from the group $\mathcal{G}$ to the complex numbers $\mathbb{C}$, so in principle we have to define it on all the elements of $\mathcal{G}$.

For example, if $\mathcal{G} = \{1, g, g^2, g^3\}$ is the cyclic group of rotations of a square in the plane, where

$$1 = \begin{bmatrix} 1 & 0 \\ 0 & 1 \end{bmatrix}, \quad g = \begin{bmatrix} 0 & -1 \\ 1 & 0 \end{bmatrix}, \quad g^2 = \begin{bmatrix} -1 & 0 \\ 0 & -1 \end{bmatrix}, \quad g^3 = \begin{bmatrix} 0 & 1 \\ -1 & 0 \end{bmatrix},$$

then the character χ of this representation is

$$\chi(1) = 2, \quad \chi(g) = 0, \quad \chi(g^2) = -2, \quad \chi(g^3) = 0.$$

9.5.4 Properties of Characters

For any two elements g_1 and g_2 of a group $\mathcal{G}$, we say they are *conjugate* if there is an element h in $\mathcal{G}$ such that $hg_1h^{-1} = g_2$. Notice that the element h that does the conjugating must be in the group $\mathcal{G}$. The relation of being conjugate in the group $\mathcal{G}$ is easily seen to be an equivalence relation and so we can speak of equivalence classes of conjugate elements called *conjugacy classes*. The point is that any character is clearly constant on all the group elements in a conjugacy class, and so usually it is defined or described on the distinct conjugacy classes of a group rather than waste time defining it for all the elements individually. The following are some easy properties of characters for the group $\mathcal{G}$.

(i) If g_1 and g_2 are conjugate in $\mathcal{G}$, then for any character $\chi(g_1)$ of $\mathcal{G}$, $\chi(g_1) = \chi(g_2)$.
(ii) If χ_1 and χ_2 are characters corresponding to equivalent representations of $\mathcal{G}$, then $\chi_1 = \chi_2$ as functions on $\mathcal{G}$.
(iii) If χ is the character corresponding the direct sum of two representations of $\mathcal{G}$ whose characters are χ_1 and χ_2, then $\chi = \chi_1 + \chi_2$.

For example, for the group and character described in Section 9.5.3, consider the 1-dimensional characters χ_1 and χ_2 defined by

$$\chi_1(1) = 1, \quad \chi_1(g) = i, \quad \chi_1(g^2) = -1, \quad \chi_1(g^3) = -i$$
$$\chi_2(1) = 1, \quad \chi_2(g) = -i, \quad \chi_2(g^2) = -1, \quad \chi_2(g^3) = i,$$

where $i^2 = -1$ as usual. Note that when the representation is 1-dimensional, the character and the representation are essentially the same. One can easily check that representations are defined by the above characters χ_1 and χ_2, that their direct sum is the representation χ defined in Section 9.5.3, and that their characters add as in the properties above. One sees also an advantage of using the complex numbers as coefficients. The representation defined in Section 9.5.3 is irreducible over the real numbers, but it is reducible over the complex numbers.

For any group $\mathcal{G}$, there is the *trivial representation* that assigns to each group element the number 1. So the character of this representation also assigns the number 1 to each group element. Although such a representation may be obvious, it must not be overlooked.

Characters form a very important tool for distinguishing between various representations. Indeed, surprisingly, characters can be used to distinguish any pair of representations as in the following theorem which shows that characters characterize representations.

Theorem 9.5.1. Characters. *Suppose that ρ_1 and ρ_2 are two (complex) representations of the finite group $\mathcal{G}$ with corresponding characters χ_1 and χ_2, respectively. Then ρ_1 is equivalent to ρ_2 if and only if $\chi_1 = \chi_2$.*

A proof of this very vital result can be found in most books on representation theory, for example, James and Liebeck (2001). There is a great deal of structure in the nature of characters, but we will just be concerned with a few basic facts here. From this result, for example, we can see that the representation in Section 9.5.3 is the direct sum of the 1-dimensional complex representations as defined above without fussing with finding the matrix P that shows the equivalence of the direct sum and the given 2-dimensional representation.

9.5.5 Permutation Representations

An example of a representation of a group G is as a group of permutation matrices as discussed in Section 9.3.3. This can be regarded as G *acting on* a finite set X, which means that G is just a set of permutations of X that form a group. The notation for this is that for every $x \in X$ and $g \in G$ $gx = g(x)$ thinking of g as a one-to-one function on the set X. We can also think of each g as an m-by-m matrix, as in Section 9.3.3, where m is the number of elements in the set X. We identify the i-th element in X as the i-th unit basis basis vector in $\mathbb{E}^m$. Then ordinary matrix multiplication corresponds to functional composition of permutation functions. Then the following follows immediately:

> **Proposition 9.5.2.** *If g is an element in a permutation group G acting on a set X, then $\chi(g)$ is the number of elements x of X such that $gx = x$.*

Here we identify G with its representation as a set of matrices.

9.5.6 The Regular Representation

One particular representation of a finite group is especially useful in that it "contains" all the irreducible representations, and, for us, appears in a very natural way in the description of certain stress matrices of certain very symmetric tensegrities.

Suppose G is a finite group with n elements $g_1, g_2, \ldots, g_n$. We regard G as a permutation group on those same elements as follows. If g_i, $i = 1 \ldots n$ is one of the elements of G, then for every j, $j = 1 \ldots n$, there is a unique $k = 1 \ldots n$ such that $g_i g_j = g_k$. In other words, each element g_i of G corresponds to a permutation of the elements of G itself, and so this describes G as a permutation group on n symbols. This, in turn, can be regarded as an n-dimensional representation by identifying each group element g_i with the standard basis vector e_i whose coordinates are all 0's except for a 1 in the i-th coordinate. This is called the *left regular representation*, and we will denote it by ρ_L.

For example, let G be the group C_{3v} (isomorphic with the dihedral group D_3), the symmetries of a regular triangle in the plane. We list the elements of G as

$$g_1 = E, \quad g_2 = C_3, \quad g_3 = C_3^2, \quad g_4 = \sigma, \quad g_5 = C_3\sigma, \quad g_6 = C_3^2\sigma,$$

where E is the identity, C_3 is rotation counterclockwise by 120 degrees, and σ is reflection about the x-axis. (For D_3, we would replace the reflection σ with C_2, a 2-fold rotation about an axis in the plane). Note that $C_3^3 = E = \sigma^2$, and $\sigma C_3\sigma = C_3^2$. These relations are sufficient to determine the product of any two elements of D_3 as written above. We associate the standard basis vector e_i with the group element g_i above, for $i = 1, \ldots, 6$. So, for example, in the left regular representation, left multiplication by g_2 is given as

$$g_2g_1 = g_2, \quad g_2g_2 = g_3, \quad g_2g_3 = g_1, \quad g_2g_4 = g_5, \quad g_2g_5 = g_6, \quad g_2g_6 = g_4.$$

So the permutation of the basis vectors is given by

$$e_1 \rightarrow e_2, \quad e_2 \rightarrow e_3, \quad e_3 \rightarrow e_1, \quad e_4 \rightarrow e_5, \quad e_5 \rightarrow e_6, \quad e_6 \rightarrow e_4.$$

The left regular representation is given by ρ_L and

$$\rho_L(g_2) = \begin{bmatrix} 0 & 0 & 1 & 0 & 0 & 0 \\ 1 & 0 & 0 & 0 & 0 & 0 \\ 0 & 1 & 0 & 0 & 0 & 0 \\ 0 & 0 & 0 & 0 & 0 & 1 \\ 0 & 0 & 0 & 1 & 0 & 0 \\ 0 & 0 & 0 & 0 & 1 & 0 \end{bmatrix}.$$

It is easy to check that $\rho_L(g_i g_j) = \rho_L(g_i)\rho_L(g_j)$ for all g_i and g_j in $\mathcal{G}$.

9.5.7 The Right Regular Representation

It will be important later to define another representation that is equivalent to the left regular representation defined in the previous section. The idea is that it corresponds to the permutations on the group elements as with the left regular representation, but the permutation associated with any given group element corresponds to multiplication on the right. But the naive method of defining such a permutation does not define a group homomorphism.

So instead, the permutation that we define on the elements of $\mathcal{G}$ corresponding to an element g_i of $\mathcal{G}$ is obtained by multiplying every element g of $\mathcal{G}$ by g_i^{-1} on the right. So the permutation is $g \to gg_i^{-1}$. We check the group homomorphism property. First we calculate the permutation given by $g_i g_j$, for g_i, g_j and g in $\mathcal{G}$,

$$g \to g(g_i g_j)^{-1} = gg_j^{-1}g_i^{-1},$$

and then the permutation given by the composition of first performing g_j and then g_i,

$$g \to gg_j^{-1} \to gg_j^{-1}g_i^{-1}.$$

So we see that we get the same element, and this defines a representation which we call ρ_R the *right regular representation*.

Continuing the example from Subsection 9.5.6, we can compare this at the level of permutation matrices by calculating the matrix $\rho_R(g_2)$, where $g_2 = C_3$ in the dihedral group as in the previous subsection:

$$g_1 g_2^{-1} = g_3, \quad g_2 g_2^{-1} = g_1, \quad g_3 g_2^{-1} = g_2, \quad g_4 g_2^{-1} = g_5, \quad g_5 g_2^{-1} = g_6, \quad g_6 g_2^{-1} = g_4.$$

So the permutation of the basis vectors is given by

$$e_1 \to e_3, \quad e_2 \to e_1, \quad e_3 \to e_2, \quad e_4 \to e_5, \quad e_5 \to e_6, \quad e_6 \to e_4.$$

The right regular representation, given by ρ_R is

$$\rho_L(g_2) = \begin{bmatrix} 0 & 1 & 0 & 0 & 0 & 0 \\ 0 & 0 & 1 & 0 & 0 & 0 \\ 1 & 0 & 0 & 0 & 0 & 0 \\ 0 & 0 & 0 & 0 & 0 & 1 \\ 0 & 0 & 0 & 1 & 0 & 0 \\ 0 & 0 & 0 & 0 & 1 & 0 \end{bmatrix}.$$

Notice that $\rho_R(g_2)$ is not the same matrix as $\rho_L(g_2)$.

9.5.8 Basics

The following are some basic facts about representations, proved in James and Liebeck (2001).

> **Theorem 9.5.3.** *Suppose that a representation ρ of a finite group $\mathcal{G}$ is the direct sum of irreducible representations in two ways, say as $\rho_1 \oplus \cdots \oplus \rho_k$ and $\sigma_1 \oplus \cdots \oplus \sigma_m$. Then $k = m$ and after appropriate rearrangement of the summands, $\rho_i = \sigma_i$ for $i = 1, \ldots, k$.*

The regular representation has a particularly interesting decomposition in terms of irreducibles.

> **Theorem 9.5.4.** *For any finite group $\mathcal{G}$, every irreducible representation ρ_i of $\mathcal{G}$ appears as a summand of the (left or right) regular representation of $\mathcal{G}$ and each ρ_i appears $\dim(\rho_i)$ times, where $\dim(\rho_i)$ is the dimension of ρ_i.*

One consequence of this is that the left and right regular representations are equivalent, and when we are only interested in a representation up to equivalence, we can simply speak of the regular representation.

For example, for the group C_{3v}, there are three distinct irreducible representations, the trivial 1-dimensional representation as defined above, the 1-dimensional representation that assigns the number 1 to each rotation (including the identity) and -1 to the three reflections, and the 2-dimensional representation that assigns the matrix corresponding to each rotation or reflection in the plane. So in the regular representation, the last 2-dimensional representation appears twice, and the two other 1-dimensional representations appear once each.

A consequence of Theorem 9.5.4 is that if n_i, for $i = 1, \ldots k$, is the dimension of the i-th distinct irreducible representation for the finite group $\mathcal{G}$, then $n_1^2 + n_2^2 + \cdots + n_k^2 = n$, the order of $\mathcal{G}$. For example, for the dihedral group C_{3v} above, $n_1 = 1, n_2 = 1, n_3 = 2$ and $1^2 + 1^2 + 2^2 = 6$, the order of C_{3v}.

So to put this last example one more way, applying Theorem 9.5.4, if ρ_1, ρ_2, ρ_3 are the three distinct representations of $\mathcal{D}_3$, and ρ is the regular representation, then ρ is equivalent to the following in block matrix form:

$$
\begin{bmatrix}
\rho_1 & 0 & 0 & 0 \\
0 & \rho_2 & 0 & 0 \\
0 & 0 & \rho_3 & 0 \\
0 & 0 & 0 & \rho_3
\end{bmatrix},
$$

and thus, there will be some matrix $\mathbf{P}_L$, and some matrix $\mathbf{P}_R$, such that, for all operations $g \in G$,

$$
\mathbf{P}_L \rho_L(g) \mathbf{P}_L^{-1} = \mathbf{P}_R \rho_R(g) \mathbf{P}_R^{-1} =
\begin{bmatrix}
\rho_1(g) & 0 & 0 & 0 \\
0 & \rho_2(g) & 0 & 0 \\
0 & 0 & \rho_3(g) & 0 \\
0 & 0 & 0 & \rho_3(g)
\end{bmatrix}.
$$

Table 9.6. Character table for C_{3v}.

$\mathcal{D}_3$ $C_{3v} = (32)$	$\{E\}$	$\{C_3, C_3^2\}$	$\{\sigma, C_3\sigma, C_3^2\sigma\}$
$A_1 = \chi_1$	1	1	1
$A_2 = \chi_2$	1	1	-1
$E = \chi_3$	2	-1	0

We can think of a representation ρ of a finite group $\mathcal{G}$ as a function from $\mathcal{G}$ to matrices. When that function is one-to-one, we say that ρ is *faithful*. We will see next that one can read off from the character table, defined below, whether a character is faithful or not.

9.5.9 Character Tables

We first introduce the idea of a conjugacy class for a group $\mathcal{G}$. Consider three elements $g_1, g_2, g_3 \in G$. We say that two elements g_1 and g_2 are *conjugate* if $g_1 = g_3^{-1} g_2 g_3$ for any $g_3 \in G$. Elements that are conjugate to one another form a *conjugacy class*. In some sense, all the elements of a conjugacy class are the same as each other, as symmetry can transform the elements into one another.

Since Theorem 9.5.1 shows that the characters completely characterize representations up to equivalence, it is very useful to have on hand a record of what those characters are. In Chapter 11 we show the characters for the groups mentioned in Section 9.4. This is traditionally done through the use of "character tables." Across the top of the table each column corresponds to a distinct conjugacy class of the group $\mathcal{G}$. Each row corresponds to a distinct irreducible representation of $\mathcal{G}$. It turns out that there are exactly as many distinct irreducible representations of $\mathcal{G}$ as there are distinct conjugacy classes of $\mathcal{G}$.

For example, for the dihedral group C_{3v} discussed in Section 9.5.6, the character table is as shown in Table 9.6.

In the top left is the name of the group, first as a permutation group (here $\mathcal{D}_3$). Then on the next line is the name in the Schoenflies notation (commonly used in applied group theory and chemistry), which is here C_3v. This is marked as being equal to the name in the Hermann–Mauguin notation commonly used in crystallography (here, 32).

Each row will be named by the appropriate "Mulliken Symbol," which are the traditional names for irreducible representations introduced by Robert S. Mulliken, who won the Nobel prize for chemistry in 1966. However, for our purposes, we will also simply number the representations as $\rho_1, \rho_2, \ldots$, and the characters as $\chi_1, \chi_2, \ldots$. The first row is traditionally reserved for the trivial representation and consists of all 1's.

Each column corresponds to a conjugacy class. Here we show that the entire conjugacy class using the Schoenflies notation (C_3, σ etc.). In our later character tables, we will indicate a conjugacy class by just one representative, and we will indicate first the notation for a group element in terms of the permutation group in question, then in Schoenflies notation.

The first column is traditionally reserved for the identity element and is the dimension of the corresponding irreducible representation.

The character table has a lot of information. For example, a representation ρ is faithful if and only if it has only the identity in its kernel. The dimension d of ρ is the first element in the row corresponding to ρ in the character table. Any other group element in the kernel will have character d as well, and any element of $\mathcal{G}$ not the kernel will not have character d. For example, in Table 9.6, the characters χ_1, χ_2 are not faithful, while the character χ_3 is faithful.

9.5.10 Tensor Products

In addition to the direct sum of representations, another important operation on representations is the tensor product, which is defined next. Suppose V and W are two finite-dimensional vector spaces with basis vectors $\mathbf{v}_1, \ldots, \mathbf{v}_n$ and $\mathbf{w}_1, \ldots, \mathbf{w}_m$, respectively. Consider the formal set of all ordered pairs of these vectors, denoted for $i = 1, \ldots, n$ and $j = 1, \ldots m$, as

$$\mathbf{v}_i \otimes \mathbf{w}_j, \tag{9.1}$$

which we call the *tensor product of* $\mathbf{v}_i$ *and* $\mathbf{w}_j$. If we take all linear combinations of the vectors of (9.1) we get the *tensor product of the vector spaces* V and W, denoted as $V \otimes W$. Note that the dimension of $V \otimes W$ is nm, the product of the dimensions of V and W.

Also there is a natural way of extending the tensor product of basis vectors to products of arbitrary vectors by

$$\mathbf{v} \otimes \mathbf{w} = \left(\sum_i \lambda_i \mathbf{v}_i \right) \otimes \left(\sum_j \lambda_j \mathbf{w}_j \right) = \sum_{i,j} \lambda_i \lambda_j \mathbf{v}_i \otimes \mathbf{w}_j.$$

It is an instructive exercise to show that this definition of tensor product of vectors does not depend on the choice of basis vectors for V and W.

A key example of tensor products is the way we have been treating configuration spaces. Let $V = \mathbb{E}^d$ be the ambient vector space, and let $\mathbb{E}^n$ be space of vectors corresponding to the vertices of a graph G. Each vector $\mathbf{e}_i$ for $i = 1, \ldots, n$, with a 1 in the i-th slot and 0's elsewhere, corresponds to the i-th vertex of G. If $\mathbf{p}_i$ is the vector corresponding to the i-th vertex of G, for $i = 1, \ldots, n$, then $\sum_i \mathbf{p}_i \otimes \mathbf{e}_i$ corresponds naturally to the configuration vector $\mathbf{p} = [\mathbf{p}_1; \ldots; \mathbf{p}_n]$ that we have been discussing frequently in previous chapters. Also, the tensor product of V and W is naturally identified as a space of n-by-m matrices, and this the way we defined the configuration matrix in Section 5.7.

Suppose that a (finite) group $\mathcal{G}$ has two representations $\rho_i : \mathcal{G} \rightarrow GL(n_i, \mathbb{C})$, for $i = 1, 2$. Defining the *tensor product* of ρ_1 and ρ_2 as

$$(\rho_1 \otimes \rho_2)(g) = \rho_1(g) \otimes \rho_2(g), \tag{9.2}$$

for each g in $\mathcal{G}$, the tensor product on the right is the Kronecker product defined in Section 5.6. It is a good exercise to check that this definition of $\rho_1 \otimes \rho_2$ is indeed a representation of $\mathcal{G}$. The following is an easy consequence of (9.2).

Proposition 9.5.5. *For two representations ρ_1 and ρ_2 of $\mathcal{G}$,*

$$\chi((\rho_1 \otimes \rho_2)(g)) = \chi(\rho_1(g))\chi(\rho_2(g)),$$

for each g in $\mathcal{G}$.

Note that the product on the right is the ordinary product of complex numbers. If we have two characters χ_i, where $\chi_i(g) = \chi(\rho_i(g))$ for $i = 1, 2$ and g in $\mathcal{G}$, then we define the character $\chi_1 \otimes \chi_2$ by $(\chi_1 \otimes \chi_2)(g) = \chi_1(g)\chi_2(g)$ for all g in $\mathcal{G}$, which is the character of the tensor product of ρ_1 and ρ_2.

For example, for the characters in Table 9.6, for the dihedral group, the character $\chi_3 \otimes \chi_3$ has the following decomposition into irreducible characters:

$$\chi_3 \otimes \chi_3 = [2, -1, 0] \otimes [2, -1, 0] = [4, 1, 0]$$
$$= [1, 1, 1] + [1, 1, -1] + [2, -1, 0] = \chi_1 + \chi_2 + \chi_3,$$

where each coordinate corresponds to the conjugacy class as in Table 9.6. Note that we can equally regard a character as a vector k-tuple, where k is the number of conjugacy classes as above, and as a complex-valued function on the elements of the group $\mathcal{G}$. Note also that once one has a decomposition of a character into a sum of irreducible characters, each of those characters corresponds to its own irreducible representation by Theorem 9.5.1.

Orthogonality Relations Between Characters

If one is given a character (and implicitly its representation), there is a simple way to determine its decomposition into irreducible terms. Define an inner product on characters by saying for any two characters χ_1 and χ_2 that

$$\langle \chi_1, \chi_2 \rangle = \frac{1}{n} \sum_{g \in \mathcal{G}} \chi_1(g)\overline{\chi_2(g)}, \tag{9.3}$$

where n is the number of elements in $\mathcal{G}$, and $\overline{\chi_2(g)}$ is the complex conjugate of $\chi_2(g)$. This inner product is the same as the ordinary dot product for real vectors (except for the factor of $1/n$), or the Hermitian inner product for complex vectors as in a standard text on linear algebra like Strang (2009).

The following basic result is shown in James and Liebeck (2001), or Bishop (1973), for example.

Proposition 9.5.6. *For any finite group, if χ_1 and χ_2 are irreducible representations, then*

$$\langle \chi_1, \chi_2 \rangle = \begin{cases} 1 & \text{if } \chi_1 = \chi_2 \\ 0 & \text{if } \chi_1 \neq \chi_2 \end{cases}$$

Proposition 9.5.6 can be used to calculate the coefficients of the irreducible characters (or representations) of a given character (or representation). For example, for the dihedral group as above,

$$\langle \chi_3 \otimes \chi_3, \chi_3 \rangle = \frac{1}{6}(1 \cdot 4 \cdot 2 + 2 \cdot 1 \cdot (-1) + 3 \cdot 0 \cdot 0) = 1,$$

where we have grouped the terms by conjugacy class, whose sizes are $1, 2, 3$ in order. So we see that the coefficient of χ_3 in $\chi_3 \otimes \chi_3$ is 1.

Another example is the regular representation χ_R for any finite group $\mathcal{G}$. Note that by Proposition 9.5.2 the number of fixed points of a permutation in a permutation representation is the character of that group element:

$$\chi_R(g) = \begin{cases} n & \text{if } g = E \\ 0 & \text{otherwise.} \end{cases}$$

For the i-th irreducible character, we calculate that

$$\langle \chi_R, \chi_i \rangle = \frac{1}{n} n n_i = n_i,$$

where n_i is the degree of the representation (and the value of the character on the identity element E). So each irreducible character (and irreducible representation) of degree n_i appears n_i times in the regular representation as stated in Theorem 9.5.4.

10

First-Order Symmetry Analysis

The present chapter will show how knowledge of the geometric symmetry of a framework, as described in Chapter 9, can be used to simplify the first order analysis of a framework described in Chapter 3.

However, before we move on to symmetry, we first recast some of the material from Chapter 3 in terms of vector spaces. Then we'll add in the symmetry within these vector spaces.

10.1 Internal and External Vector Spaces

The equilibrium and compatibility/rigidity relationship in Chapter 3 can be advantageously considered as linear mappings between two vector spaces, an "external" space X and an "internal" space I.

For any framework $(G, \mathbf{p})$, the external vector space X is defined by quantities where there is a d-dimensional vector associated with each of the n nodes of structure. In particular, we can consider X as containing all possible external loads $\mathbf{f}$ that can be applied to the framework, or alternatively all possible infinitesimal displacements $\mathbf{d}$ of a framework; however, we must take care with dimensions – the space does not have a dimension associated with it. Thus, when we write a vector with a particular physical dimension, we will consider it as a non-dimensional vector within the vector space, which is then multiplied by some suitable unit, which would be a unit of force, say newtons for $\mathbf{f}$, or units of length, say meters, for $\mathbf{d}$. We do, however, define an inner product between any vector $\mathbf{f}$ and any vector $\mathbf{d}$ as the external work, $w_E = \mathbf{f}^\mathrm{T} \mathbf{d}$. The space X can be considered as the tensor product of a linear space defined by quantities at vertices, $\mathbb{R}^n$, and a d-dimensional vector space $\mathbb{R}^d$, i.e. $X = \mathbb{R}^n \otimes \mathbb{R}^d = \mathbb{R}^{nd}$.

Similarly, for any framework $(G, \mathbf{p})$ the internal vector space I is defined by quantities where there is a scalar associated with each of the b members of the framework, i.e. $I = \mathbb{R}^b$. We could consider I as containing all possible internal forces $\mathbf{t}$, or all possible infinitesimal extensions $\mathbf{e}$ – or indeed all possible internal force coefficients ω or all possible extension coefficients $\hat{\mathbf{e}}$ for a framework. Again the comments about dimensions of these quantities applies, and we can also define an inner product between $\mathbf{t}$ and $\mathbf{e}$ (or between ω and $\hat{\mathbf{e}}$) as the internal work $w_I = \mathbf{t}^\mathrm{T} \mathbf{e}$ (or $w_I = \omega^\mathrm{T} \hat{\mathbf{e}}$).

	External	mapping $\leftrightarrow$	Internal
Forces	$\boxed{\mathbf{f}}$	equilibrium $\mathbf{f} = \mathbf{At}$	$\boxed{\mathbf{t}}$
inner product $\updownarrow$	$w_i = \mathbf{f}^\mathsf{T}\mathbf{d}$		$w_e = \mathbf{t}^\mathsf{T}\mathbf{e}$
Displacements	$\boxed{\mathbf{d}}$	compatibility $\mathbf{Cd} = \mathbf{e}$	$\boxed{\mathbf{e}}$

Figure 10.1 The relationships between the internal and external vector spaces describing forces and infinitesimal deformations of a framework. For a framework in equilibrium, for any compatible deformation, $w_i = w_e$.

Note that, for this chapter, we have changed our notation from $\mathbb{E}$ (for "Euclidean") to $\mathbb{R}$ (for "real"). This doesn't really signify anything, except that earlier on we wanted to emphasize the Euclidean nature of the space with which we are dealing, whereas in this chapter it will be important to emphasize the real nature of the spaces, to contrast with the complex spaces that it is necessary to introduce later, This also matches the terminology used in our main reference text, James and Liebeck (2001).

Figure 10.1 summarizes the relationships between quantities in the various vector spaces.

10.2 Decomposition of Internal and External Vector Spaces

The equilibrium/compatibility relationship, or equivalently the rigidity relationship, provides a natural decomposition of internal I and external X vector spaces for a framework. The basic scheme is illustrated in Table 10.1. Later in the chapter (Section 10.5) we will then consider the equivalent decomposition when symmetry is also incorporated, when I and X will be considered as $\mathbb{R}\mathcal{G}$-modules.

We use the 'natural' basis for the internal I and external X vector spaces described in the previous section, and Section 3.2. An illustration of the way that the equilibrium matrix $\mathbf{A}$ then provides a linear transformation from the internal to the external vector space, is shown in Figure 10.2. The four fundamental subspaces of $\mathbf{A}$ shown in the figure (see, e.g. Strang, 2009) provide a natural decomposition of I and X.

The row space I_r subspace of I: extensions $\mathbf{e}$ that are compatible with some infinitesimal nodal displacement $\mathbf{d}$. Equivalently, internal forces that are orthogonal to all states of self-stress.

The nullspace S subspace of I: all possible states of self-stress – internal forces $\mathbf{t}$ that are in equilibrium with zero applied load. Equivalently, extensions that are orthogonal with all compatible extensions.

Table 10.1. The decomposition of internal and external vector spaces.

	Internal space I	$\overset{\mathbf{A}}{\underset{\mathbf{A}^\mathrm{T}}{\rightleftharpoons}}$	External space X
Decomposition into orthogonal vector subspaces	$I = I_r + S$		$X = X_r + M_i + M_r$
Dimensions of these vector spaces	$b = r + s$		$nd = r + m + \binom{d+1}{2}$

The column space X_r subspace of X: Nodal forces $\mathbf{f}$ that can be equilibrated by some internal set of forces $\mathbf{t}$. Equivalently, all displacements that are orthogonal to all mechanisms.

The row space M subspace of X: all possible infinitesimal mechanisms – infinitesimal nodal displacements that are compatible with zero extensions. Equivalently, all forces that are orthogonal to the equilibratable forces.

The space of infinitesimal mechanisms can be further subdivided,

$$M = M_i \oplus M_R, \tag{10.1}$$

where M_i is the space of internal infinitesimal mechanisms, and M_R is the space of rigid-body displacements. By definition M_i and M_R are orthogonal spaces.

10.2.1 Dimensions of Internal and External Vector Spaces

The internal vector space I has dimension b (the number of members of the structure), while the external vector space X has dimension nd (the number of vertices times the underlying dimension of the space); all vectors in X can be considered as vectors in $\mathbb{R}^{nd} = \mathbb{R}^n \otimes \mathbb{R}^d$.

Consider that the equilibrium matrix A has rank r. Then the fundamental theorem of Linear Algebra states that

Theorem 10.2.1. *Both I_r and X_r have dimension r. S has dimension $e - r$ and M has dimension $nd - r$*

We define the dimension of S to be s, often (loosely) called the number of states of self-stress, and thus we can write

$$s = b - r. \tag{10.2}$$

There are $\binom{d+1}{2}$ independent rigid-body infinitesimal displacements available for a non-degenerate framework, and hence M_R has dimension $\binom{d+1}{2}$. We define the dimension of M_i to be m, often (loosely) called the number of internal mechanisms, and thus we can write for a non-degenerate framework,

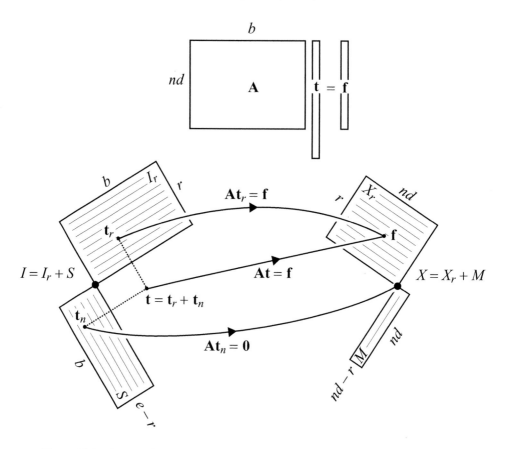

Figure 10.2 The mapping between internal and external vector spaces provided by the equilibrium matrix $\mathbf{A}$. A vector $\mathbf{t} \in I$ (a set of internal forces) can be split into a component $\mathbf{t}_r$ in the row-space I_r of the matrix, and an orthogonal component $\mathbf{t}_n$ that is a self-stress, i.e. $\mathbf{A}\mathbf{t}_n = \mathbf{0}$. A similar mapping in reverse would be provided by the compatibility matrix $\mathbf{C} = \mathbf{A}^\mathrm{T}$. Figure conceptually borrowed from (Strang, 2009).

$$m + \binom{d+1}{2} = nd - r. \tag{10.3}$$

We can now revisit the Maxwell counting rule, described in Section 3.4.1.

The following theorem, first explicitly stated for $d = 2$ and 3 by Calladine (1978), relates dimensions of the vector spaces defined, without requiring any knowledge of anything beyond the counting the number of components of the framework.

Theorem 10.2.2 (Calladine/Maxwell). *For a non-degenerate framework,*

$$m - s = nd - b - \binom{d+1}{2}. \tag{10.4}$$

Proof. Eliminating the rank r between (10.2) and (10.3) leads directly to (10.4). ☐

Maxwell (1864b) first described a counting rule for frameworks, where he wrote that a necessary condition for $m = 0$ in $d = 3$ was $d \geq 3n - 6$.

10.2.2 Singular Value Decomposition of the Equilibrium Matrix

The singular value decomposition (SVD) of the equilibrium matrix (Pellegrino, 1993) provides a convenient and numerically robust way of finding basis vectors for the subspaces I_r, S, X_r, and M.

Any real matrix, but specifically here the equilibrium matrix, $\mathbf{A}$, of dimension nd by b, can be factored into

$$\mathbf{A} = \mathbf{Q}_1 \mathbf{\Sigma} \mathbf{Q}_2^{\mathsf{T}}. \tag{10.5}$$

The matrices $\mathbf{Q}_1$ and $\mathbf{Q}_2$ are both orthogonal. The columns of $\mathbf{Q}_1$ (nd by nd) are the eigenvectors of $\mathbf{A}\mathbf{A}^{\mathsf{T}}$. The columns of $\mathbf{Q}_2$ (b by b) are the eigenvectors of $\mathbf{A}^{\mathsf{T}}\mathbf{A}$. The r singular values are on the diagonal of $\mathbf{\Sigma}$ (nd by b). The singular values are the square roots of the non-zero eigenvalues of both $\mathbf{A}\mathbf{A}^{\mathsf{T}}$ and $\mathbf{A}^{\mathsf{T}}\mathbf{A}$.

The first r columns of $\mathbf{Q}_1$ give an orthonormal basis for X_r, and the last $nd - r$ columns give an orthonormal basis for M. Similarly, the first r columns of $\mathbf{Q}_2$ give an orthonormal basis for I_r, and the last $e - r$ columns give an orthonormal basis for S. Furthermore, the columns of $\mathbf{Q}_1$ and $\mathbf{Q}_2$ are chosen in a special way: if $\mathbf{A}$ multiplies a column of $\mathbf{Q}_2$, it gives a multiple of the corresponding column of $\mathbf{Q}_1$; similarly, if $\mathbf{A}^{\mathsf{T}}$ multiplies a column of $\mathbf{Q}_1$, it gives a multiple of the corresponding column of $\mathbf{Q}_2$.

10.3 Internal and External Vector Spaces as $\mathbb{R}\mathcal{G}$-Modules

Now we introduce symmetry. We consider the vector spaces in the presence of symmetry — which allows us to consider vector spaces as $\mathbb{R}\mathcal{G}$-modules.

10.3.1 Transformation Operators and Matrices

Every framework $(G, \mathbf{p})$ has a geometric symmetry group $\mathcal{G}$, even if this symmetry group is only $\mathcal{C}_1$, the trivial group containing only the identity operation. Thus, if V is a vector space, and $g \in \mathcal{G}$ is one of the symmetry operations of the group, there will be a *transformation operator* for the operation g, $\mathbf{O}_V(g)$, that gives the effect of g on V; it maps any vector $\mathbf{v}$ in V to another vector $\mathbf{O}_V(g)\mathbf{v}$ in V:

$$g : \mathbf{v} \rightarrow \mathbf{O}_V(g)\mathbf{v}.$$

When we write the vector $\mathbf{v}$ in a particular coordinate system, this transformation operator is described by a transformation matrix. In Chapter 3, vectors in X or I were defined using particular coordinate systems, that we will now define as being the *natural coordinate systems* for the framework. Using the natural coordinate system, we now define a transformation matrix $\mathbf{D}_V(g)$ for each symmetry operation $g \in \mathcal{G}$ that maps any vector $\mathbf{v} \in V$ to another vector $\mathbf{D}_V(g)\mathbf{v} \in V$:

$$g : \mathbf{v} \rightarrow \mathbf{D}_V(g)\mathbf{v}.$$

Note that $\mathbf{D}_V(g)$ will always be an orthogonal matrix, and $\mathbf{D}_V(E)$ will be the identity matrix.

We only need to consider two sets of transformation matrices, one of which we call the external transformation matrices $\mathbf{D}_X(g)$ and the other of which we call the internal transformation matrices, $\mathbf{D}_I(g)$.

As an example, Figures 10.3 and 10.4 show the effect of all symmetry operations $g \in C_{4v}$ on, respectively, a particular external vector and a particular internal vector, both associated with the example structure shown in Figure 3.2(c). The external vector is given by

$$\mathbf{v}_X = [0; 0; 0; -1; 2; 1; -2; 0], \tag{10.6}$$

and the internal vector is given by

$$\mathbf{v}_I = [0; 1; 2; 0; 0; 0]. \tag{10.7}$$

The vectors $\mathbf{v}_X$ and $\mathbf{v}_I$ would be have to be multiplied by units of force to be considered as a vector $\mathbf{f}$ and $\mathbf{t}$, respectively, and would have to be multiplied by units of displacement to be considered as a vector $\mathbf{d}$ and $\mathbf{e}$, respectively. In fact, such vectors considered as forces would be in equilibrium (related by the equilibrium matrix $\mathbf{A}$ in Table 3.2(c)), but considered as deformations, they would not be compatible (i.e. not related by the compatibility matrix $\mathbf{C}$ in Table 3.1(c)).

Tables 10.2 and 10.3 show the transformation matrices $\mathbf{D}_X(g)$ and $\mathbf{D}_I(g)$ for each of the symmetry operations $g \in C_{4v}$ for the example framework.

10.3.2 Transformation Matrices and Group Representations

Note that both sets of transformation matrices defined in Section 10.3.1, $\mathbf{D}_X(g)$ and $\mathbf{D}_I(g)$, where $g \in \mathcal{G}$ are the symmetry operations of the symmetry group of a framework, form a matrix group. This is straightforward to show. Consider initially the orthogonal matrices $\mathbf{D}_X(g)$. Matrix multiplication is associative, and the identity $\mathbf{D}_X(E)$ exists as an identity matrix. Closure of the group is guaranteed as $\mathbf{D}_X(g_1 g_2) = \mathbf{D}_X(g_1)\mathbf{D}_X(g_2)$, and the inverse is given by $\mathbf{D}_X(g^{-1}) = \mathbf{D}_X^{-1}(g)$. Thus the matrices $\mathbf{D}_X(g)$, and similarly $\mathbf{D}_I(g)$, form a group.

A key point here is that we have thus shown that the mappings $g \rightarrow \mathbf{D}_X(g)$ and $g \rightarrow \mathbf{D}_I(g)$, each form a representation of the group $\mathcal{G}$.

10.3.3 $\mathbb{R}\mathcal{G}$-Modules and Their Character

A vector space V, defined over the real numbers $\mathbb{R}$, for which there is a transformation operator $\mathbf{O}_V(g)$ for all $g \in \mathcal{G}$ is called an $\mathbb{R}\mathcal{G}$-*module* – $\mathbb{R}$ from the real numbers, and $\mathcal{G}$ from the group. Thus, the vector spaces described in Section 10.1, X and I, are both $\mathbb{R}\mathcal{G}$-modules.

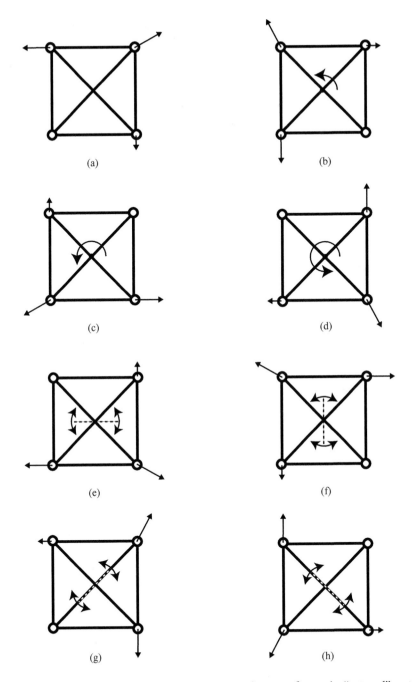

Figure 10.3 The effect of each of the symmetry operations $g \in C_{4v}$ on the "external" vector $\mathbf{v}_X$ defined in (10.6):(a) E; (b) C_4; (c) C_4^2; (d) C_4^3; (e) σ_{v1}; (f) σ_{v2}; (g) σ_{d1}; (h) σ_{d2}. The axes $v1$, $v1$, $d1$, and $d2$ are shown in Figure 9.1.

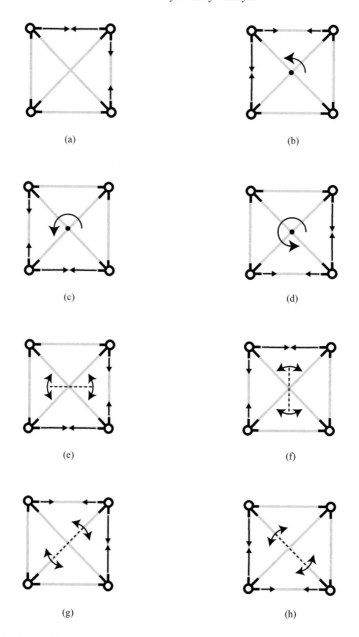

(a)					(b)

(c)					(d)

(e)					(f)

(g)					(h)

Figure 10.4 The effect of each of the symmetry operations $g \in \mathcal{C}_{4v}$ on the "internal" vector $\mathbf{v}_I$ defined in (10.7):(a) E; (b) C_4; (c) C_4^2; (d) C_4^3; (e) σ_{v1}; (f) σ_{v2}; (g) σ_{d1}; (h) σ_{d2}. The axes $v1$, $v1$, $d1$, and $d2$ are shown in Figure 9.1.

There is a clear link between $\mathbb{R}\mathcal{G}$-modules and representations of $\mathcal{G}$. For any given coordinate system, the transformation operator $\mathbf{O}_V(g)$ will be written as a set of transformation matrices $\mathbf{D}_V(g)$ for all $g \in \mathcal{G}$, and the mapping from g to $\mathbf{D}_V(g)$ forms a representation of $\mathcal{G}$. The character of this representation does not depend on the particular

Table 10.2. The transformation matrices $D_X(g)$ for each of the symmetry operations $g \in C_{4v}$ for the example framework:(a) $D_X(E)$; (b) $D_X(C_4)$; (c) $D_X(C_4^2)$; (d) $D_X(C_4^3)$; (e) $D_X(\sigma_{v1})$; (f) $D_X(\sigma_{v2})$; (g) $D_X(\sigma_{d1})$; (h) $D_X(\sigma_{d2})$. In each case, 0 is a 2×2 zero matrix. The character of X, $\chi_X(E), C_4, C_4^2, C_4^3, \sigma_{v1}, \sigma_{v2}, \sigma_{d1}, \sigma_{d2}) = (8, 0, 0, 0, 0, 0, 0, 0)$.

$$
\text{(a)} \qquad \begin{bmatrix}
\begin{bmatrix} 1 & 0 \\ 0 & 1 \end{bmatrix} & 0 & 0 & 0 \\
0 & \begin{bmatrix} 1 & 0 \\ 0 & 1 \end{bmatrix} & 0 & 0 \\
0 & 0 & \begin{bmatrix} 1 & 0 \\ 0 & 1 \end{bmatrix} & 0 \\
0 & 0 & 0 & \begin{bmatrix} 1 & 0 \\ 0 & 1 \end{bmatrix}
\end{bmatrix}
$$

$$
\text{(b)} \qquad \begin{bmatrix}
0 & 0 & 0 & \begin{bmatrix} 0 & -1 \\ 1 & 0 \end{bmatrix} \\
\begin{bmatrix} 0 & -1 \\ 1 & 0 \end{bmatrix} & 0 & 0 & 0 \\
0 & \begin{bmatrix} 0 & -1 \\ 1 & 0 \end{bmatrix} & 0 & 0 \\
0 & 0 & \begin{bmatrix} 0 & -1 \\ 1 & 0 \end{bmatrix} & 0
\end{bmatrix}
$$

$$
\text{(c)} \qquad \begin{bmatrix}
0 & 0 & \begin{bmatrix} -1 & 0 \\ 0 & -1 \end{bmatrix} & 0 \\
0 & 0 & 0 & \begin{bmatrix} -1 & 0 \\ 0 & -1 \end{bmatrix} \\
\begin{bmatrix} -1 & 0 \\ 0 & -1 \end{bmatrix} & 0 & 0 & 0 \\
0 & \begin{bmatrix} -1 & 0 \\ 0 & -1 \end{bmatrix} & 0 & 0
\end{bmatrix}
$$

$$
\text{(d)} \qquad \begin{bmatrix}
0 & \begin{bmatrix} 0 & 1 \\ -1 & 0 \end{bmatrix} & 0 & 0 \\
0 & 0 & \begin{bmatrix} 0 & 1 \\ -1 & 0 \end{bmatrix} & 0 \\
0 & 0 & 0 & \begin{bmatrix} 0 & 1 \\ -1 & 0 \end{bmatrix} \\
\begin{bmatrix} 0 & 1 \\ -1 & 0 \end{bmatrix} & 0 & 0 & 0
\end{bmatrix}
$$

$$
\text{(e)} \qquad \begin{bmatrix}
0 & 0 & 0 & \begin{bmatrix} 1 & 0 \\ 0 & -1 \end{bmatrix} \\
0 & 0 & \begin{bmatrix} 1 & 0 \\ 0 & -1 \end{bmatrix} & 0 \\
0 & \begin{bmatrix} 1 & 0 \\ 0 & -1 \end{bmatrix} & 0 & 0 \\
\begin{bmatrix} 1 & 0 \\ 0 & -1 \end{bmatrix} & 0 & 0 & 0
\end{bmatrix}
$$

$$
\text{(f)} \qquad \begin{bmatrix}
0 & \begin{bmatrix} -1 & 0 \\ 0 & 1 \end{bmatrix} & 0 & 0 \\
\begin{bmatrix} -1 & 0 \\ 0 & 1 \end{bmatrix} & 0 & 0 & 0 \\
0 & 0 & 0 & \begin{bmatrix} -1 & 0 \\ 0 & 1 \end{bmatrix} \\
0 & 0 & \begin{bmatrix} -1 & 0 \\ 0 & 1 \end{bmatrix} & 0
\end{bmatrix}
$$

$$
\text{(g)} \qquad \begin{bmatrix}
\begin{bmatrix} 0 & 1 \\ 1 & 0 \end{bmatrix} & 0 & 0 & 0 \\
0 & 0 & 0 & \begin{bmatrix} 0 & 1 \\ 1 & 0 \end{bmatrix} \\
0 & 0 & \begin{bmatrix} 0 & 1 \\ 1 & 0 \end{bmatrix} & 0 \\
0 & \begin{bmatrix} 0 & 1 \\ 1 & 0 \end{bmatrix} & 0 & 0
\end{bmatrix}
$$

$$
\text{(h)} \qquad \begin{bmatrix}
0 & 0 & \begin{bmatrix} 0 & -1 \\ -1 & 0 \end{bmatrix} & 0 \\
0 & \begin{bmatrix} 0 & -1 \\ -1 & 0 \end{bmatrix} & 0 & 0 \\
\begin{bmatrix} 0 & -1 \\ -1 & 0 \end{bmatrix} & 0 & 0 & 0 \\
0 & 0 & 0 & \begin{bmatrix} 0 & -1 \\ -1 & 0 \end{bmatrix}
\end{bmatrix}
$$

choice of coordinate system, and thus we can naturally define the *character* $\chi_V(g)$ of an $\mathbb{R}\mathcal{G}$-module V as

$$
\chi_V(g) = \mathrm{tr}(\mathbf{D}_V(g))
$$

for any choice of coordinate system for the transformation matrices.

Table 10.3. The transformation matrices $D_I(g)$ for each of the symmetry operations $g \in C_{4v}$ for the example framework. All other entries in the matrices are zero. The character of I, $\chi_I(E, C_4, C_4^2, C_4^3, \sigma_{v1}, \sigma_{v2}, \sigma_{d1}, \sigma_{d2}) = (6, 0, 0, 0, 2, 2, 2, 2)$.

$$\mathbf{D}_I(E) = \begin{bmatrix} 1 & & & & & \\ & 1 & & & & \\ & & 1 & & & \\ & & & 1 & & \\ & & & & 1 & \\ & & & & & 1 \end{bmatrix}$$

(a)

$$\mathbf{D}_I(C_4) = \begin{bmatrix} & & 1 & & & \\ 1 & & & & & \\ & 1 & & & & \\ & & & & & 1 \\ & & & & 1 & \\ & & & 1 & & \end{bmatrix}$$

(b)

$$\mathbf{D}_I(C_4^2) = \begin{bmatrix} & 1 & & & & \\ & & 1 & & & \\ 1 & & & & & \\ & & & & 1 & \\ & & & & & 1 \\ & & & 1 & & \end{bmatrix}$$

(c)

$$\mathbf{D}_I(C_4^3) = \begin{bmatrix} & & 1 & & & \\ & 1 & & & & \\ 1 & & & & & \\ & & & & & 1 \\ & & & & 1 & \\ & & & 1 & & \end{bmatrix}$$

(d)

$$\mathbf{D}_I(\sigma_{v1}) = \begin{bmatrix} & & 1 & & & \\ & 1 & & & & \\ 1 & & & & & \\ & & & 1 & & \\ & & & & & 1 \\ & & & & 1 & \end{bmatrix}$$

(e)

$$\mathbf{D}_I(\sigma_{v2}) = \begin{bmatrix} 1 & & & & & \\ & & 1 & & & \\ & 1 & & & & \\ & & & & 1 & \\ & & & 1 & & \\ & & & & & 1 \end{bmatrix}$$

(f)

$$\mathbf{D}_I(\sigma_{d1}) = \begin{bmatrix} & & 1 & & & \\ & 1 & & & & \\ 1 & & & & & \\ & & & & 1 & \\ & & & 1 & & \\ & & & & & 1 \end{bmatrix}$$

(g)

$$\mathbf{D}_I(\sigma_{d2}) = \begin{bmatrix} & 1 & & & & \\ 1 & & & & & \\ & & 1 & & & \\ & & & 1 & & \\ & & & & 1 & \\ & & & & & 1 \end{bmatrix}$$

(h)

Many calculations on the impact of symmetry on a first-order analysis of a framework can actually be done just working with the characters of the various $\mathbb{R}\mathcal{G}$-modules – and next we will see that these can be easily found.

10.3.4 Simple Evaluation of the Characters of an $\mathbb{R}\mathcal{G}$-Module

The calculation of characters doesn't require the construction of an entire representation. Characters are traces of matrices, and thus only the entries on the diagonal matter. For a permutation matrix, for example, an entry on a diagonal implies that an element is unshifted by that operation, and the trace will just count how many elements are unshifted.

Consider the internal vector space I, which is an $\mathbb{R}\mathcal{G}$-module. For the natural coordinate system, the transformation matrices $\mathbf{D}_I(g)$ will form a set of permutation matrices, where an entry on the diagonal will imply that, for that particular symmetry operation, that bar of the framework is unshifted. If we count how many bars are unshifted by the operation, that counts the trace of the matrix, and thus gives the character $\chi_I(g)$, which must be equal to the permutation representation of the bars of the framework $\chi_e(g)$. We thus write the characters as equal:

$$\chi_I = \chi_e. \tag{10.8}$$

As an example, consider the framework shown in Figure 9.1:

- Under the identity $g = E$, all bars are unshifted, and hence $\chi_I(E) = 6$;
- Under the 2-fold rotation $g = C_2$, the two bars crossing the centre transform into themselves, but every other bar moves, and hence $\chi_I(C_2) = 2$;
- For either of the 4-fold rotations all bars move, and so $\chi_I(C_4) = \chi_I(C_4^{-1}) = 0$;
- For the reflections σ_{v1} and σ_{v2}, only the two bars perpendicular to the line of reflection transform into themselves, and so $\chi_I(\sigma_{v1}) = \chi_I(\sigma_{v2}) = 2$;
- For the reflections σ_{d1} and σ_{d2}, both the diagonal bar along the reflection axis and the diagonal bar perpendicular to the line of reflection transform into themselves, while no other does, so $\chi_I(\sigma_{d1}) = \chi_I(\sigma_{d2}) = 2$.

In summary, for the example,

$$\chi_I(E, C_4, C_4^2, C_4^3, \sigma_{v1}, \sigma_{v2}, \sigma_{d1}, \sigma_{d2}) = (6, 0, 0, 0, 2, 2, 2, 2).$$

It is straightforward to check that this agrees with the calculation made by finding the trace of the transformation matrices given in Table 10.3, where each entry of 1 on the diagonal of a matrix corresponds to an unshifted bar described in the previous paragraph.

The position is slightly more complicated for the external vector space X. Here, the transformation matrices $\mathbf{D}_X(g)$ will not be permutation matrices, as can be seen in the example in Table 10.2. However, using the natural coordinate system for X, we can write each of the transformation matrices using a Kronecker product,

$$\mathbf{D}_X(g) = \mathbf{D}_v(g) \otimes \mathbf{D}_T, \tag{10.9}$$

where $\mathbf{D}_v(g)$ is a permutation matrix for the nodes (v for vertices of the underlying graph), $\mathbf{D}_T(g)$ is a $d \times d$ rotation/reflection matrix, with all eigenvalues having magnitude 1, and $\otimes$ being the Kronecker product. For the example framework, the matrices $\mathbf{D}_v(g)$ and $\mathbf{D}_T(g)$ are given in Table 10.4. $\mathbf{D}_v(g)$ describes how nodes move around, while $\mathbf{D}_T(g)$ describes how the vectors attached to the node are transformed.

The character of $\mathbf{D}_v(g)$ is straightforward to find – it simply counts the number of nodes unshifted by each symmetry operation. The character of $\mathbf{D}_T(g)$ is straightforward to calculate from first principles, or can simply be looked up in a book of Group Theory tables such as Atkins et al. (1970) or Altmann and Herzig (1994). Then the character of X can simply

Table 10.4. The nodal permutation matrices $D_V(g)$ for each of the symmetry operations $g \in \mathcal{C}_{4v}$ for the example framework. Also shown are the transformation matrix for an orthonormal pair of vectors $D_V(g)$ for each of the symmetry operations $g \in \mathcal{C}_{4v}$. The character of $D_V(g)$, $\chi_V(E, C_4, C_4^2, C_4^3, \sigma_{v1}, \sigma_{v2}, \sigma_{d1}, \sigma_{d2}) = (4, 0, 0, 0, 0, 0, 2, 2)$, and the character of $D_T(g)$, $\chi_T(E, C_4, C_4^2, C_4^3, \sigma_{v1}, \sigma_{v2}, \sigma_{d1}, \sigma_{d2}) = (2, 0, -2, 0, 0, 0, 2, 2)$.

$$D_v(E) = \begin{bmatrix} 1 & & & \\ & 1 & & \\ & & 1 & \\ & & & 1 \end{bmatrix} ;$$

$$D_T(E) = \begin{bmatrix} 1 & 0 \\ 0 & 1 \end{bmatrix}$$

(a)

$$D_v(C_4) = \begin{bmatrix} & & & 1 \\ 1 & & & \\ & 1 & & \\ & & 1 & \end{bmatrix} ;$$

$$D_T(C_4) = \begin{bmatrix} 0 & -1 \\ 1 & 0 \end{bmatrix}$$

(b)

$$D_v(C_4^2) = \begin{bmatrix} & & 1 & \\ & & & 1 \\ 1 & & & \\ & 1 & & \end{bmatrix} ;$$

$$D_T(C_4^2) = \begin{bmatrix} -1 & 0 \\ 0 & -1 \end{bmatrix}$$

(c)

$$D_v(C_4^3) = \begin{bmatrix} & 1 & & \\ & & 1 & \\ & & & 1 \\ 1 & & & \end{bmatrix} ;$$

$$D_T(C_4^3) = \begin{bmatrix} 0 & 1 \\ -1 & 0 \end{bmatrix}$$

(d)

$$D_v(\sigma_{v1}) = \begin{bmatrix} & & 1 & \\ & 1 & & \\ 1 & & & \\ & & & 1 \end{bmatrix} ;$$

$$D_T(\sigma_{v1}) = \begin{bmatrix} 1 & 0 \\ 0 & -1 \end{bmatrix}$$

(e)

$$D_v(\sigma_{v2}) = \begin{bmatrix} 1 & & & \\ & & 1 & \\ & 1 & & \\ & & & 1 \end{bmatrix} ;$$

$$D_T(\sigma_{v2}) = \begin{bmatrix} -1 & 0 \\ 0 & 1 \end{bmatrix}$$

(f)

$$D_v(\sigma_{d1}) = \begin{bmatrix} 1 & & & \\ & & 1 & \\ & 1 & & \\ & & & 1 \end{bmatrix} ;$$

$$D_T(\sigma_{d1}) = \begin{bmatrix} 0 & 1 \\ 1 & 0 \end{bmatrix}$$

(g)

$$D_v(\sigma_{d2}) = \begin{bmatrix} 1 & & & \\ & & 1 & \\ & 1 & & \\ & & & 1 \end{bmatrix} ;$$

$$D_T(\sigma_{d2}) = \begin{bmatrix} 0 & -1 \\ -1 & 0 \end{bmatrix}$$

(h)

be found by multiplying the two characters for each operation together – it follows from (10.9) directly that

$$\chi_X(g) = \chi_v(g)\chi_T(g) \quad \forall g \in \mathcal{G}, \tag{10.10}$$

which we write as

$$\chi_X = \chi_v \times \chi_T. \tag{10.11}$$

Thus for the example where

$$\chi_v(E, C_4, C_4^2, C_4^3, \sigma_{v1}, \sigma_{v2}, \sigma_{d1}, \sigma_{d2}) = (4, 0, 0, 0, 0, 0, 2, 2)$$

and

$$\chi_T(E, C_4, C_4^2, C_4^3, \sigma_{v1}, \sigma_{v2}, \sigma_{d1}, \sigma_{d2}) = (2, 0, -2, 0, 0, 0, 2, 2),$$

then

$$\chi_X(E, C_4, C_4^2, C_4^3, \sigma_{v1}, \sigma_{v2}, \sigma_{d1}, \sigma_{d2}) = \chi_v \times \chi_T = (8, 0, 0, 0, 0, 0, 0, 0),$$

in agreement with the results shown in Table 10.2.

10.3.5 $\mathbb{R}\mathcal{G}$-Submodules

If, for some $\mathbb{R}\mathcal{G}$-module V, there is a subspace of V that is preserved by all of the symmetry operations in the group, then that subspace is defined as an $\mathbb{R}\mathcal{G}$-submodule. More formally, if there is a subspace V_i of V, where the transformation operators $\mathbf{O}_V(g)$ for each symmetry operation $g \in \mathcal{G}$ (note that the transformation operators are defined for the parent space V) are such that, for any $\mathbf{v} \in V_i$,

$$\mathbf{v} \to \mathbf{O}_V(g)\mathbf{v} \in V_i,$$

then V_i is an $\mathbb{R}\mathcal{G}$-submodule of the $\mathbb{R}\mathcal{G}$-module V.

We have already seen a number of examples of $\mathbb{R}\mathcal{G}$-submodules for the simple example structures shown in Figure 3.2. Figure 10.5 reproduces Figures 3.3, 3.4, and 3.9 to show these.

- Figure 10.5(a) shows a basis vector for the 1-dimensional subspace of states of self-stress for the structure. This is a subspace of the internal vector space I, and further, the subspace is preserved by all possible symmetry operations (the structure has symmetry C_{4v}, and the symmetry operations are shown in Figure 10.3). Thus the self-stress shown is a basis for a 1-dimensional $\mathbb{R}C_{4v}$-submodule of the $\mathbb{R}C_{4v}$-module I.

- Figure 10.5(b) shows an infinitesimal mechanism for a structure which also has symmetry C_{4v}. This internal mechanism is a basis for the 1-dimensional subspace of all possible internal infinitesimal mechanisms, and is a subspace of the external vector space X. The mechanism shown is unchanged by any rotation operation, and is reversed (every component multiplied by -1) by every reflection operation. The mechanism is said to be *symmetric* with respect to rotations, and *anti-symmetric* with respect to reflections. Thus although the basis vector is reversed by some operations, the subspace of internal infinitesimal mechanisms itself is preserved by every symmetry operation, and is hence a 1-dimensional $\mathbb{R}C_{4v}$-submodule of the $\mathbb{R}C_{4v}$-module X.

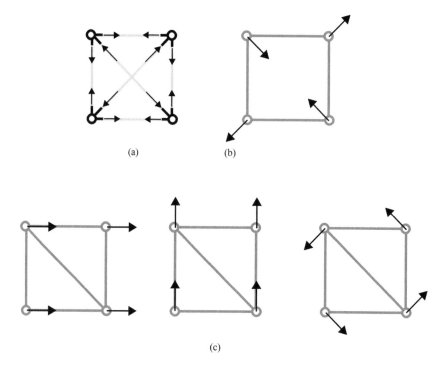

(a)   (b)

(c)

Figure 10.5 Examples of basis vectors for $\mathbb{R}\mathcal{G}$-submodules: (a) a state of self-stress; (b) an internal infinitesimal mechanism; (c) infinitesimal rigid-body displacements and rotations.

- Figure 10.5(c) shows basis vectors for the rigid-body translations and rigid-body rotations of a structure. This structure has symmetry C_{2v}. The rigid-body rotations in this example form a 1-dimensional $\mathbb{R}C_{2v}$-submodule of the $\mathbb{R}C_{2v}$-module X. The rigid-body displacements form a 2-dimensional $\mathbb{R}C_{2v}$-submodule of the $\mathbb{R}C_{2v}$-module X – 2-dimensional because some symmetry operations will transform one basis vector into the other.

In fact, we shall see in Section 10.4 that the vector subspace of all states of self-stress and the vector subspace of all infinitesimal internal mechanisms and the vector subspace of all infinitesimal rigid-body displacements and the vector subspace of all infinitesimal rigid-body rotations will always form $\mathbb{R}\mathcal{G}$ submodules.

10.4 Symmetry Operations, Equilibrium and Compatibility – $\mathbb{R}\mathcal{G}$-Homomorphisms

In this section we will examine the effect of symmetry operations on the equilibrium and compatibility relationships.

10.4.1 Equilibrium

Consider a framework $(G, \mathbf{p})$ with symmetry group $\mathcal{G}$, carrying a vector of internal forces, $\mathbf{t}$, in equilibrium with a vector of external forces $\mathbf{f}$. We describe this with (3.16), using the equilibrium matrix $\mathbf{A}$:

$$\mathbf{f} = \mathbf{At}. \tag{10.12}$$

Now consider the effect of a symmetry operation, $g \in \mathcal{G}$:

$$\mathbf{f} \rightarrow \mathbf{D}_X(g)\mathbf{f}$$

$$\mathbf{t} \rightarrow \mathbf{D}_I(g)\mathbf{t}.$$

The geometric symmetry of the framework implies that these transformed vectors will also be in equilibrium, and so

$$\mathbf{D}_X(g)\mathbf{f} = \mathbf{AD}_I(g)\mathbf{t}. \tag{10.13}$$

Substituting from the original equilibrium equation (10.12) into (10.13) gives an equation that is valid for all $\mathbf{t}$:

$$\mathbf{D}_X(g)\mathbf{At} = \mathbf{AD}_I(g)\mathbf{t},$$

and hence

$$\mathbf{D}_X(g)\mathbf{A} = \mathbf{AD}_I(g) \quad \forall g \in \mathcal{G}. \tag{10.14}$$

The transformation matrices are orthogonal matrices, where $\mathbf{D}_X^{-1}(g) = \mathbf{D}_X^{\mathsf{T}}(g)$ and $\mathbf{D}_I^{-1}(g) = \mathbf{D}_I^{\mathsf{T}}(g)$, and so (10.14) can equivalently be written as

$$\mathbf{A} = \mathbf{D}_X^{\mathsf{T}}(g)\mathbf{AD}_I(g) \quad \forall g \in \mathcal{G} \tag{10.15}$$

and as

$$\mathbf{D}_X(g)\mathbf{AD}_i^{\mathsf{T}}(g) = \mathbf{A} \quad \forall g \in \mathcal{G}. \tag{10.16}$$

Equation (10.14) or (10.15) or (10.16) can be considered as being essentially what we mean when we say that the equilibrium relations for a framework are symmetric with respect to the group $\mathcal{G}$.

10.4.2 Compatibility

An equivalent development for the compatibility matrix $\mathbf{C}$ or the rigidity matrix $\mathbf{R}$ will give

$$\mathbf{D}_I(g)\mathbf{C} = \mathbf{CD}_X(g) \quad \forall g \in \mathcal{G} \tag{10.17}$$

or

$$\mathbf{D}_I(g)\mathbf{R} = \mathbf{RD}_X(g) \quad \forall g \in \mathcal{G}. \tag{10.18}$$

10.4.3 $\mathbb{R}\mathcal{G}$-Homomorphisms

We define an $\mathbb{R}\mathcal{G}$-*homomorphism* as follows. Let U and V be $\mathbb{R}\mathcal{G}$-modules, where $\mathbf{D}_U(g)$ and $\mathbf{D}_V(g)$ are the transformation matrices for the symmetry operation g for U and V respectively in the chosen coordinate system. Let $\mathbf{X}$ be a matrix that maps from U to V in the chosen coordinate system. Then $\mathbf{X}$ represents an $\mathbb{R}\mathcal{G}$-homomorphism if

$$\mathbf{D}_V(g)\mathbf{X}\mathbf{u} = \mathbf{X}\mathbf{D}_U(g)\mathbf{u} \quad \forall u \in U, g \in \mathcal{G}, \tag{10.19}$$

i.e. if $\mathbf{v} = \mathbf{X}\mathbf{u}$, then $\mathbf{D}_V(g)\mathbf{v} = \mathbf{X}\mathbf{D}_U(g)\mathbf{u}$.

Theorem 10.4.1. *If* $\mathbf{X}$ *represents an* $\mathbb{R}\mathcal{G}$-*homomorphism from* U *to* V, *then* $\mathbf{X}^T$ *represents an* $\mathbb{R}\mathcal{G}$-*homomorphism from* V *to* U

Proof. If $\mathbf{X}$ represents an $\mathbb{R}\mathcal{G}$-homomorphism from U to V, then, from (10.19),

$$\mathbf{D}_V(g)\mathbf{X} = \mathbf{X}\mathbf{D}_U(g) \quad \forall g \in \mathcal{G}.$$

Taking the transpose,

$$\mathbf{X}^T\mathbf{D}_V^T(g) = \mathbf{D}_U^T(g)\mathbf{X}^T \quad \forall g \in \mathcal{G},$$

and $\mathbf{D}_V^T(g) = \mathbf{D}_V(g^{-1})$, $\mathbf{D}_U^T(g) = \mathbf{D}_U(g^{-1})$, and so,

$$\mathbf{X}^T\mathbf{D}_V(g^{-1}) = \mathbf{D}_U(g^{-1})\mathbf{X}^T \quad \forall g^{-1} \in \mathcal{G},$$

and thus $\mathbf{X}^T$ represents an $\mathbb{R}\mathcal{G}$-homomorphism from V to U. $\qquad\square$

We are interested in $\mathbb{R}\mathcal{G}$-homomorphisms because it is clear that both the equilibrium and compatibility matrices are $\mathbb{R}\mathcal{G}$-homomorphisms.

10.4.4 Fundamental Subspaces are $\mathbb{R}\mathcal{G}$-Submodules

$\mathbb{R}\mathcal{G}$-homomorphisms naturally give rise to $\mathbb{R}\mathcal{G}$-submodules through the following theorem:

Theorem 10.4.2. *The four fundamental subspaces of the matrix* $\mathbf{X}$, *where* $\mathbf{X}$ *represents an* $\mathbb{R}\mathcal{G}$-*homomorphism, are* $\mathbb{R}\mathcal{G}$-*submodules.*

Proof. Consider that $\mathbf{X}$ is an $\mathbb{R}\mathcal{G}$-homomorphism from U to V, and that transformation matrices $\mathbf{D}_U(g)$ and $\mathbf{D}_V(g)$ represent each symmetry operation g in U and V respectively.

Consider initially a vector $\mathbf{u}$ in the nullspace of $\mathbf{X}$ (a subspace of U), i.e. $\mathbf{X}\mathbf{u} = \mathbf{0}$. Then, for any $g \in \mathcal{G}$,

$$\mathbf{D}_V(g)\mathbf{X}\mathbf{u} = \mathbf{D}_V(g)\mathbf{0} = \mathbf{0}.$$

From (10.19), $\mathbf{D}_V(g)\mathbf{X}\mathbf{u} = \mathbf{X}\mathbf{D}_U(g)\mathbf{u}$, and so,

$$\mathbf{X}\mathbf{D}_U(g)\mathbf{u} = \mathbf{0},$$

and hence $\mathbf{D}_U(g)\mathbf{u}$ is also in the nullspace of $\mathbf{X}$, for any symmetry operation $g \in G$. Thus the nullspace of $\mathbf{X}$ is an $\mathbb{R}\mathcal{G}$-submodule of U.

Now consider a vector $\mathbf{v}$ in the column space of $\mathbf{X}$ (a subspace of V), i.e. $\mathbf{Xu} = \mathbf{v}$ for some $\mathbf{u} \in U$. Then, for any $g \in \mathcal{G}$,

$$\mathbf{D}_V(g)\mathbf{Xu} = \mathbf{D}_V(g)\mathbf{v}.$$

From (10.19), $\mathbf{D}_V(g)\mathbf{Xu} = \mathbf{XD}_U(g)\mathbf{u}$, and so,

$$\mathbf{XD}_U(g)\mathbf{u} = \mathbf{D}_V(g)\mathbf{v},$$

and hence $\mathbf{D}_V(g)\mathbf{v}$ is also in the column space of $\mathbf{X}$, for any symmetry operation $g \in G$. Thus the column space of $\mathbf{X}$ is an $\mathbb{R}\mathcal{G}$-submodule of V.

The same proofs can also be applied to the nullspace and column space of $\mathbf{X}^T$ (also an $\mathbb{R}\mathcal{G}$-homomorphism, from Theorem 10.4.1), and hence the left-nullspace of $\mathbf{X}$ is an $\mathbb{R}\mathcal{G}$-submodule of V and the rowspace of $\mathbf{X}$ is an $\mathbb{R}\mathcal{G}$-submodule of U. Thus the theorem is proved. $\square$

10.4.5 Isomorphic $\mathbb{R}\mathcal{G}$-Modules

Consider U and V to be $\mathbb{R}\mathcal{G}$-modules, where $\mathbf{D}_U(g)$, and $\mathbf{D}_V(g)$ are the transformation matrices for the symmetry operation g for U and V respectively in the chosen coordinate system. Let $\mathbf{X}$ be a matrix that maps from U to V in the chosen coordinate system. Then, if $\mathbf{X}$ represents an $\mathbb{R}\mathcal{G}$-homomorphism, and $\mathbf{X}$ is invertible, then $\mathbf{X}$ represents an *isomorphic $\mathbb{R}\mathcal{G}$-homomorphism*, and U and V are *isomorphic $\mathbb{R}\mathcal{G}$-modules*.

Isomorphic $\mathbb{R}\mathcal{G}$-submodules can almost be considered to be identical to one another. In particular, they will have the same character under any symmetry operation.

10.4.6 The Column Space and Row Space are Isomorphic $\mathbb{R}\mathcal{G}$-Modules

Theorem 10.4.3. *The row space and the column space of the matrix $\mathbf{X}$, where $\mathbf{X}$ represents an $\mathbb{R}\mathcal{G}$-homomorphism, are isomorphic $\mathbb{R}\mathcal{G}$-submodules.*

Proof. Consider the relation $\mathbf{Xa} = \mathbf{b}$, where $\mathbf{a}$ is in the row space of $\mathbf{X}$, and $\mathbf{b}$ is in the column space of $\mathbf{X}$. Then every $\mathbf{a}$ gives a unique $\mathbf{b}$, and every $\mathbf{b}$ gives a unique $\mathbf{a}$ – the matrix $\mathbf{X}$ defines an invertible function between the row space and the column space. $\square$

10.4.7 Orthogonal $\mathbb{R}\mathcal{G}$-Modules

Two $\mathbb{R}\mathcal{G}$-modules U and V are orthogonal if, for any vector $\mathbf{u} \in U$ and any vector $\mathbf{v} \in V$, $\mathbf{u}^T\mathbf{v} = 0$.

10.4.8 Infinitesimal Mechanisms and States of Self-Stress as $\mathbb{R}\mathcal{G}$-Submodules

It follows directly from Theorem 10.4.2 that, as the equilibrium matrix $\mathbf{A}$ is an $\mathbb{R}\mathcal{G}$-homomorphism, then the nullspace of $\mathbf{A}$, i.e. the space of all possible states of self-stress, is an $\mathbb{R}G$-submodule of the vector space I. Similarly, the left-nullspace of $\mathbf{A}$, i.e. the space of all infinitesimal mechanisms, is an $\mathbb{R}G$-submodule of X.

Consider now the infinitesimal rigid-body motions of a framework $(G, \mathbf{p})$. We distinguish between displacements (where every node moves in the same direction by the same amount) and rotations (any rigid-body motion that is orthogonal to all displacements). The rotations will leave one point unshifted, the centre of rotation, and that point will be the centroid $\bar{\mathbf{p}}$, where

$$\bar{\mathbf{p}} = \frac{\sum_{i=1}^{n} \mathbf{p}_i}{n}.$$

Clearly the d-dimensional space of infinitesimal rigid-body displacements, and the space of infinitesimal rigid-body rotations (whose dimension depends on the space spanned by $\mathbf{p}$), each form $\mathbb{R}G$-submodules of X.

Examples illustrating all of the above have already been shown in Figure 10.5.

10.5 Decomposition of Internal and External $\mathbb{R}G$-Modules

The basic development in Section 10.2 can be now be followed, including symmetry, by considering I and X as $\mathbb{R}G$-modules. The key difference is that we now consider the character χ of the $\mathbb{R}G$-modules rather than just the dimension of the linear spaces – remember that the character χ is a vector containing a scalar value $\chi(g)$ for each operation $g \in G$. However, note that the character under the identity, $\chi(E)$, is equal to the dimension of the linear space, and thus considering characters automatically subsumes the consideration of dimensions in Section 10.2.

The basic scheme followed in the current section is illustrated in Table 10.5.

A difficulty of notation arises here, because what we are here calling the "character" is called the "representation" in papers and books on applied group theory, and is usually given the symbol Γ; and the term character and the symbol χ is reserved for the scalar value under a particular symmetry operation. However, here we follow mathematical representation

Table 10.5. The decomposition of internal and external $\mathbb{R}G$-modules.

	Internal $\mathbb{R}G$-module I $\overset{\mathbf{A}}{\underset{\mathbf{A}^\mathsf{T}}{\rightleftharpoons}}$ External $\mathbb{R}G$ module-X	
Decomposition into orthogonal $\mathbb{R}G$-submodules	$I = I_r + S$	$X = X_r + M_i + M_R + M_T$
Character of $\mathbb{R}G$-submodules	$\chi_e = \chi_r + \chi_s$	$\chi_v \times \chi_T = \chi_r + \chi_m + \chi_R + \chi_T$

theory, and will reserve the term representation for the mapping from a symmetry group g to a representation matrix $\mathbf{\Gamma}(g)$; translating what follows to the terms used by applied group theorists simply requires the substitution of the word "character" with "representation", and the symbol "χ" with "Γ".

Our starting point is to note that, from Theorem 10.4.2, I_r and S are $\mathbb{R}G$-submodules of I, and X_r and M are submodules of X. Further, from Theorem 10.4.3, I_r and X_r are isomorphic.

The next step is to break down M into three separate submodules: M_T, M_R, and M_i. The orthogonal $\mathbb{R}G$-modules M_T and M_R have already been introduced in Section 10.4.8, and contain all possible rigid body translations, and rotations, respectively. Any mechanisms that are orthogonal to both M_T and M_R must, by definition, be internal infinitesimal mechanisms (see Section 3.2.2), and it is a simple exercise to show that all such mechanisms form an $\mathbb{R}G$-module, which we label M_i. Clearly $M = M_T + M_R + M_i$.

For the internal space we can thus write

$$I = I_r + S, \tag{10.20}$$

and for the external space

$$X = X_r + M_i + M_R + M_T. \tag{10.21}$$

10.5.1 Symmetry-Adapted Counting for Frameworks

Fowler and Guest (2000) introduced the use of characters to count mechanisms and states of self-stress in a framework.

Theorem 10.5.1 (Symmetry Calladine/Maxwell). *For a framework,*

$$\chi_m - \chi_s = \chi_v \otimes \chi_T - \chi_e - \chi_T - \chi_R, \tag{10.22}$$

where:

χ_m, χ_s *are the characters of the internal mechanisms and states of self-stress, respectively;*

χ_v, χ_e *are the characters of the vertices and edges of the framework, respectively;*

χ_T, χ_R *are the characters of the rigid-body displacements and rigid-body displacement, respectively, considered in the appropriate dimension for the framework.*

Proof. We start from the decomposition of the internal I and external X $\mathbb{R}G$-modules expressed in equations (10.20) and (10.21). This decomposition into orthogonal submodules shows that representations $\mathbf{D}$ of symmetry operations in I and X can be written in a block-diagonal form using the representations in the sub-modules, i.e. that with the right choice of coordinate system, for any symmetry operation $g \in G$,

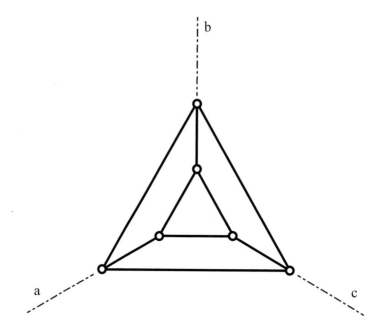

Figure 10.6 A 2-dimensional framework with C_{3v} symmetry. Symmetry operations that leave the framework unchanged are the identity (E), rotation by $2\pi/3$ and $4\pi/3$ (C_3, C_3^2), and reflection in the lines a, b, and c $(\sigma_a, \sigma_b, \sigma_c)$. Symmetry-adapted counting shows that this framework cannot be isostatic.

$$\mathbf{D}_I(g) = \mathbf{D}_{I_r}(g) \oplus \mathbf{D}_S(g) \tag{10.23}$$

$$\mathbf{D}_X(g) = \mathbf{D}_{X_r}(g) \oplus \mathbf{D}_{M_i}(g) \oplus \mathbf{D}_{M_R}(g) \oplus \mathbf{D}_{M_T}(g), \tag{10.24}$$

where $\oplus$ represents a block-diagonal composition of representations for submodules.

Now consider the characters of the representations in (10.23) and (10.23), incorporating the simple evaluation of characters from (10.8) and (10.11),

$$\chi_e = \chi_I = \chi_r + \chi_s \tag{10.25}$$

$$\chi_v \times \chi_T = \chi_X = \chi_r + \chi_m + \chi_R + \chi_T, \tag{10.26}$$

where χ_r is the character of both the isomorphic $\mathbb{R}\mathcal{G}$-modules I_r and X_r.

Eliminating χ_r (which, in general, can only be found from numerical analysis of a particular framework) allows us to find (10.22) for any framework. $\qquad\square$

Consider as an example the framework shown in Figure 10.6. With $d = 2$, $n = 6$, and $e = 9$, Calladine/Maxwell counting (10.4) gives $m - s = 0$. But in this case, symmetry counting can tell us more. We will show this by giving the calculation in (10.22) in the form of a tabular calculation:

	E	C_3	C_3^2	σ_a	σ_b	σ_c
χ_v	6	0	0	2	2	2
$\otimes \chi_T$	2	-1	-1	0	0	0
$=$	12	0	0	0	0	0
$-\chi_e$	-9	0	0	-3	-3	-3
$-\chi_T$	-2	1	1	0	0	0
$-\chi_R$	-1	-1	-1	1	1	1
$= \chi_m - \chi_s$	0	0	0	-2	-2	-2

Each of the rows of the tabular calculation gives the character of one of the components of (10.22), and the columns show the components of each character under each of the symmetry operations of the group. The numbers corresponding to the characters on the RHS of (10.22) can all be straightforwardly found, according to the scheme described in Section 10.3.4: χ_v and χ_e are the number of vertices and edges, respectively, unshifted by each symmetry operation, and χ_T and χ_R can be straightforwardly calculated, or looked up in a book of group theory tables. (Note that books of tables are typically designed to be used for chemical calculations on molecules in three dimensions, and assume χ_T and χ_R are in a 3-dimensional setting – so a little care needs to be taken using these for a 2-dimensional framework.)

The result of the calculation shows that neither the vector space of self-stress S, nor the space of internal mechanisms M_i, can be empty – if they were, the character under every symmetry operation would be 0. We know that the dimension of each of these spaces must be equal from Calladine/Maxwell counting (equation (10.4), reproduced by the column of the tabular calculation under the identity operation E), but further characterization requires a little more theory, coming in Section 10.6. Before we do that, we will look at more detail of what we can say, if we require that the characters χ_m and χ_s be equal under every symmetry operation for a framework.

10.5.2 Isostatic Symmetric Frameworks

If a framework has neither an infinitesimal mechanism nor a state of self-stress (i.e. the framework is isostatic), then the characters χ_m and χ_s, and hence $(\chi_m - \chi_s)$ must all be zero. Thus, the symmetry Calladine/Maxwell condition in (10.22) can be used to extend the simple Maxwell counting for isostatic structures. Here we reproduce the consideration of this in 2D given in Connelly et al. (2009) (the paper goes on to also consider frameworks in 3D).

For 2D, the possible symmetry operations are: the identity (E), rotation by $2\pi/p$ about a point (C_p), and reflection in a line (σ). In Table 10.6, we give a general tabular formulation of the symmetry Calladine/Maxwell condition (10.22). Each column shows the value of

Table 10.6. Calculations of characters for the 2D symmetry Calladine/Maxwell equation (10.22).

	E	$C_{n>2}$	C_2	σ
χ_v	n	n_c	n_c	n_σ
$\otimes \chi_T$	2	$2\cos(2\pi/p)$	-2	0
$=$	$2n$	$2j_c\cos(2\pi/p)$	$-2n_c$	0
$-\chi_e$	$-b$	0	$-b_2$	$-b_\sigma$
$-\chi_T$	-2	$-2\cos(2\pi/p)$	2	0
$-\chi_R$	-1	-1	-1	1
$= \chi_m - \chi_s$	$2n-b-3$	$2(n_c-1)\cos(2\pi/p)-1$	$-2n_c-b_2+1$	$-b_\sigma+1$

each character for one of four possible symmetry operations (four rather than three because we distinguish the C_2 from the C_p operation with $p > 2$).

To treat all 2-dimensional cases in a single table, we need a notation that keeps track of the fate of structural components under the various operations, which in turn depends on how the joints and bars are placed with respect to the symmetry elements. The notation used in Table 10.6 is as follows.

n is the total number of joints;
n_c is the number of joints lying on the point of rotation ($C_{p>2}$ or C_2). For simplicity here, we will only consider cases where all joints have distinct locations, so $n_c = 0$ or 1);
n_σ is the number of joints lying on a given mirror line;
b is the total number of bars;
b_2 is the number of bars left unshifted by a C_2 operation (see Figure 10.7(a) and note that C_p with $p > 2$ shifts all bars);
b_σ is the number of bars unshifted by a given mirror operation (see Figure 10.7(b); the unshifted bar may lie in, or perpendicular to, the mirror line).

Each of the counts refers to a particular symmetry element and any structural component may therefore contribute to one or more count, for instance, a joint counted in j_c also contributes to j_σ for each mirror line present.

From Table 10.6, the symmetry treatment of the 2D Maxwell equation reduces to scalar equations of four types. If $\Gamma(m) - \Gamma(s) = 0$, then

E: $2n - b = 3$ (10.27)

C_2: $2n_c + b_2 = 1$ (10.28)

σ: $b_\sigma = 1$ (10.29)

$C_{p>2}$: $2(n_c - 1)\cos\left(\dfrac{2\pi}{p}\right) = 1,$ (10.30)

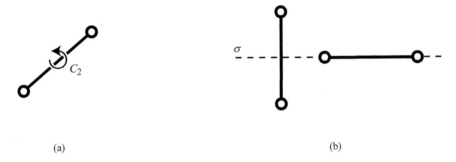

(a)   (b)

Figure 10.7 Possible placements of a bar with respect to a symmetry element in two dimensions, such that it is unshifted by the associated symmetry operation: (a) a C_2 centre of rotation; (b) a σ mirror line, where the bar might be placed perpendicular to the line, or along the line.

where a given equation applies when the corresponding symmetry operation is present in $\mathcal{G}$. Some observations on 2D isostatic frameworks, arising from this set of equations are:

(i) All frameworks have the identity element and (10.27) simply restates the scalar Maxwell counting rule (10.4) with $m - s = 0$ and dimension $d = 2$.

(ii) A C_2 element imposes limitations on the placement of bars and joints. As both j_c and b_2 must be non-negative integers, (10.28) has the unique solution $b_2 = 1$, $j_c = 0$. In other words, an isostatic 2D framework with a C_2 element of symmetry has no joint on the point of rotation, but exactly one bar centred at that point.

(iii) Similarly, the presence of a mirror line implies, by (10.29), that $b_\sigma = 1$ for that line, but places no restriction on the number of joints in the same line, and hence allows this bar to lie either in, or perpendicular to, the mirror.

(iv) Deduction of the condition imposed by a rotation of higher order $C_{p>2}$ proceeds as follows. Equation (10.30) with $\phi = 2\pi/p$ implies

$$(j_c - 1) \cos\left(\frac{2\pi}{n}\right) = \frac{1}{2}, \tag{10.31}$$

and as j_c is either 0 or 1, this implies that $j_c = 0$ and $n = 3$. Thus, a 2D isostatic framework cannot have a C_p rotational element with $p > 3$, and when either a C_2 or a C_3 rotational element is present, no joint may lie at the centre of rotation.

In summary, a 2D isostatic framework may have only symmetry operations drawn from the list $\{E, C_2, C_3, \sigma\}$, and hence the possible symmetry groups $\mathcal{G}$ are $C_1, C_2, C_3, C_s, C_{2v}$, and C_{3v}. Group by group, the conditions necessary for a 2D framework to be isostatic are then as follows.

C_1: $b = 2n - 3$.

C_2: $b = 2n - 3$ with $b_2 = 1$ and $n_c = 0$, and as all other bars and joints occur in pairs, n is even and b is odd.

C_3: $b = 2n - 3$ with $n_c = 0$, and hence all joints and bars occur in sets of three.

C_s: $b = 2n - 3$ with $b_\sigma = 1$ and all other bars occurring in pairs. Symmetry does not restrict j_σ.

C_{2v}: $b = 2n - 3$ with $n_c = 0$ and $b_2 = b_\sigma = 1$. A central bar lies in one of the two mirror lines, and perpendicular to the other. Any additional bars must lie in the general position, and hence occur in sets of four, with joints in sets of two and four. Hence b is odd and n is even.

C_{3v}: $b = 2n - 3$ with $n_c = 0$ and $b_\sigma = 1$ for each of the three mirror lines.

Figure 10.8 gives examples of small 2D isostatic frameworks for each of the possible groups, including cases where bars lie in, and perpendicular to, mirror lines.

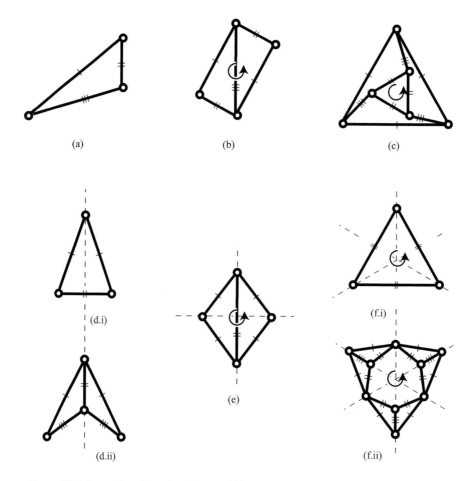

Figure 10.8 Examples, for each of the possible groups, of small 2D isostatic frameworks, with bars which are equivalent under symmetry marked with the same symbol: (a) C_1; (b) C_2; (c) C_3; (d) $C_s \equiv C_{1v}$; (e) C_{2v}; (f) C_{3v}. Mirror lines are shown dashed, and rotation axes are indicated by a circular arrow. For each of C_s and C_{3v}, two examples are given: (i) where each mirror has a bar centred at, and perpendicular to, the mirror line; (ii) where a bar lies in each mirror line. For C_{2v}, the bar lying at the centre must lie in one mirror line, and perpendicular to the other.

So far in this section, we have only considered necessary conditions for symmetric frameworks to be isostatic. But with sufficient additional conditions, these conditions can also be necessary. Firstly, we need to specify that the joints are placed as generally as possible given the symmetry conditions. Then, Schulze has shown that the conditions given above are both necessary and sufficient for C_2 and C_s (Schulze, 2010b), and for C_3 (Schulze, 2010c). For the dihedral groups C_{2v} and C_{3v}, the sufficiency condition remains a conjecture.

10.6 Irreducible Submodules

Section 10.5 showed the advantage of decomposing $\mathbb{R}\mathcal{G}$-modules into orthogonal $\mathbb{R}\mathcal{G}$-submodules. Here we will take that process to its limit, where an $\mathbb{R}\mathcal{G}$-submodule can be decomposed no further – we say that the $\mathbb{R}\mathcal{G}$-submodule is then *irreducible*.

However, to go further we are best to acknowledge that much of representation theory works better if we work with complex numbers, and hence $\mathbb{C}\mathcal{G}$-modules rather than $\mathbb{R}\mathcal{G}$-submodules. This is a consequence of Schur's Lemma, below, which concerns $\mathbb{C}\mathcal{G}$-modules rather than $\mathbb{R}\mathcal{G}$-modules. However, this doesn't cause much of a problem: all the previous definitions for $\mathbb{R}\mathcal{G}$-modules have a natural counterpart in $\mathbb{C}\mathcal{G}$-modules, and as for our calculations we ultimately want real results, and it is not difficult to work with complex numbers (which inevitably appear alongside their complex conjugate) and then to revert to real numbers at the end of the calculation.

> **Theorem 10.6.1** (Schur's Lemma). *Consider U and V to be irreducible $\mathbb{C}\mathcal{G}$-modules, where $\mathbf{D}_U(g)$, and $\mathbf{D}_V(g)$ are the transformation matrices for the symmetry operation $g \in \mathcal{G}$ for U and V respectively in the chosen coordinate system.*
>
> > (i) *Let $\mathbf{X}$ be a matrix that maps from U to V in the chosen coordinate system. If $\mathbf{X}$ represents a $\mathbb{C}\mathcal{G}$-homomorphism, then either $\mathbf{X}$ is invertible, and U and V are isomorphic, or $\mathbf{X} = \mathbf{0}$*
> > (ii) *If $\mathbf{X}\mathbf{D}_U(g) = \mathbf{D}_U(g)\mathbf{X}$ for all $g \in \mathcal{G}$ then $\mathbf{X}$ is a complex scalar multiple of the identity matrix.*

The consequence of Schur's Lemma is that, for each symmetry group, there is a unique set of possible irreducible $\mathbb{C}\mathcal{G}$-modules, and up to the choice of coordinate system for each module, a unique set of irreducible representations. Any $\mathbb{R}\mathcal{G}$-module must be made up of a unique direct sum of irreducible $\mathbb{C}\mathcal{G}$-modules, and any reducible representation must be made up the corresponding direct sum of irreducible representations.

To work with 2- and 3-dimensional structures, there is a limited set of groups that might be of interest – and for finite structures, these are listed in Table 9.5. For these groups, the once-and-forever calculation of characters can be found in character tables published in useful reference works, for instance Atkins et al. (1970); Altmann and Herzig (1994) or Bishop (1973).

We will now complete this chapter by showing how the symmetry-adapted counting in Section 10.5.1 can be considered in terms of the irreducible representations of the group rather than the characters. Each of the terms shown in (10.22) can be considered as a sum of irreducible representation (hence the notation "representation" that is used in applied theory). The decomposition into irreducible representation is easily done by projection using equation (9.3), and this and many other practical techniques will be described in applied texts such as Bishop (1973).

Consider revisiting the tabular calculation for the structure in Figure 10.6. We have reproduced this here, but have also added, for each line of the calculation, the decomposition into irreducible representations using the labeling shown in Table 9.6. (For compactness, we have also grouped the columns into conjugacy classes, noting the number of operations in each conjugacy class in the labeling of each column.) Unfortunately the symbol E is having to do double duty here within one table – it is both the identity (as a column heading) and the 2-dimensional irreducible representation in the final column.

	E	$2C_3$	$3\sigma_a$	
χ_v	6	0	2	$2A_1 + 2E$
$\otimes \chi_T$	2	−1	0	E
=	12	0	0	$2A_1 + 2A_2 + 4E$
$-\chi_e$	−9	0	−3	$-3A_1 - 3E$
$-\chi_T$	−2	1	0	$-E$
$-\chi_R$	−1	−1	1	$-A_2$
$= \chi_m - \chi_s$	0	0	−2	$A_2 - A_1$

Thus the calculation can be written as

$$
\begin{aligned}
\chi_m - \chi_s &= \chi_v \otimes \chi_T & - \chi_e & \qquad - \chi_T - \chi_R \\
&= (2A_1 + 2E) \otimes E - (3A_1 + 3E) - E \ - A_2 \\
&= A_2 - A_1,
\end{aligned}
$$

where multiplication of irreducible representations can be looked up in reference books of character tables, or worked out from characters, and in particular here $A_1 \otimes E = E$ and $E \otimes E = A_1 + A_2 + E$.

One final step is to note that the decomposition of a representation will always yield a positive coefficient for each irreducible representation. Thus we can guarantee that in the final result calculated from (10.22), any terms with positive coefficients must be part of χ_m, and any terms with negative coefficients must be a part of χ_s.

For the example structure, where we have $\chi_m - \chi_s = A_2 - A_1$, we can guarantee that χ_m must contain A_2, and χ_s must contain A_1. Symmetry counting alone cannot, of course, guarantee that there will not be equisymmetric mechanisms and states of self-stress that would cancel in $\chi_m - \chi_s$. However, numerical calculation shows for the example that the calculation here gives a complete characterization for the example structure, with the mechanism and state of self-stress illustrated in Figure 10.9. It can be seen that the symmetry

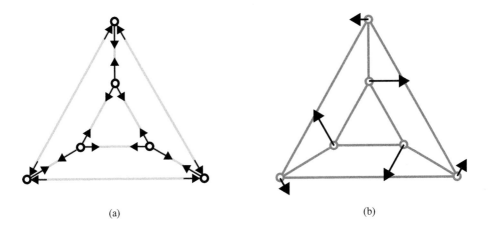

Figure 10.9 For the example structure: (a) the state of self-stress, with irreducible representation $\chi_s = A_1$; (b) the infinitesimal mechanism, with irreducible representation $\chi_m = A_2$.

nature of the state of self-stress, with character $\chi_s = A_1$, is fully captured by the totally symmetric irreducible representation A_1. Similarly the symmetry nature of the infinitesimal mechanisms, with character $\chi_m = A_2$, is fully captured by the irreducible representation A_2 – it is preserved by any proper symmetry operation, and reversed by any improper symmetry operation.

Working with coordinates systems where the internal and external spaces are decomposed into irreducible submodules naturally leads to the block-diagonalization of the rigidity and equilibrium matrices defined in Chapter 3. A number of authors have considered some of the implications and practical uses of this, such as Kangwai and Guest (2000); Schulze (2010a); Owen and Power (2010); and Schulze and Tanigawa (2015).

10.6.1 Guaranteeing a Flex

Schulze (2010d), building on work from Kangwai and Guest (1999) and Guest and Fowler (2007), used the block-diagonalization of the rigidity matrix to consider when it is possible to use symmetry to show that an infinitesimal mechanism extends to a finite flex. The key step is to consider systems that are *symmetry-regular* with a given symmetry, which we define as meaning that the block of the rigidity matrix that corresponds to the fully symmetric irreducible representation is full rank. Then, if there is a fully symmetric infinitesimal mechanism, it will extend to a fully symmetric finite mechanism. The key point of the proof is that the flex does not alter the symmetry of the structure, and therefore does not alter the block-diagonal form of the rigidity matrix, which reflects that for the totally symmetric $\mathbb{C}\mathcal{G}$-submodules, the "external" space has more degrees of freedom than the "internal" space. Schulze (2010d) also shows that, if the nodes are placed as generically as possible, given the symmetry, that this will guarantee a symmetry-regular configuration.

Table 10.7. Character table for C_{6v}

C_{6v}	$\{E\}$	$\{C_6, C_6^{-1}\}$	$\{C_6^2, C_6^{-2}\}$	$\{C_6^3\}$	$\{3\sigma_v\}$	$\{3\sigma_d\}$
A_1	1	1	1	1	1	1
A_2	1	1	1	1	−1	−1
B_1	1	−1	1	−1	1	−1
B_2	1	−1	1	−1	−1	1
E_1	2	1	−1	−2	0	0
E_2	2	−1	−1	2	0	0

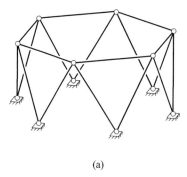

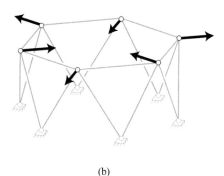

(a)   (b)

Figure 10.10 (a) A framework in 3D that satisfies the relevant Maxwell count that the number of bars is three times the number of non-foundation bars, but which symmetry counting shows has an infinitesimal mechanism. (b) Because a symmetry analysis (for group C_{3v}) shows a totally symmetric mechanism without an equisymmetric state of self-stress, the infinitesimal mechanism shown must extend to a flex.

An interesting example structure is the ring shown in Figure 10.10(a), originally analysed by Tarnai (1980) and then by Kangwai and Guest (1999) and Guest and Fowler (2007). The structure has C_{6v} symmetry in 3D (a 6-fold rotation axis, and two sets of three reflection planes, one set passing through foundation nodes (labelled σ_d in Table 10.7), and one set passing through non-foundation nodes (labelled σ_v). As the structure is pinned to the foundation, we will analyse it using a modified form of (10.22) in which rigid body modes do not need to be subtracted, and for which we only count the non-foundation nodes,

$$\chi_m - \chi_s = \chi_v \otimes \chi_T - \chi_e. \tag{10.32}$$

Again, we write this in both tabular form, and in terms of the decomposition into irreducible representations given in the character table in Table 10.7.

	E	C_6, C_6^{-1}	C_6^2, C_6^{-2}	C_6^3	$3\sigma_v$	$3\sigma_d$	
χ_v	6	0	0	0	2	0	$2A_1 + B_1 + E_1 + 2E_2$
$\otimes \chi_T$	3	2	0	-1	1	1	$A_1 + E_1$
$=$	18	0	0	0	2	0	$2A_1 + A_2 + 2B_1 + B_2 + 3E_1 + 3E_2$
$-\chi_e$	-18	0	0	0	0	-2	$-2A_1 - A_2 - B_1 - 2B_2 - 3E_1 - 3E_2$
$= \chi_m - \chi_s$	0	0	0	0	2	-2	$B_1 - B_2$

We have $\chi_m - \chi_s = B_1 - B_2$, which shows that the detected infinitesimal mechanism, shown in Figure 10.10(b), is not totally symmetric in C_{6v}. However, we can follow the irreducible representations through the descent in symmetry to C_{3v} (Guest and Fowler, 2007), or redo the calculation in this group. This shows that, when considered in C_{3v}, the infinitesimal mechanism is now totally symmetric, while the state of self-stress is not. It is straightforward to show that the configuration is C_{3v}-regular, and hence that the infinitesimal mechanism will extend to a flex with C_{3v} symmetry.

11

Generating Stable Symmetric Tensegrities

11.1 Symmetric Tensegrities

Here we introduce and explain our catalogue, which is a computer program that can show you over a hundred different tensegrities that are "highly symmetric" and for which the user can choose parameters to change its shape. Here we explain some of the group theory that one needs to understand in order to use the program as well as to understand the symmetry of the tensegrity. Some examples of the catalogue can be seen in Subsection 11.6.4. One can access the program at:

https://robertconnelly.github.io/symmetric-tensegrity/

Now we consider a configuration $\mathbf{p} = [\mathbf{p}_1; \ldots; \mathbf{p}_n]$ in d-dimensional space, where there is a group $\mathcal{G}$ of symmetries on the set of points of $\mathbf{p}$. We only insist, though, that $\mathcal{G}$ be a subgroup of all of the symmetries of the configuration. One way of saying this is that $\mathcal{G}$ *acts* on the configuration $\mathbf{p}$. For example, in Figure 11.1 the dihedral group $\mathcal{D}_3$ acts on the six points indicated. But we could just as well consider the cyclic group $\mathcal{C}_3$ as acting on the same six points.

If the configuration is part of a tensegrity, then we also insist that the cables are transformed to cables, struts to struts, and bars to bars under the group operations.

The notation for this situation is that if $\mathbf{p}_i$ is a point in the configuration $\mathbf{p}$, and g is a group element in $\mathcal{G}$, then $g\mathbf{p}_i$ is the image of the point $\mathbf{p}_i$ under the action of the group element g. If $\{g_i, g_j\}$ represents a cable (or a strut or bar), then under this convention, the action of the group element g on the cable is the unordered set $\{gg_i, gg_j\}$ representing the cable (strut or bar).

11.2 Some Group Definitions

When a group $\mathcal{G}$ acts on a configuration of points $\mathbf{p}$, or any set for that matter, it is helpful to make a few definitions that help us understand the situation. We say that $\mathcal{G}$ acts *freely* on $\mathbf{p}$ if for all $\mathbf{p}_i$ in $\mathbf{p}$, $g\mathbf{p}_i = \mathbf{p}_i$ implies that the group element g is the identity.

In the example with Figure 11.1, the dihedral group $\mathcal{D}_3$ acts freely on the configuration of six points – as does the cyclic group $\mathcal{C}_3$.

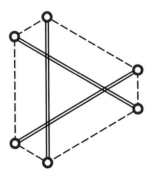

Figure 11.1 A symmetric tensegrity in the plane.

We say that a subset T of the configuration $\mathbf{p}$ is a *transitivity class* if $T = \{g\mathbf{p}_i|g \text{ in } \mathcal{G}\}$ for some $\mathbf{p}_i$ of the configuration. A transitivity class is sometimes called an *orbit*. Note that transitivity classes form a partition of the points of the configuration – two transitivity classes are either disjoint, or the same. Similarly we define transitivity classes of cables and struts. If there is only one transitivity class (of vertices, say) then we say that $\mathcal{G}$ is *transitive* (on the vertices) or *acts transitively*.

In the example of Figure 11.1, $\mathcal{D}_3$ acts transitively, and there are two transitivity classes of cables, and one transitivity class of struts.

11.2.1 Highly Symmetric Tensegrities

One interesting special class of tensegrities is those where a group $\mathcal{G}$ of symmetries acts transitively and freely. This means that there is a one-to-one correspondence between the elements of $\mathcal{G}$ and the vertices $\mathbf{p} = [\mathbf{p}_1; \ldots ; \mathbf{p}_n]$ of the tensegrity. Indeed, choose any vertex $\mathbf{p}_1$ and identify it with the identity element 1 of $\mathcal{G}$. If we enumerate the elements of $\mathcal{G} = (g_1 = 1, g_2, \ldots, g_n)$, then g_i is identified with $g_i\mathbf{p}_1 = \mathbf{p}_i$, the i-th vertex of the configuration $\mathbf{p}$. For these tensegrities, we will show here that we can write the stress matrix in terms of the right regular representation permutation matrix.

First we need to find a symmetric self-stress. Suppose that $\boldsymbol{\omega} = [\ldots ; \omega_{ij}; \ldots]$ is a proper self-stress for the tensegrity $(G, \mathbf{p})$, and $\mathcal{G}$ is a finite group acting freely and transitively on the vertices of $\mathbf{p}$, as described above. Further, we will assume that the associated stress matrix $\boldsymbol{\Omega}$ is positive semi-definite of rank $n - d - 1$. Now, without changing the rank or the positive semi-definiteness of the stress matrix, we can replace $\boldsymbol{\omega}$ and thus $\boldsymbol{\Omega}$ with the average stress $\Sigma_{g\in G}g\boldsymbol{\omega}$, where $g\omega_{ij} = \omega_{kl}$ when the image of the member $\{ij\}$ is the member $\{kl\}$ under the action of g, and $g\boldsymbol{\omega} = [\ldots ; g\omega_{ij}; \ldots]$. Since the sum of positive semi-definite quadratic forms is positive semi-definite, we have not introduced any negative eigenvalues in the stress matrix, and since the rank cannot decrease in the averaging process, we have not changed the rank either. But the action of $\mathcal{G}$ is now invariant on the force coefficients.

In other words, $g\omega_{ij} = \omega_{ij}$, for all members $\{ij\}$ of G. So from now on we will assume that G is invariant on the force coefficients ω and, of course, on the stress matrix Ω.

Consider one transitivity class of, say, cables. It has a representative of the form $\{1, c\}$ for c in G and we can identify the group element c with that transitivity class of cables. So every cable in that transitivity class appears in the group $\{g, gc\}$ as g varies over the elements of G. Note the class will include the cable $\{c^{-1}, c^{-1}c\} \equiv \{1, c^{-1}\}$, so we an also identify the group element c^{-1} with the same transitivity class of cables.

Let $\rho_R(c)$ denote the permutation matrix that corresponds to right multiplication by c^{-1} in the right regular representation. This implies that $\rho_R(c)e_i = e_j$ if and only if $g_i c^{-1} = g_j$, for the standard basis vectors e_i, $i = 1, \ldots, n$ of Euclidean n-space. In other words, the (j, i) entry of $\rho_R(c)$, which is $e_j^T \rho_R(c)e_i$, is 1 if and only if $g_i c^{-1} = g_j$, otherwise it is 0.

Note that $\rho_R(c)$ is not necessarily a symmetric matrix. However, if $g_i c^{-1} = g_j$ and $g_j c^{-1} = g_i$, then substituting we get $g_i c^{-1} c^{-1} = g_i$, which implies that $c^2 = 1$ in G. And if $c^2 = 1$ in G, $\rho_R(c)$ is symmetric, otherwise it is not. Furthermore, no diagonal entry of $\rho_R(c)$ is 1 unless $c = 1$.

Consider two possibilities. Suppose $c^2 = 1$ in G, and there is a cable from p_1 to p_2, where $p_2 = cp_1$, c in G. Then there is a cable from p_i to p_j if and only if $g_i c^{-1} = g_j$ (or equivalently $g_i c = g_j$), and there is a 1 in the (i, j) and (j, i) entries of $\rho_R(c)$, and all the other entries are 0.

Alternatively, suppose $c^2 \ne 1$ in G, and there is a cable from p_1 to p_2, where $p_2 = cp_1$, c in G. Then there is a cable from from p_i to p_j if and only if $g_i c = g_j$ (or equivalently $g_i = g_j c^{-1}$) if and only if there is a 1 in the (i, j) entry of $\rho_R(c)$. So the matrix $\rho_R(c) + \rho_R(c^{-1}) = \rho_R(c) + \rho_R(c)^T$ has a 1 in the (i, j) entry and the (j, i) entry if and only if there is a cable between p_i and p_j.

Now consider the matrix

$$\Omega(c) = \begin{cases} I - \rho_R(c) & \text{if } c^2 = 1; \\ 2I - (\rho_R(c) + \rho_R(c^{-1})) = 2I - (\rho_R(c) + \rho_R(c)^T) & \text{otherwise} \end{cases} \quad (11.1)$$

where I is the n-by-n identity matrix. For this matrix it is clear that it is symmetric, the row and column sums are 0 (since this is true for $I - \rho_R(c)$, $\rho_R(c)$ being a permutation matrix), and the (i, j) and (j, i) entries are -1 if and only if there is cable between p_i and p_j. So $\Omega(c)$ is a stress matrix with a force coefficient of 1 on the cables associated to the group element c.

Note that the same definition and properties hold if c is a strut instead of a cable.

We now describe the stress matrix for a tensegrity that has a group G operating freely and transitively on it. Choose one vertex, say p_1, in the configuration and identify that vertex with the identity 1 in G. Then consider $c_1, c_2, \ldots, c_a$ in G that correspond to the transitivity classes of cables in the tensegrity, and $s_1, s_2, \ldots, s_b$ that correspond to the transitivity classes of struts in the tensegrity. So, for example, there is a cable corresponding to the transitivity class c_k between p_i and p_j if and only if $g_i c_k = g_j$ or $g_i c_k^{-1} = g_j$.

Define $\omega_k = \omega_{1i} > 0$, for $k = 1, \ldots, a$, where $\{p_1, p_i\}$ corresponds to the transitivity class of cables given by c_k, and ω is the starting equilibrium stress for the configuration **p**. Similarly, define $\omega_{-k} < 0$ for $k = 1, \ldots, b$ for the struts.

The result of the definitions and the discussion above is that

$$\mathbf{\Omega} = \sum_{k=1}^{a} \omega_k \mathbf{\Omega}(c_k) + \sum_{k=1}^{b} \omega_{-k} \mathbf{\Omega}(s_k). \tag{11.2}$$

11.3 Irreducible Components

Once the right regular representation of the group $\mathcal{G}$ is obtained, each of the terms in (11.2) is an element of the right regular representation. Let $\mathbf{P}_R$ be the matrix that conjugates the right regular representation to the direct sum of irreducible representations. In other words $\mathbf{P}_R \rho_R \mathbf{P}_R^{-1}$ is the direct sum of irreducible representations as described in Theorem 9.5.4. By restricting to the subspace corresponding to each summand of the decomposition of Theorem 9.5.4, we can replace each term ρ_R in the definition of $\mathbf{\Omega}(c)$ and in Equation (11.2) with any irreducible representation. If ρ_i, $i = 1, \ldots, m$ are the irreducible representations of the group $\mathcal{G}$, then we define

$$\mathbf{\Omega}_i(c) = \begin{cases} I - \rho_i(c) & \text{if } c^2 = 1; \\ 2I - (\rho_i(c) + \rho_i(c^{-1})) = 2I - (\rho_i(c) + \rho_i(c)^{\mathsf{T}}) & \text{otherwise,} \end{cases} \tag{11.3}$$

and similar to Equation 11.2 we define the *local stress matrix* for representation i as

$$\mathbf{\Omega}_i = \sum_{k=1}^{a} \omega_k \mathbf{\Omega}_i(c_k) + \sum_{k=1}^{b} \omega_{-k} \mathbf{\Omega}_i(s_k). \tag{11.4}$$

We have effectively block-diagonalized the stress matrix, with d blocks for each d-dimensional irreducible representation, which could thus be written as

$$\mathbf{P}_R \mathbf{\Omega} \mathbf{P}_R^{-1} = \mathbf{\Omega}_1 \oplus \sum_{j=1}^{\dim(\rho_2)} \mathbf{\Omega}_2 \oplus \cdots \oplus \sum_{j=1}^{\dim(\rho_m)} \mathbf{\Omega}_m.$$

This brings us to one of the main points of using representation theory. Instead of computing whether $\mathbf{\Omega}$ is positive semi-definite of the appropriate rank directly, we see that it is enough to compute whether each of the components $\mathbf{\Omega}_i$ is positive definite or positive semi-definite. Furthermore, it is possible to keep track of the rank of $\mathbf{\Omega}$ by keeping track of the rank of each $\mathbf{\Omega}_i$. This holds the possibility of greatly reducing the amount of computation, and allows the possibility of first deciding on the stress with some desired properties, and then calculating the configuration by determining the kernel of $\mathbf{\Omega}$, and even more easily by calculating the kernel of each $\mathbf{\Omega}_i$.

Let us label the trivial representation, which takes all the group elements into the identity, as the first representation ρ_1. Then we see that $\mathbf{\Omega}_1 = 0$. Suppose that only one other irreducible representation, say ρ_2, is such that Ω_2 is singular with a 1-dimensional kernel, and that all the other representations Ω_i for $i = 3, \ldots, m$ are positive semi-definite. If the dimension of ρ_2 is d, then by Theorem 9.5.4, the ρ_2 representation appears exactly d times in the right regular representation ρ_R. So the kernel of Ω is $d + 1$-dimensional, and it is positive semi-definite. Under these conditions, it means that the associated tensegrity framework is super stable in all dimensions up to affine motions.

11.3.1 Example of the Method

Let us take $\mathcal{G}$ to be the dihedral group $\mathcal{D}_3$ with six elements, and the group elements corresponding to cables to be $c_1 = C_3\sigma$, $c_2 = C_3^2 S$, and $s_1 = \sigma$, using the notation of Subsection 9.5.9:

$$\rho_3(C_3) = \begin{bmatrix} -1/2 & -\sqrt{3}/2 \\ \sqrt{3}/2 & -1/2 \end{bmatrix}, \quad \rho_3(s_1) = \rho_2(\sigma) = \begin{bmatrix} 1 & 0 \\ 0 & -1 \end{bmatrix}.$$

We then calculate

$$\rho_3(c_1) = \rho_3(C_3\sigma) = \begin{bmatrix} -1/2 & \sqrt{3}/2 \\ \sqrt{3}/2 & 1/2 \end{bmatrix}, \quad \rho_3(c_2) = \rho_2(C_3^2\sigma) = \begin{bmatrix} -1/2 & -\sqrt{3}/2 \\ -\sqrt{3}/2 & 1/2 \end{bmatrix}.$$

Using Equation 11.3 we get

$$\Omega_3(c_1) = \begin{bmatrix} 3/2 & -\sqrt{3}/2 \\ -\sqrt{3}/2 & 1/2 \end{bmatrix}, \quad \Omega_3(c_2) = \begin{bmatrix} 3/2 & \sqrt{3}/2 \\ \sqrt{3}/2 & 1/2 \end{bmatrix}, \quad \Omega_3(s_1) = \begin{bmatrix} 0 & 0 \\ 0 & 2 \end{bmatrix}.$$

Then the local stress matrix for the third representation is

$$\Omega_3 = \begin{bmatrix} \frac{3}{2}(\omega_1 + \omega_2) & \frac{\sqrt{3}}{2}(-\omega_1 + \omega_2) \\ \frac{\sqrt{3}}{2}(-\omega_1 + \omega_2) & \frac{1}{2}(\omega_1 + \omega_2) + 2\omega_{-1} \end{bmatrix}.$$

The second representation is 1-dimensional, so we can use the character χ_2 as in the character table in Subsection 9.5.9 to calculate the following:

$$\rho_2(c_1) = \begin{bmatrix} -1 \end{bmatrix}, \quad \rho_2(c_2) = \begin{bmatrix} -1 \end{bmatrix}, \quad \rho_2(s_1) = \begin{bmatrix} -1 \end{bmatrix}.$$

Using Equation 11.3 we get

$$\Omega_2(c_1) = \begin{bmatrix} 2 \end{bmatrix}, \quad \Omega_2(c_2) = \begin{bmatrix} 2 \end{bmatrix}, \quad \Omega_2(s_1) = \begin{bmatrix} 2 \end{bmatrix}.$$

Then the local stress matrix for the second representation is

$$\Omega_2 = \begin{bmatrix} 2\omega_1 + 2\omega_2 + 2\omega_{-1} \end{bmatrix}.$$

We calculate the determinants of the local stress matrices, which we call the i-th stress determinant $\Delta_i = \Delta_i(\omega_1, \ldots, \omega_a, \omega_{-1}, \ldots, \omega_{-b})$, for representation i:

$$\Delta_3 = \det(\Omega_3) = 3\omega_1\omega_2 + 3(\omega_1 + \omega_2)\omega_{-1}, \quad \Delta_2 = \det(\Omega_3) = 2\omega_1 + 2\omega_2 + 2\omega_{-1}.$$

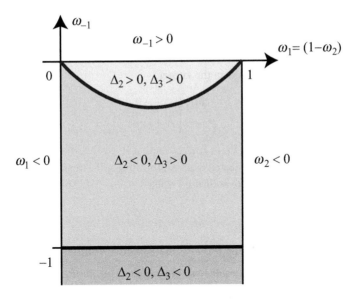

Figure 11.2 A determinant plot for the tensegrity shown in Figure 11.1. The vertical axis is the strut force coefficient ω_{-1}, and the horizontal axis shows the cable force coefficients ω_1 and ω_2, normalized so that $\omega_1 + \omega_2 = 1$. The shaded region shows where the force coefficients are proper, i.e. $\omega_{-1} < 0$, $\omega_1 > 0$, $\omega_2 > 0$. The shaded region is split into different regions depending on the signs of the stress determinants Δ_2 and Δ_3. The quadratic line shows where $\Delta_2 = 0$, and the straight line shows where $\Delta_3 = 0$ ($\Delta_1 = 0$ everywhere, by definition). We are interested in points where some stress determinants are zero but the rest are all positive: for any value of ω_1 this will be found by descending from the horizontal axis until the first $\Delta_i = 0$ line is crossed, defining the value of ω_{-1} for equilibrium to be satisfied.

We can normalize the stresses so that $\omega_1 + \omega_2 = 1$. Then $\omega_2 = 1 - \omega_1$, and for $0 < \omega_1 < 1$ and $0 < 1 - \omega_1 = \omega_2 < 1$. Note that when ω_1, ω_2, and ω_{-1} are all positive Ω_2 and Ω_3 are both positive definite. Then allow to ω_{-1} to decrease and become negative while fixing ω_1 and $\omega_2 = 1 - \omega_1$. The polynomial Δ_3 first becomes 0, changing from positive to negative, when $\omega_{-1} = -\omega_1(1 - \omega_1) > -1$. The polynomial Δ_2 first becomes 0, changing from positive to negative, when $\omega_{-1} = -1$ – see Figure 11.2. Thus when $\omega_{-1} = -\omega_1(1 - \omega_1)$, Ω_2 is positive definite, Ω_3 is positive semi-definite with a 1-dimensional kernel, and $\Omega_1 = 0$ but is 1-dimensional. Thus in the full right regular representation Ω is positive semi-definite with a 3-dimensional kernel, since the 2-dimensional representation appears twice in the direct sum. Thus the associated tensegrity is (universally) rigid and prestress stable in all dimensions, since there are at least three distinct stressed directions.

Notice that we started with the graph of the framework only, not the actual configuration itself. Then we found the force coefficients corresponding to an equilibrium self-stress for some configuration. To find the configuration, we first find a vector in the kernel of the third local stress matrix Ω_3,

$$\Omega_3 = \begin{bmatrix} \frac{3}{2} & \frac{\sqrt{3}}{2}(1 - 2\omega_1) \\ \frac{\sqrt{3}}{2}(1 - 2\omega_1) & \frac{1}{2} - 2\omega_1(1 - \omega_1) \end{bmatrix}.$$

For example, the following vector is in the kernel,

$$\begin{bmatrix} \frac{\sqrt{3}}{2}(1 - 2\omega_1) \\ -\frac{3}{2} \end{bmatrix} = \begin{bmatrix} x_1 \\ y_1 \end{bmatrix} = \mathbf{p}_1.$$

From the definition of equation (11.1) it is clear that if we let $p_i = \rho_3(g_i)p_1$, then the configuration $\mathbf{p} = [\mathbf{p}_1; \ldots; \mathbf{p}_n]$ will have an equilibrium self-stress with a 3-dimensional kernel, and be positive semi-definite as desired when $0 < \omega_1 < 1$. This gives a configuration as shown in Figure 11.1.

11.4 Groups for 3-Dimensional Examples

As with any representation of a group, one must decide on the initial description of the group that will be represented into the group of matrices. For tensegrities in three-space, a natural choice is to use certain permutation groups as the initial groups. One reason for this is that permutations are unbiased as far as pointing to any particular representation. This is helpful since the process that we describe here will be such that different representations will be chosen for the configuration that displays the stress which has a stress matrix of maximal rank as well as being positive semi-definite. But an even more relevant reason for the permutation description is that the group multiplication is particularly convenient and efficient to calculate. Also, properties of the underlying graph of the tensegrity can be read off easily from the permutation description of the cable and strut generators c_i and s_i.

A number of the groups that we will use are formed from the direct product of a permutation group with the permutation group S_2. For two groups G_1 and G_2, their *direct product*, written as $G_1 \times G_2$, is the set of pairs (g_1, g_2), and the group multiplication is given by $(g_1, g_2)(g_1', g_2') = (g_1 g_1', g_2 g_2')$, where g_1, g_1' are in G_1, and g_2, g_2' are in G_2. It is easy to check that the required properties of a group are satisfied by $G_1 \times G_2$. It might be useful to note that the permutation group S_2 is isomorphic with $\mathbb{Z}_2$, which is the group $\{1, -1\}$ with group multiplication being the multiplication of real numbers.

We use the following six groups plus the dihedral groups, which will be described in later subsections. See Table 9.5 for further notes about these groups.

(i) The alternating group on four symbols A_4. This is the group of even permutations of the symbols $\{1, 2, 3, 4\}$. A permutation is *even* if it can be written as an even number of transpositions, where a *transposition* interchanges exactly two symbols, leaving all the others fixed. If a permutation is not even, then it is called an *odd permutation*. It is a nice exercise to show that the even permutations form a group. The order of A_4 is 12.

(ii) The symmetric group on four symbols S_4. This is the group of all permutations of the symbols $\{1, 2, 3, 4\}$, and has order 24.

(iii) The alternating group on five symbols $\mathcal{A}_5$. This is the group of all even permutations of the symbols $\{1, 2, 3, 4, 5\}$ and has order 60.

(iv) The group $\mathcal{A}_4 \times \mathcal{S}_2$. It has order 24. Note that this group has the same order as $\mathcal{S}_4$, but it is not isomorphic to $\mathcal{S}_4$, since, for example, $\mathcal{S}_4$ has an element of order 3, whereas $\mathcal{A}_4 \times \mathcal{S}_2$ does not.

(v) The group $\mathcal{S}_4 \times \mathcal{S}_2$. It has order 48.

(vi) The group $\mathcal{A}_5 \times \mathcal{S}_2$. It has order 120.

11.5 Presentation of Groups

Another method that can be used to define a group is what is called a *presentation* of a group $\mathcal{G}$. This is a list of generators $a, b, \dots$ for $\mathcal{G}$, together with what are called *relations* $r_1, r_2, \dots$. These are finite products of the generators and their inverses, called *words*, such that each word is equal to the identity in $\mathcal{G}$. The group $\mathcal{G}$ is defined by this presentation in the sense that if any other group $\mathcal{H}$ has the property that it has corresponding generators satisfying the same relations, then $\mathcal{H}$ is homomorphic to $\mathcal{G}$. The presentation is written as $\mathcal{G} = \{a, b, \dots \mid r_1 = r_2 = \dots = 1\}$.

As an example, the cyclic group $\mathcal{C}_n$ can be defined by the presentation with one generator a, and one relation $a^n = 1$, so $\mathcal{C}_n = \{a \mid a^n = 1\}$. For another example the alternating group $\mathcal{A}_5$ has the presentation, $\mathcal{A}_5 = \{a, b \mid a^2 = b^3 = (ab)^5 = 1\}$.

11.6 Representations for Groups of Interest

To generate tensegrities, we need all the irreducible representation of the group we are working with, even though we may be interested only in the 3-dimensional (or possibly the 2-dimensional) representations. We give some of the ideas here that we use to generate those irreducible representations.

Here we will give concrete examples of matrices $\rho_i(g)$ that form an irreducible representation for the dihedral groups, and for each of the groups listed in Section 11.4. In order to compute any element $\rho_i(g)$ for g in $\mathcal{G}$ it is enough to do it for the generators $g_1, \dots$ We have included the character tables for all of the groups that we are considering, and this can be used to check the correctness of the choices that we will describe next for the g_i.

11.6.1 Dihedral Groups

The dihedral group $\mathcal{D}_n$ can be defined as the full group of rotational symmetries of the regular polygon in space. It is not too hard to see that $\mathcal{D}_n$ has the presentation $\mathcal{D}_n = \{r, s \mid r^n = s^2 = (sr)^2 = 1\}$. Rotation by $2\pi/n$, denoted as $\mathcal{C}_n$, corresponds to r, and rotation by π about some suitable axis in the plane of the polygon C_2 corresponds to s.

The following is the character table for $\mathcal{D}_n$. It is usual to distinguish the case when n is odd, and when n is even. In both cases the elements r and s generate $\mathcal{D}_n$.

Groups $\mathcal{D}_n$, n odd n $\geq$ 3

$\mathcal{D}_n$	1	$r^k(k = 1, \ldots, \frac{n-1}{2})$	s
$\mathcal{D}_n = (n2)$	E	C_n^k	C_2
$A_1 = \chi_1$	1	1	1
$A_2 = \chi_2$	1	1	-1
$E_j = \psi_j$	2	$2\cos\left(\frac{2\pi jk}{n}\right)$	0
$\left(j = 1, \ldots, \frac{n-1}{2}\right)$			

There are $(n + 1)/2$ conjugacy classes and irreducible representations. It is clear that r and s generate $\mathcal{D}_n$, and that $r^a s^b$; $a = 0, 1, \ldots, n - 1$; $b = 0, 1$ enumerate $\mathcal{D}_n$.

Since the characters χ_1 and χ_2 are 1-dimensional, we can regard them as representations. Let $\rho_j, j = 1, \ldots, \frac{n-1}{2}$ denote the representation corresponding to the character ψ_j. Then we define

$$\rho_j(r) = \begin{bmatrix} \cos\left(\frac{2\pi j}{n}\right) & -\sin\left(\frac{2\pi j}{n}\right) \\ \sin\left(\frac{2\pi j}{n}\right) & \cos\left(\frac{2\pi j}{n}\right) \end{bmatrix}, \quad \rho_j(s) = \begin{bmatrix} 1 & 0 \\ 0 & -1 \end{bmatrix}.$$

It is easy to check that the trace of each of these matrices is the character of the corresponding group element.

Groups $\mathcal{D}_n$, n even n $\geq$ 4

$\mathcal{D}_n$	1	$r^k(k = 1, \ldots, \frac{n}{2} - 1)$	$r^{n/2}$	s	rs
$\mathcal{D}_n = (n22)$	E	C_n^k	$C_n^{n/2}$	C_2'	C_2''
$A_1 = \chi_1$	1	1	1	1	1
$A_2 = \chi_2$	1	1	1	-1	-1
$B_1 = \chi_3$	1	$(-1)^k$	$(-1)^{n/2}$	1	-1
$B_2 = \chi_4$	1	$(-1)^k$	$(-1)^{n/2}$	-1	1
$E_j = \psi_j$	2	$2\cos\left(\frac{2\pi jk}{n}\right)$	$2(-1)^{n/2}$	0	0
$\left(j = 1, \ldots, \frac{n}{2} - 1\right)$					

There are $n/2 + 3$ conjugacy classes and irreducible representations. Again r and s generate $\mathcal{D}_n$, and $r^a s^b$; $a = 0, 1, \ldots, n - 1$; $b = 0, 1$ enumerate $\mathcal{D}_n$.

Since the characters $\chi_1, \chi_2, \chi_3, \chi_4$ are 1-dimensional, we can regard them as representations. Let $\rho_j, j = 1, \ldots, \frac{n}{2} - 1$ denote the representation corresponding to the character ψ_j. Then, as before, we define

$$\rho_j(r) = \begin{bmatrix} \cos\left(\frac{2\pi j}{n}\right) & -\sin\left(\frac{2\pi j}{n}\right) \\ \sin\left(\frac{2\pi j}{n}\right) & \cos\left(\frac{2\pi j}{n}\right) \end{bmatrix}, \quad \rho_j(s) = \begin{bmatrix} 1 & 0 \\ 0 & -1 \end{bmatrix}.$$

It is again easy to check that the trace of each of these matrices is the character of the corresponding group element.

11.6.2 Basic Non-Dihedral Representations

In each of the cases below, we have indicated that products of the generators $g_1^a g_2^b g_3^c \ldots$ can be used to enumerate the group $\mathcal{G}$. For example for the group S_4, each g_i permutes the set $\{1, \ldots, i\}$. So the same is true for $g_1^a g_2^b \ldots g_i^c$. So if we are given any g in $\mathcal{G}$ as a permutation, choose the integer c so that g_i^{-c} permutes the symbol i to the same symbol as g. Then c is the exponent of g_i in the decomposition $g = g_1^a g_2^b \ldots g_3^c$. Continuing this way one can easily determine all the exponents $a, b, \ldots c$. A similar idea works for the other groups A_4 and A_5 as well.

Group S_4

S_4	1	(123)	(12)(34)	(1234)	(12)
$O = (432)$	E	$8C_3$	$3C_2$	$6C_4$	$6C_4'$
$T_d = (43m)$	E	$8C_3$	$3C_2$	$6S_4$	$6\sigma_d$
$A_1 = \chi_1$	1	1	1	1	1
$A_2 = \chi_2$	1	1	1	-1	-1
$E = \chi_3$	2	-1	2	0	0
$T_1 = \chi_4$	3	0	-1	1	-1
$T_2 = \chi_5$	3	0	-1	-1	1

We describe all the elements of S_4 so that they can be enumerated easily with the following generators. Let

$$g_1 = (12), \quad g_2 = (123), \quad g_3 = (1234),$$

in disjoint cycle notation. Then it is easy to check that $g_1^a g_2^b g_3^c$, for $a = 0, 1$, $b = 0, 1, 2$, $c = 0, 1, 2, 3$, are all distinct. Thus these are the $2 \cdot 3 \cdot 4 = 24$ elements of S_4.

There are two irreducible 3-dimensional representations of S_4. They both can be thought of as symmetries of the unit 3-dimensional cube. One representation ρ_4 is the set of rotations that are symmetries of the cube. Use the numbers $1, 2, 3, 4$ to label the vertices of the cube, where opposite vertices have the same label:

$$1 \to \pm \begin{bmatrix} 1 \\ 1 \\ 1 \end{bmatrix}, \quad 2 \to \pm \begin{bmatrix} -1 \\ 1 \\ 1 \end{bmatrix}, \quad 3 \to \pm \begin{bmatrix} 1 \\ -1 \\ 1 \end{bmatrix}, \quad 4 \to \pm \begin{bmatrix} 1 \\ 1 \\ -1 \end{bmatrix}.$$

It is easy to check that any rotation of the cube corresponds to a permutation of these labels, and that this correspondence is a group isomorphism from $\mathcal{S}_4$ to the rotations of the cube. We calculate the matrices implied by the above correspondence as follows:

$$\rho_4(g_1) = \begin{bmatrix} -1 & 0 & 0 \\ 0 & 0 & 1 \\ 0 & 1 & 0 \end{bmatrix}, \quad \rho_4(g_2) = \begin{bmatrix} 0 & 1 & 0 \\ 0 & 0 & -1 \\ -1 & 0 & 0 \end{bmatrix}, \quad \rho_4(g_3) = \begin{bmatrix} 0 & 0 & -1 \\ 0 & 1 & 0 \\ 1 & 0 & 0 \end{bmatrix}.$$

For the representation ρ_5 we make the correspondence between permutation elements and a subset of the vertices of the cube that form a regular tetrahedron. Then ρ_5 corresponds to the full symmetry group of the tetrahedron. To make definite the correspondence we define the following correspondence:

$$1 \to \begin{bmatrix} 1 \\ 1 \\ 1 \end{bmatrix}, \quad 2 \to \begin{bmatrix} 1 \\ -1 \\ -1 \end{bmatrix}, \quad 3 \to \begin{bmatrix} -1 \\ 1 \\ -1 \end{bmatrix}, \quad 4 \to \begin{bmatrix} -1 \\ -1 \\ 1 \end{bmatrix}.$$

We calculate the matrices implied by the above correspondence as follows:

$$\rho_5(g_1) = \begin{bmatrix} 1 & 0 & 0 \\ 0 & 0 & -1 \\ 0 & -1 & 0 \end{bmatrix}, \quad \rho_5(g_2) = \begin{bmatrix} 0 & 1 & 0 \\ 0 & 0 & -1 \\ -1 & 0 & 0 \end{bmatrix}, \quad \rho_5(g_3) = \begin{bmatrix} 0 & 0 & 1 \\ 0 & -1 & 0 \\ -1 & 0 & 0 \end{bmatrix}.$$

From these correspondences it is possible to look at any particular symmetry operation, and be able to determine the corresponding permutation in $\mathcal{S}_4$.

Now consider the 2-dimensional irreducible representation ρ_3 of $\mathcal{S}_4$. It is possible to check that the following correspondence will work:

$$\rho_3(g_1) = \begin{bmatrix} -1 & 0 \\ 0 & 1 \end{bmatrix}, \quad \rho_3(g_2) = \begin{bmatrix} -1/2 & -\sqrt{3}/2 \\ \sqrt{3}/2 & -1/2 \end{bmatrix}, \quad \rho_3(g_3) = \begin{bmatrix} 1/2 & -\sqrt{3}/2 \\ -\sqrt{3}/2 & -1/2 \end{bmatrix}.$$

(The group elements $g_1 = (12)$, $g_2 = (123)$ generate the symmetric group S_3, which has a 2-dimensional representation which are the transformation matrices corresponding to symmetry operations acting on a triangle. From the character table, we can determine that $\rho_3((12)(34)) = 1$ and so $\rho_3((12)) = \rho_3((34))$, and thus $\rho_3((1234)) = \rho_3((123)(34)) = \rho_3((123))\rho_3((34)) = \rho_3((123))\rho_3((12))$.)

The non-trivial 1-dimensional representation ρ_2 of $\mathcal{S}_4$, can be read from the character table, observing that it is [1] for even permutations, and [−1] for odd permutations:

$$\rho_2(g_1) = \begin{bmatrix} -1 \end{bmatrix}, \quad \rho_2(g_2) = \begin{bmatrix} 1 \end{bmatrix}, \quad \rho_2(g_3) = \begin{bmatrix} -1 \end{bmatrix}.$$

Group $\mathcal{A}_4$

$\mathcal{A}_4$	1	(123)	(132)	(13)(24)
$T = (23)$	E	$4C_3$	$4C_3^2$	$3C_2$
$A = \chi_1$	1	1	1	1
$^1E = \chi_2$	1	$\frac{-1+i\sqrt{3}}{2}$	$\frac{-1-i\sqrt{3}}{2}$	1
$^2E = \chi_3$	1	$\frac{-1-i\sqrt{3}}{2}$	$\frac{-1+i\sqrt{3}}{2}$	1
$T = \chi_4$	3	0	0	-1

Since the alternating group of even permutations on four symbols $\mathcal{A}_4$ is a subgroup of $\mathcal{S}_4$, most of the work for finding the irreducible representations is done. One new feature is that this group has two non-trivial irreducible 1-dimensional complex representations. When

$$g_1 = (123), \ g_2 = (12)(34), \ g_3 = (13)(24),$$

then $g_1^a g_2^b g_3^c$ for $a = 0, 1, 2, \ b = 0, 1, \ c = 0, 1$ enumerate all the elements of $\mathcal{A}_4$, similar to the case for $\mathcal{S}_4$.

For the 3-dimensional irreducible representation ρ_4 of $\mathcal{A}_4$, we can simply restrict to the subset of either irreducible 3-dimensional representation of $\mathcal{S}_4$. So we get the following:

$$\rho_4(g_1) = \begin{bmatrix} 0 & 1 & 0 \\ 0 & 0 & -1 \\ -1 & 0 & 0 \end{bmatrix}, \ \rho_4(g_2) = \begin{bmatrix} 1 & 0 & 0 \\ 0 & -1 & 0 \\ 0 & 0 & -1 \end{bmatrix}, \ \rho_4(g_3) = \begin{bmatrix} -1 & 0 & 0 \\ 0 & 1 & 0 \\ 0 & 0 & -1 \end{bmatrix}.$$

The other two irreducible representations ρ_2 and ρ_3, which are complex, are the following:

$$\rho_2(g_1) = \begin{bmatrix} -1/2 + i\sqrt{3}/2 \end{bmatrix}, \ \rho_2(g_2) = \begin{bmatrix} 1 \end{bmatrix}, \ \rho_2(g_3) = \begin{bmatrix} 1 \end{bmatrix}$$

and

$$\rho_3(g_1) = \begin{bmatrix} -1/2 - i\sqrt{3}/2 \end{bmatrix}, \ \rho_3(g_2) = \begin{bmatrix} 1 \end{bmatrix}, \ \rho_3(g_3) = \begin{bmatrix} 1 \end{bmatrix}.$$

Note that when these complex representations appear in the stress matrix, the imaginary parts vanish because each time a group element appears, so does its inverse with the same real coefficient and the representation for the inverse is the complex conjugate.

Group $\mathcal{A}_5$

$\mathcal{A}_5$	1	(12345)	(13524)	(123)	(12)(34)
I	E	$12C_5$	$12C_5^2$	$20C_3$	$15C_2$
$A = \chi_1$	1	1	1	1	1
$T_1 = \chi_2$	3	τ	τ'	0	-1
$T_2 = \chi_3$	3	τ'	τ	0	-1
$G = \chi_4$	4	-1	-1	1	0
$H = \chi_5$	5	0	0	-1	1

We use the numbers $\tau = (1 + \sqrt{5})/2$ (the golden ratio) and $\tau' = (1 - \sqrt{5})/2$. Note that $\tau\tau' = -1$, $\tau + \tau' = 1$, $\tau^2 - \tau - 1 = (\tau')^2 - \tau' - 1 = 0$. It is also worth noting that $\cos(2\pi/5) = -\tau'/2$ and $\cos(4\pi/5) = -\tau/2$.

Similar to the previous cases, we define the following group elements:

$$g_1 = (123), \quad g_2 = (12)(34), \quad g_3 = (13)(24), \quad g_4 = (12345).$$

Then $g_1^a g_2^b g_3^c g_4^d$, for $a = 0, 1, 2$, $b = 0, 1$, $c = 0, 1$, $d = 0, 1, 2, 3, 4$, enumerate $\mathcal{A}_5$.

To find the 3-dimensional representations of the permutation group $\mathcal{A}_5$, following the ideas used for $\mathcal{S}_4$, we partition the vertices of the regular dodecahedron into five sets of four vertices, where each of those sets forms a regular tetrahedron, and the tetrahedra are permuted by any rotation of the dodecahedron.

The following are one of two choices for the vertices of the dodecahedron partitioned into the five tetrahedra:

$$T_1 = \left\{ \begin{bmatrix} 1 \\ -1 \\ 1 \end{bmatrix}, \begin{bmatrix} 0 \\ \tau \\ -\tau' \end{bmatrix}, \begin{bmatrix} -\tau' \\ 0 \\ -\tau \end{bmatrix}, \begin{bmatrix} -\tau \\ \tau' \\ 0 \end{bmatrix} \right\}, T_2 = \left\{ \begin{bmatrix} -1 \\ 1 \\ 1 \end{bmatrix}, \begin{bmatrix} 0 \\ -\tau \\ -\tau' \end{bmatrix}, \begin{bmatrix} \tau' \\ 0 \\ -\tau \end{bmatrix}, \begin{bmatrix} \tau \\ -\tau' \\ 0 \end{bmatrix} \right\},$$

$$T_3 = \left\{ \begin{bmatrix} 1 \\ 1 \\ -1 \end{bmatrix}, \begin{bmatrix} 0 \\ -\tau \\ \tau' \end{bmatrix}, \begin{bmatrix} -\tau' \\ 0 \\ \tau \end{bmatrix}, \begin{bmatrix} -\tau \\ -\tau' \\ 0 \end{bmatrix} \right\}, T_4 = \left\{ \begin{bmatrix} -1 \\ -1 \\ -1 \end{bmatrix}, \begin{bmatrix} 0 \\ \tau \\ \tau' \end{bmatrix}, \begin{bmatrix} \tau' \\ 0 \\ \tau \end{bmatrix}, \begin{bmatrix} \tau \\ \tau' \\ 0 \end{bmatrix} \right\},$$

$$T_5 = \left\{ \begin{bmatrix} 1 \\ 1 \\ 1 \end{bmatrix}, \begin{bmatrix} -1 \\ -1 \\ 1 \end{bmatrix}, \begin{bmatrix} 1 \\ -1 \\ -1 \end{bmatrix}, \begin{bmatrix} -1 \\ 1 \\ -1 \end{bmatrix} \right\}.$$

We identify each tetrahedron T_i with i, for $i = 1, 2, 3, 4, 5$, and any rotation of the dodecahedron corresponds uniquely to an even permutation of the set $\{1, 2, 3, 4, 5\}$. Then it is easy to calculate that the corresponding representation on the generators g_j defined above:

$$\rho_3(g_1) = \begin{bmatrix} 0 & 1 & 0 \\ 0 & 0 & 1 \\ 1 & 0 & 0 \end{bmatrix}, \quad \rho_3(g_2) = \begin{bmatrix} -1 & 0 & 0 \\ 0 & -1 & 0 \\ 0 & 0 & 1 \end{bmatrix},$$

$$\rho_3(g_3) = \begin{bmatrix} 1 & 0 & 0 \\ 0 & -1 & 0 \\ 0 & 0 & -1 \end{bmatrix}, \quad \rho_3(g_4) = \frac{1}{2}\begin{bmatrix} 1 & \tau' & \tau \\ -\tau' & -\tau & -1 \\ \tau & 1 & \tau' \end{bmatrix}.$$

We can check that this is the representation ρ_3 in the character table by computing traces of the appropriate matrices above. In particular $\chi_3(g_4) = \tau'$.

For the representation ρ_2, we observe that if we interchange τ and τ' everywhere (which corresponds to the other possible choice for partitioning the vertices between tetrahedra), then we get another representation as described by the following generators:

$$\rho_2(g_1) = \begin{bmatrix} 0 & 1 & 0 \\ 0 & 0 & 1 \\ 1 & 0 & 0 \end{bmatrix}, \quad \rho_2(g_2) = \begin{bmatrix} -1 & 0 & 0 \\ 0 & -1 & 0 \\ 0 & 0 & 1 \end{bmatrix},$$

$$\rho_2(g_3) = \begin{bmatrix} 1 & 0 & 0 \\ 0 & -1 & 0 \\ 0 & 0 & -1 \end{bmatrix}, \quad \rho_2(g_4) = \frac{1}{2}\begin{bmatrix} 1 & \tau & \tau' \\ -\tau & -\tau' & -1 \\ \tau' & 1 & \tau \end{bmatrix}.$$

It is clear that this is the representation corresponding to χ_2, since $\chi_2(g_4) = \tau$.

To find the 4-dimensional representation ρ_4, we can consider the transformation matrices of a regular 4-dimensional simplex (consisting of five vertices) in four-space corresponding to even permutations of the vertices. We choose the following vectors for the vertices of the 4-simplex (centred on the origin):

$$1 \to \begin{bmatrix} -1 \\ -1 \\ 1 \\ -1/\sqrt{5} \end{bmatrix}, \quad 2 \to \begin{bmatrix} 1 \\ -1 \\ -1 \\ -1/\sqrt{5} \end{bmatrix}, \quad 3 \to \begin{bmatrix} -1 \\ 1 \\ -1 \\ -1/\sqrt{5} \end{bmatrix}, \quad 4 \to \begin{bmatrix} 1 \\ 1 \\ 1 \\ -1/\sqrt{5} \end{bmatrix}, \quad 5 \to \begin{bmatrix} 0 \\ 0 \\ 0 \\ 4/\sqrt{5} \end{bmatrix}.$$

We then calculate the following matrices for the generators above:

$$\rho_4(g_1) = \begin{bmatrix} 0 & 0 & 1 & 0 \\ 1 & 0 & 0 & 0 \\ 0 & 1 & 0 & 0 \\ 0 & 0 & 0 & 1 \end{bmatrix}, \quad \rho_4(g_2) = \begin{bmatrix} -1 & 0 & 0 & 0 \\ 0 & 1 & 0 & 0 \\ 0 & 0 & -1 & 0 \\ 0 & 0 & 0 & 1 \end{bmatrix},$$

$$\rho_4(g_3) = \begin{bmatrix} 1 & 0 & 0 & 0 \\ 0 & -1 & 0 & 0 \\ 0 & 0 & -1 & 0 \\ 0 & 0 & 0 & 1 \end{bmatrix}, \quad \rho_4(g_4) = \frac{1}{4}\begin{bmatrix} -3 & 1 & 1 & -\sqrt{5} \\ 1 & 1 & -3 & -\sqrt{5} \\ -1 & 3 & -1 & \sqrt{5} \\ \sqrt{5} & \sqrt{5} & \sqrt{5} & -1 \end{bmatrix}.$$

Note that if we rescale the last coordinate of each of the vectors in the 4-simplex above, by replacing $\sqrt{5}$ by 1, then the corresponding representation will have all rational entries, but will not be into the orthogonal group. This implies that some of the polynomials calculated later will eventually have only rational coefficients.

For the irreducible 5-dimensional representation ρ_5, we can calculate the following matrices for the g_i's:

$$\rho_5(g_1) = \begin{bmatrix} -\frac{1}{2} & -\frac{1}{2} & 0 & 0 & 0 \\ \frac{3}{2} & -\frac{1}{2} & 0 & 0 & 0 \\ 0 & 0 & 0 & 0 & 1 \\ 0 & 0 & 1 & 0 & 0 \\ 0 & 0 & 0 & 1 & 0 \end{bmatrix}, \quad \rho_5(g_2) = \begin{bmatrix} 1 & 0 & 0 & 0 & 0 \\ 0 & 1 & 0 & 0 & 0 \\ 0 & 0 & 1 & 0 & 0 \\ 0 & 0 & 0 & -1 & 0 \\ 0 & 0 & 0 & 0 & -1 \end{bmatrix},$$

$$\rho_5(g_3) = \begin{bmatrix} 1 & 0 & 0 & 0 & 0 \\ 0 & 1 & 0 & 0 & 0 \\ 0 & 0 & -1 & 0 & 0 \\ 0 & 0 & 0 & -1 & 0 \\ 0 & 0 & 0 & 0 & 1 \end{bmatrix}$$

$$\rho_5(g_4) = \begin{bmatrix} \frac{1-3\sqrt{5}}{16} & \frac{-1-\sqrt{5}}{16} & \frac{1+\sqrt{5}}{8} & -\frac{1}{4} & \frac{1-\sqrt{5}}{8} \\ \frac{-3-3\sqrt{5}}{16} & \frac{-1+3\sqrt{5}}{16} & \frac{-3+\sqrt{5}}{8} & -\frac{\sqrt{5}}{4} & \frac{3+\sqrt{5}}{8} \\ \frac{-3-3\sqrt{5}}{8} & \frac{3-\sqrt{5}}{8} & -\frac{1}{2} & 0 & -\frac{1}{2} \\ -\frac{3}{4} & -\frac{\sqrt{5}}{4} & 0 & \frac{1}{2} & \frac{1}{2} \\ \frac{-3+3\sqrt{5}}{8} & \frac{-3-\sqrt{5}}{8} & -\frac{1}{2} & -\frac{1}{2} & 0 \end{bmatrix}.$$

One warning about this representation is that not all the matrices are orthogonal. (In particular $\rho_5(g_4)$ is not orthogonal.)

The representation ρ_5 was constructed by taking first what is called the tensor product $\rho_2 \otimes \rho_2$ and then calculating what is called the symmetric component. This turns out to be equivalent to the 6-dimensional representation that is the sum $\rho_5 + \rho_1$ of the 5-dimensional representation ρ_5 we are looking for and the trivial representation ρ_1. We can project this 6-dimensional representation onto ρ_5 easily.

To check that the matrices have been calculated correctly, one can observe that $\mathcal{A}_5 = \{a, b \mid a^2 = b^2 = (ab)^5 = 1\}$ is a presentation of $\mathcal{A}_5$. If one has a group $\mathcal{G}$, which is not the identity, and the relations in the presentation above are satisfied by some pair of the group elements a and b in $\mathcal{G}$, which generate all of $\mathcal{G}$, then $\mathcal{G}$ is isomorphic to the group $\mathcal{A}_5$. (It turns out that $\mathcal{A}_5$ has no non-trivial homomorphic group images, i.e. it is *simple*, so we know that any such group is isomorphic to $\mathcal{A}_5$.) In our case, we can take $a = (12)(34)$ and $b = (235)$ in disjoint cycle notation. Then $a = g_2$, $b = g_2 g_4$, $ab = g_4$. So we can verify that our matrices are chosen properly by checking the relations for the corresponding matrices. The characters are also easily checked.

Groups $A_4 \times S_2$, $S_4 \times S_2$, and $A_5 \times S_2$

For the direct product with S_2, there is the generator of order 2 that commutes with all of the elements of the group, which can be multiplied with all the other elements easily.

$A_4 \times S_2$	1	(123)	(132)	(13)(24)	−1	−(132)	−(123)	−(13)(24)
$T_h = (m3)$	E	$4C_3$	$4C_3^2$	$3C_2$	i	$4S_6$	$4S_6^2$	$3\sigma_d$
$A_g = \chi_1$	1	1	1	1	1	1	1	1
$E_g = \chi_2$	1	$\frac{-1+i\sqrt{3}}{2}$	$\frac{-1-i\sqrt{3}}{2}$	1	1	$\frac{-1-i\sqrt{3}}{2}$	$\frac{-1+i\sqrt{3}}{2}$	1
$E_g = \chi_3$	1	$\frac{-1-i\sqrt{3}}{2}$	$\frac{-1+i\sqrt{3}}{2}$	1	1	$\frac{-1+i\sqrt{3}}{2}$	$\frac{-1-i\sqrt{3}}{2}$	1
$T_g = \chi_4$	3	0	0	−1	3	0	0	−1
$A_u = \chi_5$	1	1	1	1	−1	−1	−1	−1
$E_u = \chi_6$	1	$\frac{-1+i\sqrt{3}}{2}$	$\frac{-1-i\sqrt{3}}{2}$	−1	−1	$-\frac{-1-i\sqrt{3}}{2}$	$-\frac{-1+i\sqrt{3}}{2}$	1
$E_u = \chi_7$	1	$\frac{-1-i\sqrt{3}}{2}$	$\frac{-1+i\sqrt{3}}{2}$	−1	−1	$-\frac{-1+i\sqrt{3}}{2}$	$-\frac{-1-i\sqrt{3}}{2}$	1
$T_u = \chi_8$	3	0	0	−1	−3	0	0	1

$S_4 \times S_2$	1	(123)	(12)	(1234)	(12)(34)	−1	−(1234)	−(123)	−(12)(34)	−(12)
$O_h = (m3m)$	E	$8C_3$	$6C_2$	$6C_4$	$3C_2$	i	$6S_4$	$8S_6$	$3\sigma_h$	$6\sigma_d$
$A_{1g} = \chi_1$	1	1	1	1	1	1	1	1	1	1
$A_{2g} = \chi_2$	1	1	−1	−1	1	1	−1	1	1	−1
$E_g = \chi_3$	2	−1	0	0	2	2	0	−1	2	0
$T_{1g} = \chi_4$	3	0	−1	1	−1	3	1	0	−1	−1
$T_{2g} = \chi_5$	3	0	1	−1	−1	3	−1	0	−1	1
$A_{1u} = \chi_6$	1	1	1	1	1	−1	−1	−1	−1	−1
$A_{2u} = \chi_7$	1	1	−1	−1	1	−1	1	−1	−1	1
$E_u = \chi_8$	2	−1	0	0	2	−2	0	1	−2	0
$T_{1u} = \chi_9$	3	0	−1	1	−1	−3	−1	0	1	1
$T_{2u} = \chi_{10}$	3	0	1	−1	−1	−3	1	0	1	−1

$A_5 \times S_2$	1	(12345)	(13524)	(123)	(12)(34)	-1	$-(13524)$	$-(12345)$	$-(123)$	$-(12)(34)$
I_h	E	$12C_5$	$12C_5^2$	$20C_3$	$15C_2$	i	$12S_{10}$	$12S_{10}^3$	$20S_6$	15σ
$A_g = \chi_1$	1	1	1	1	1	1	1	1	1	1
$T_{1g} = \chi_2$	3	τ	τ'	0	-1	3	τ'	τ	0	-1
$T_{2g} = \chi_3$	3	τ'	τ	0	-1	3	τ	τ'	0	-1
$G_g = \chi_4$	4	-1	-1	1	0	4	-1	-1	1	0
$H_g = \chi_5$	5	0	0	-1	1	5	0	0	-1	1
$A_u = \chi_6$	1	1	1	1	1	-1	-1	-1	-1	-1
$T_{1u} = \chi_7$	3	τ	τ'	0	1	-3	$-\tau'$	$-\tau$	0	1
$T_{2u} = \chi_8$	3	τ'	τ	0	-1	-3	$-\tau$	$-\tau'$	0	1
$G_u = \chi_9$	4	-1	-1	1	0	-4	1	1	-1	0
$H_u = \chi_{10}$	5	0	0	-1	1	-5	0	0	1	-1

Some notation here does double duty. For example, when the symbol i appears as one of the numbers in the character table, it represents the complex number whose square is -1. When it appears as the name of group element, it denotes *inversion*, which is a group operation that multiplies each coordinate by -1. (It corresponds to $-I$, the negative of the identity matrix.) The symbols T (with subscripts), when they name a particular representation, are not to be confused with our labelling of tetrahedra. The symbols C_n denote rotation of $2\pi/n$ about a line in three-space as well as cyclic groups of order n. The symbol σ is used to signify reflection about a plane in three-space, and it should not be confused with C_2, which is not a reflection, although both have order 2 as group elements. The symbol S with subscripts corresponds to reflection about a plane, followed by a rotation about a line perpendicular to that plane.

Each of these groups $\mathcal{G} \times S_2$ has twice the number of elements as $\mathcal{G}$, twice the number of conjugacy classes as $\mathcal{G}$, and twice the number of irreducible representations as $\mathcal{G}$. For each conjugacy class say $[g]$ of $\mathcal{G}$, $[(g, 1)]$ and $[(g, -1)]$ are distinct conjugacy classes of $\mathcal{G} \times S_2$. For each representation ρ of $\mathcal{G}$ and g in $\mathcal{G}$, $\rho_1((g, 1)) = \rho(g), \rho_1((g, -1)) = \rho(g)$, and $\rho_2((g, 1)) = \rho(g), \rho_2((g, -1)) = -\rho(g)$ defines two other corresponding irreducible representations of $\mathcal{G} \times S_2$. In the enumeration generators of the elements of $\mathcal{G}$, we can add one extra element, say $g_0 = -1$. Then $g_0^a g_1^b, \ldots$, where $a = 0, 1, b = \ldots$, enumerate $\mathcal{G}$.

This completes the description of our six basic groups.

11.6.3 Determinant Plots

We are now in a position to follow the method shown in Section 11.3 (illustrated by the example in Section 11.3.1) for the basic groups described above, as we have an explicit set of matrices that form each irreducible representation of each of our six groups.

Suppose we have chosen the group $\mathcal{G}$ and two group elements c_1 and c_2 in $\mathcal{G}$ that generate it (c_1 and c_2 will correspond to two transitivity classes of cables in the final tenesgrity we construct). This means that every element of $\mathcal{G}$ can be written as a product of some number of these two group elements in some order, and is necessary if we wish the final stress matrix that we construct to be positive semi-definite, given that the tensegrity is connected by struts and cables. We also choose s_1 in $\mathcal{G}$ not equal to any of $c_1, c_1^{-1}, c_2, c_2^{-1}$, which will correspond to the transitivity class of the strut.

From the definition in (11.4), for each irreducible representation ρ_i, we form the matrix $\mathbf{\Omega}_i = \mathbf{\Omega}_i(\omega_1, \omega_2, \omega_{-1})$ in terms of the real variables $\omega_1, \omega_2, \omega_{-1}$.

$$\mathbf{\Omega}_i = \omega_1 \mathbf{\Omega}_i(c_1) + \omega_2 \mathbf{\Omega}_i(c_2) + \omega_{-k} \mathbf{\Omega}_i(s_k).$$

When $\omega_1 > 0, \omega_2 > 0, \omega_{-1} \geq 0$ it is easy to see that $\mathbf{\Omega}_i$ is positive definite for ρ_i not the trivial representation. This is because the graph of the tensegrity that it defines is connected and effectively has only cables. So the only kernel vector comes from the trivial representation.

We now define the following, which we call the *i-th determinant polynomial* corresponding to the non-trivial irreducible representation ρ_i of $\mathcal{G}$,

$$\Delta_i(\omega_1, \omega_2, \omega_{-1}) = \det(\mathbf{\Omega}_i(\omega_1, \omega_2, \omega_{-1})).$$

Notice that Δ_i is implicitly a function of the choice of c_1, c_2, and s_1, and when $\omega_1 > 0$, $\omega_2 > 0, \omega_{-1} \geq 0$, then $\Delta_i(\omega_1, \omega_2, \omega_{-1}) > 0$ for all $i \neq 1$. Since we are ultimately interested in when $\omega_1 > 0, \omega_2 > 0$, we can normalize the force coefficients by dividing all the ω_i's by $\omega_1 + \omega_2$, and this has the effect that we may assume that $\omega_1 + \omega_2 = 1$. Thus we are essentially considering the two variable polynomial $\Delta_i(\omega_1, 1 - \omega_1, \omega_{-1})$.

Define the following region in (ω_1, ω_{-1}) space:

$$R = R(c_1, c_2, s_1) = \{(\omega_1, \omega_{-1}) \mid 0 < \omega_1 < 1, \ \Delta_i(\omega_1, 1 - \omega_1, \omega_{-1}) > 0, \text{ for all } i \neq 1\}.$$

Each region R is convex, because the sum of positive definite matrices is positive definite. This means that if (ω_1, ω_{-1}) and $(\omega_1', \omega_{-1}')$ are in R, then so is any point on the line segment connecting them $((t\omega_1 + (1 - t)\omega_1, t\omega_{-1}' + (1 - t)\omega_{-1}')$, where $0 \leq t \leq 1$).

Next fix $\omega_1 > 0$ and $\omega_2 = 1 - \omega_1 > 0$, and vary ω_{-1}. As $\omega_{-1} \to -\infty$ eventually $\mathbf{\Omega}$ must have a negative eigenvalue. So every vertical ray from the horizontal axis, starting in the interval from 0 to 1, must eventually leave R. So that point on the lower boundary of R corresponds to a force coefficient $\omega_{-1} < 0$ where, for some i, $\mathbf{\Omega}_i$ is singular, but still positive semi-definite (indeed, every point on the boundary of R corresponds to a positive semi-definite $\mathbf{\Omega}$).

The zero set of each of the polynomials Δ_i corresponds to values of the force coefficients where $\mathbf{\Omega}_i$ is singular, and the boundary of R must be part of that zero set. The i for which $\Delta_i(\omega_1, 1 - \omega_1, \omega_{-1}) = 0$ indicates which representation ρ_i corresponds to the singular $\mathbf{\Omega}_i(\omega_1, \omega_2, \omega_{-1})$. So the result of this analysis is that for fixed ω_1, ω_2, the critical force coefficient ω_{-1}, and the representation ρ_i to go with it, can be found by finding the smallest magnitude negative root of the polynomials $\Delta_i(\omega_1, \omega_2, \omega_{-1})$. We refer to that representation ρ_i as the *winner*.

Typical cases are shown later in Figures 11.3 to 11.9. Each figure shows, in (a) a perspective view of the resultant tensegrity, and in (b) the determinant plot showing the choice of stresses to generate the tensegrity by a cross, on the boundary of R. For each of the cases shown, the winning representation is 3-dimensional, giving a positive-definite stress

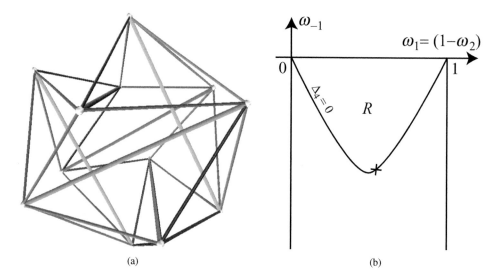

(a)   (b)

Figure 11.3 An $\mathcal{A}_4$ tensegrity with cables $\{(134),(234)\}$ and strut$\{(14)(23)\}$ – in this case the force coefficient is unequal in the two classes of cables. (a) The physical configuration. (b) The determinant plot, showing the choice of stresses by a cross, on the boundary of the region R, which ensures that the stress matrix is positive semi-definite. In this case, ρ_4 is the winner, which is a faithful representation. As ρ_4 is 3-dimensional representation, the stress matrix is rank-deficient by 4.

matrix that is rank-deficient by 4. In each tensegrity, the cables are shown in red and blue (blue carrying the force coefficient ω_1, red carrying ω_2). Examples are shown for the groups $\mathcal{A}_4, \mathcal{A}_4 \times \mathcal{S}_2, \mathcal{S}_4, \mathcal{S}_4 \times \mathcal{S}_2, \mathcal{A}_5, \mathcal{A}_5 \times \mathcal{S}_2$.

For each Figures 11.3–11.9, the choice of struts and cables, c_1, c_2, s_1, is given. However, to understand how those choices can be made, we need to define equivalence classes of tensegrity Cayley graphs, as described in the next section.

11.6.4 Cayley Graphs

We have seen that if we are given the elements c_1, c_2, s_1 in one of our groups $\mathcal{G}$, it is possible to determine those geometric representations of a tensegrity, where c_1 and c_2 correspond to transitivity classes of cables, and s_1 corresponds to a transitivity class of struts. In principle, we could enumerate all such triples of elements of $\mathcal{G}$ and do the calculations for each of them, but that would involve several redundant cases, since several pairs of such tensegrities would be essentially identical. With this in mind, for any finite group $\mathcal{G}$ we define the *tensegrity Cayley graph* $\Gamma = \Gamma(c_1, \ldots, s_1, \ldots)$ corresponding to any finite number of elements $c_1, c_2, \ldots$ and $s_1, \ldots$ of $\mathcal{G}$ as follows. The vertices of Γ are the elements of $\mathcal{G}$. For any pair of elements g_1, g_2 of $\mathcal{G}$, there is an unoriented edge between them, labelled as a cable, if there is a c_i such that $g_1 c_i = g_2$ or $g_2 c_i = g_1$. Similarly, for any pair of elements g_1, g_2 of $\mathcal{G}$, there is an unoriented edge between them, labelled as a strut, if there is a s_i such that $g_1 s_i = g_2$ or $g_2 s_i = g_1$.

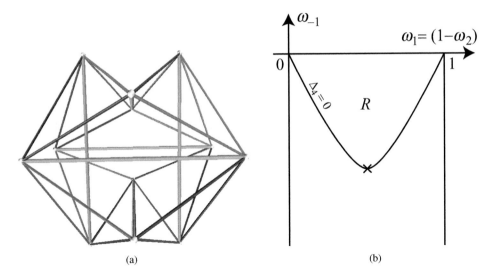

(a)   (b)

Figure 11.4 The same choice of group, cables, and struts as in Figure 11.3, but now with the force coefficient in the cables equal, which creates an additional reflection symmetry. This tensegrity is used as a baby toy called the "Sqwish."

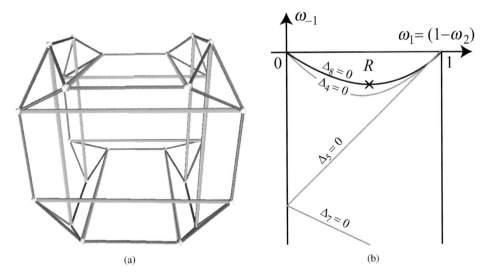

(a)   (b)

Figure 11.5 An $\mathcal{A}_4 \times \mathcal{S}_2$ tensegrity with cables $\{(124), -(13)(24)\}$ and strut $\{-(12)(34)\}$ and its determinant plot. In this case, ρ_8 is the winner, which is a faithful representation. As ρ_8 is 3-dimensional representation, the stress matrix is rank-deficient by 4.

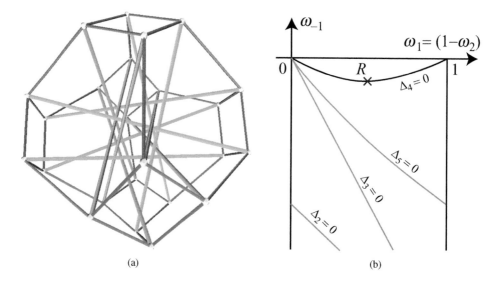

(a)   (b)

Figure 11.6 An $\mathcal{S}_4$ tensegrity with cables $\{(1423),(13)\}$ and strut$\{(12)\}$ and its determinant plot. In this case, ρ_4 (faithful, 3-dimensional) is the winner.

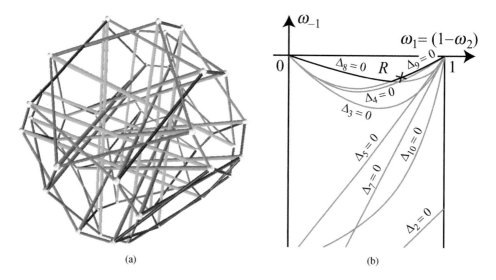

(a)   (b)

Figure 11.7 An $\mathcal{S}_4 \times \mathcal{S}_2$ tensegrity with cables $\{-(14),(1234)\}$ and strut $\{(34)\}$ and its determinant plot. In this case, ρ_9 (faithful, 3-dimensional) is the winner. Note that in this case, the force coefficient ω_1 (in the blue cables) has to be sufficient to ensure that the 3-dimensional ρ_9 is the winner – ρ_8 is 2-dimensional, and hence does not give a 3-dimensional configuration.

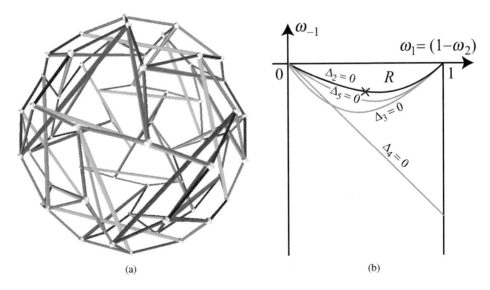

Figure 11.8 An $\mathcal{A}_5$ tensegrity with cables $\{(15243),(15)(34)\}$ and strut $\{(13)(24)\}$, together with its determinant plot. The cable graph is that of the soccer ball, a truncated icosahedron, although the hexagons are distorted. In this case, ρ_2 (faithful, 3-dimensional) is the winner.

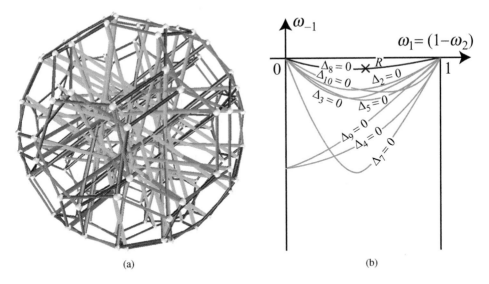

Figure 11.9 An $\mathcal{A}_5 \times \mathcal{S}_2$ tensegrity with cables $\{-(15)(24),(14532)\}$ and strut $\{(13)(24)\}$ and its determinant plot. In this case, ρ_8 (faithful, 3-dimensional) is the winner. In this case, as the force coefficient ω_2 (in the red cables) is increased, the red cables shorten and the configuration moves towards a regular icosahedron where sets of 10 vertices converge.

The standard definition of a Cayley graph does not usually make a distinction between cables and struts, but this is natural for us. This can be regarded as a sort of colouring of the edges of the standard Cayley graph.

An isomorphism $\alpha : \mathcal{G} \to \mathcal{G}$ of $\mathcal{G}$ to itself is called an *automorphism*. For example, for any g_0 in $\mathcal{G}$ the function $g \to g_0 g g_0^{-1}$, which is conjugation by g_0, is an automorphism of $\mathcal{G}$ called an *inner automorphism*. If an automorphism of a group is not an inner automorphism, it is called an *outer automorphism*.

The following are some easy consequences of our definitions.

(i) If any cable or strut is replaced by its inverse, the corresponding tensegrity Cayley graph is the same.

(ii) If $\alpha : \mathcal{G} \to \mathcal{G}$ is an automorphism of $\mathcal{G}$, then the tensegrity Cayley graph $\Gamma(c_1, \ldots, s_1, \ldots)$ is the same as the tensegrity Cayley graph of $\Gamma(\alpha(c_1), \ldots, \alpha(s_1), \ldots)$.

(iii) The elements $c_1, c_2, \ldots$ generate $\mathcal{G}$ if and only if the cable subgraph of Γ is connected.

We can regard an automorphism of the group $\mathcal{G}$ as a way of relabeling the vertices of the associated Cayley graph Γ that is consistent with the group structure. The group $\mathcal{G}$ acts on Γ as a group of symmetries in the sense that left multiplication by any element of $\mathcal{G}$ takes Γ to itself.

For Cayley graphs, it is usual to assume that the elements that are used to define it actually generate it, so the associated Cayley graph is connected. For us, it will be natural, also, to assume that the subgraph of Γ, determined by the edges labeled cables, be connected.

11.6.5 Automorphisms

In light of the discussion in the previous subsection, we will determine all the automorphisms of the dihedral groups and the six groups that we have considered. One other definition is useful. We say that a subgroup $\mathcal{H}$ of the group $\mathcal{G}$ is *normal* (or equivalently *self-conjugate*) if for all g in $\mathcal{G}$, $g \mathcal{H} g^{-1} = \mathcal{H}$, where $g \mathcal{H} g^{-1} = \{ ghg^{-1} \mid h \text{ in } \mathcal{H} \}$. For example, the group of even permutations of n symbols $\mathcal{A}_n$ is a normal subgroup of $\mathcal{S}_n$ the group of all permutations of n symbols, for $n = 2, 3, \ldots$. It is a well-known fact, which can be read off from the character table, that $\mathcal{A}_5$ has no normal subgroups other than itself and the trivial group consisting of just the identity element.

If a group $\mathcal{H}$ is a normal subgroup of a larger group $\mathcal{G}$, then it is easy to see that conjugation of $\mathcal{H}$ by any element of $\mathcal{G}$, $h \to ghg^{-1}$ is an automorphism of $\mathcal{H}$. So for the groups that we considered, $\mathcal{A}_4$ and $\mathcal{A}_5$ are normal subgroups of $\mathcal{S}_4$ and $\mathcal{S}_5$, respectively, and so conjugation by elements in the larger symmetric group provides outer automorphisms.

Conversely, suppose that we identify $\mathcal{G}$ as a subgroup of $GL(n, \mathbb{R})$, the group of all n-by-n non-singular matrices, for some $n = 1, 2, \ldots$. We can regard the identity map as a representative of one class of equivalent representations. Then any automorphism α of $\mathcal{G}$ is also a representation of $\mathcal{G}$ of the same dimension. Of course, if α is obtained by conjugation by an element of $\mathcal{G}$ (an inner automorphism) or even conjugation by an element of $GL(n, \mathbb{R})$, α will be equivalent to a representation to the identity representation. But if there is an automorphism that does not arise from conjugation by an element of $GL(n, \mathbb{R})$, then it will determine an inequivalent representation and this can be determined in the character table for $\mathcal{G}$.

Since we are dealing largely with permutation groups, it is helpful to describe the operation of conjugation by an element of S_n. We wish to describe $\alpha(g) = g_0 g g_0^{-1}$. First regard each of these elements as functions, i.e. permutations, of the symbols $\{1, \ldots, n\}$. Then $\alpha(g)(g_0(i)) = g_0(g(i))$, and so if we write $g = (a, b, c, \ldots)$ in disjoint cycle notation, then $\alpha(g) = (g_0(a), g_0(b), \ldots)$ in disjoint cycle notation. For example, if $g_0 = (12)$ and $g = (12345)$ in disjoint cycle notation, then $\alpha(g) = (21345)$ in disjoint cycle notation. This is useful to simplify some of the calculations.

We now describe the automorphisms of the dihedral groups and our other six groups.

$\mathcal{D}_n$, n odd, $n \geq 3$

We can describe any automorphism α of $\mathcal{D}_n$ by the value of $\alpha(r)$ and $\alpha(s)$. Since the order of r is n, $\alpha(r)$ must also have order n. But the only possibilities are $\alpha(r) = r^j$, where j relatively prime to n. Similarly, since the order of s is 2, $\alpha(s) = r^k s$, for some $k = 1, \ldots n$. It is easy to check that r^j and $r^k s$ generate $\mathcal{D}_n$, and that $(r^k s) r^j (r^k s) = (r^j)^{-1}$. Thus such any such α for j relatively prime to n determines an automorphism of $\mathcal{D}_n$.

$\mathcal{D}_n$, n even $n \geq 4$

Again we can describe any automorphism α of $\mathcal{D}_n$ by the value of $\alpha(r)$ and $\alpha(s)$. Since r has order n, $\alpha(r)$ must also have order n, which implies $\alpha(r) = r^j$, where j is relatively prime to n as before. But now there are more elements that have order 2 that could be images of s. The element $r^{n/2}$ cannot be the image of s, since r^j and $r^{n/2}$ do not generate all of $\mathcal{D}_n$. On the other hand $\alpha(s) = r^k s$ for any $k = 1, \ldots$, and $\alpha(r) = r^j$, j relatively prime to n does define an automorphism of $\mathcal{D}_n$ as before.

$\mathcal{A}_4$

Recall that the irreducible 3-dimensional representation of $\mathcal{A}_4$ assigns the linear extension of the corresponding permutation of the vertices of a regular tetrahedron centred at the origin. Any automorphism of the image of this representation in $GL(3, \mathbb{R})$ arising from the conjugation by an element of $GL(3, \mathbb{R})$, that is a matrix with positive determinant, will be an inner automorphism. If the matrix has a negative determinant, it will correspond to conjugation by an element in S_4, since the elements of S_4 correspond to arbitrary permutations of the vertices of the regular tetrahedron. Thus conjugation $\mathcal{A}_4$ by elements in the larger group S_4 describe all the automorphisms of $\mathcal{A}_4$.

S_4

There are two distinct 3-dimensional irreducible representations of S_4, but their images in $Gl(3, \mathbb{R})$ are distinct. One image contains some matrices with negative determinant, and the other only those with positive determinant. So there are only inner automorphisms of S_4.

$$\mathcal{A}_5$$

This is similar to $\mathcal{A}_4$ in that the only automorphisms are those coming from conjugation by elements in the larger group S_5. This can be seen by considering the irreducible
4-dimensional representation, or if one looks at the two irreducible 3-dimensional representations, their images can be taken to be the rotations of the regular dodecahedron (or
equivalently the icosahedron). In addition to the inner automorphisms of $\mathcal{A}_5$, conjugation by
an odd permutation of S_5 (as described above), takes an element of order 5 in one conjugacy
class to the other conjugacy class, and this permutes the two irreducible 3-dimensional
representations.

$$\mathcal{A}_4 \times S_2$$

The group of symmetries of a cube that induce an even permutation on the long diagonals
of the cube is the image of the only 3-dimensional irreducible representation of $\mathcal{A}_4 \times S_2$. So
the only automorphisms of $\mathcal{A}_4 \times S_2$ are restrictions of symmetries of cube. These are then
the automorphisms of the group $\mathcal{A}_4$ (conjugation by an element of S_4 not changing the sign
of S_2. In other words $\alpha(g, z) = (g_0 g g_0^{-1}, z)$, where g is in $\mathcal{A}_4$, g_0 is in S_4, and z is in S_2.

   Note that here and later we regard the group $S_2 = \{1, -1\}$, where 1 is the identity element,
and -1 is the other element.

$$S_4 \times S_2$$

This group is isomorphic to the full group of symmetries of the cube. There are two inequivalent irreducible one-to-one 3-dimensional representations of this group as can be seen from
the character table. The automorphism that takes one representation to the other is given
by $\alpha(g, z) = (g_0 g g_0^{-1}, \theta(g)z)$, where g is in S_4, g_0 is in S_4, z is in S_2 and $\theta: S_4 \rightarrow S_2$
is the group homomorphism that assigns $+1$ to an even permutation and -1 to an odd
permutation. So all automorphisms of $S_4 \times S_2$ are described as inner automorphisms or one
of the automorphisms above. For example, the pair of elements $-(123), (1234)$ are taken to
the corresponding pair $-(213), -(2134)$ by an automorphism of the first type.

$$\mathcal{A}_5 \times S_2$$

This group is isomorphic to the full group of symmetries of the regular dodecahedron (or
icosahedron). But again this group is the image of two distinct irreducible one-to-one 3-
dimensional representations. But then it is easy to see that the only automorphisms are
given as the product of conjugation on the $\mathcal{A}_5$ factor by an element of S_5 and the identity
on the S_2 factor.

11.6.6 Enumeration of Tensegrity Cayley Graphs

From the discussion in Section 11.6.4 we see what the conditions are for the group elements
c_1, c_2, s_1 to define the same tensegrity Cayley graph $\Gamma(c_1, c_2, s_1)$. With this in mind we define
the following set, which we call the *defining set*:

$$D = D(c_1, c_2, s_1) = (\{\{c_1, c_1^{-1}\}, \{c_2, c_2^{-1}\}\}, \{s_1, s_1^{-1}\}).$$

Here we are using set notation, where $\{x, y\} = \{y, x\}$, but $(x, y) \neq (y, x)$, unless $x = y$. We only create a defining set D when all three of the sets $\{c_1, c_1^{-1}\}, \{c_2, c_2^{-1}\}, \{s_1, s_1^{-1}\}$ are distinct. With this notation, we see that D remains the same if any of the elements are replaced by their inverse, or the roles of c_1 and c_2 are reversed. Note, also, that if an element is replaced by its inverse, the set it defines collapses to a singleton. From the discussion to this point we have the following.

Proposition 11.6.1. Cayley defining sets. *For a group G, two tensegrity Cayley graphs $\Gamma(c_1, c_2, s_1)$ and $\Gamma(c_1', c_2', s_1')$ are the same if and only if there is an automorphism α of G such that $D(\alpha(c_1), \alpha(c_2), \alpha(s_1)) = D(c_1', c_2', s_1')$.*

We also insist that c_1 and c_2 generate G in order that the cable graph be connected, which in turn is needed if there is any chance for the tensegrity to be super stable.

Our catalogue is then organized so that for each group G there is one entry for each equivalence class of triples of elements in G, where two triples are equivalent if they define the same defining set. Furthermore it is natural to collect those triples together that define the same cable graph. For each pair of elements c_1 and $c2$ in G, define the *cable defining set* as

$$C = C(c_1, c_2) = \{\{c_1, c_1^{-1}\}, \{c_2, c_2^{-1}\}\}.$$

Then we first consider the C-equivalence classes, where (c_1, c_2) is C-equivalent to (c_1', c_2') if there is an automorphism α of G such that $C(\alpha(c_1), \alpha(c_2)) = C(c_1', c_2')$. Then for each of these C-equivalence classes, there will be several equivalence classes of triples, each defining a distinct tensegrity Cayley graph. Keep in mind, though, that c_1 and c_2 must generate G.

For example, for the group A_4 there are only two C-equivalence classes, which are represented by $\{(134), (243)\}$ and $\{(124), (14)(23)\}$. This amounts to saying that any two generators of A_4 that are of order 3 are C-equivalent, and any two generators, one of order 3, the other of order 2, are C-equivalent. Furthermore, these are the only pairs of generators of A_4. Each of these C-equivalence classes has three tensegrity Cayley graphs, making six in all.

For the dihedral groups, and from the discussion in Section 11.6.5, we see that the only C-equivalence class for $D_n, n = 3, 4, \ldots$ is represented by $\{r, s\}$, since any two elements that generate D_n are an automorphic image of r and s.

Table 11.1 gives the C-equivalence classes for the six non-dihedral groups we have considered.

11.6.7 Algorithm

We have all the tools to describe the process that goes into creating the catalogue of symmetric tensegrities. The goal and restrictions are as follows.

    (i) The tensegrity has a symmetry group G isomorphic to one of $D_n, n = 3, 4, \ldots, A_4$, $S_4, A_5, A_4 \times S_2, S_4 \times S_2$, or $A_5 \times S_2$.

    (ii) The group G acts transitively on the vertices of the tensegrity.

(iii) The group $\mathcal{G}$ acts freely on the vertices of the tensegrity. That is, the only group element that fixes any vertex is the identity.

(iv) The tensegrity has only two transitivity classes of cables, and one transitivity class of strut.

Table 11.1. C-equivalence classes for the six non-dihedral groups we have considered. Bear in mind that for each choice of cable graph, there are several choices of struts that give distinct tensegrity Cayley graphs. The symbol {}* indicates that the cable Cayley graph is planar. It turns out that this implies that the representation is almost always determined by the cable graph only.

Group	$\mathcal{A}_4$	$\mathcal{S}_4$	$\mathcal{A}_5$
C- equivalence classes	$\{(134),(243)\}^*$ $\{(124),(14)(23)\}^*$	$\{(1423),(234)\}^*$ $\{(12),(143)\}^*$ $\{(1423),(13)\}^*$ $\{(1423),(1324)\}$	$\{(134),(23)(45)\}^*$ $\{(15)(24),(14532)\}$ $\{(12354),(143)\}^*$ $\{(12354),(145)\}$ $\{(12354),(14235)\}$ $\{(15243),(15)(34)\}^*$ $\{(142),(354)\}$ $\{(14325),(14253)\}$

Group	$\mathcal{A}_4 \times \mathcal{S}_2$	$\mathcal{S}_4 \times \mathcal{S}_2$	$\mathcal{A}_5 \times \mathcal{S}_2$
C- equivalence classes	$\{(124),-(13)(24)\}^*$ $\{(142),-(134)\}$ $\{-(243),(12)(34)\}$ $\{-(234),-(12)(34)\}$ $\{-(234),-(134)\}$	$\{-(1342),(34)\}$ $\{-(132),-(14)\}$ $\{-(1423),(1234)\}$ $\{-(1423),-(142)\}$	$\{-(13425),(23)(45)\}$ $\{(23)(45),-(152)\}$ $\{-(13425),-(23)(45)\}$ $\{(124),-(23)(45)\}$ $\{-(23)(45),(15342)\}$ $\{-(23)(45),-(13245)\}$ $\{(23)(45),-(14523)\}$ $\{(15243),-(142)\}$ $\{-(14)(25),(13524)\}$ $\{-(132),-(154)\}$ $\{-(132),(12354)\}$ $\{-(25)(34),-(123)\}$ $\{(354),-(123)\}$ $\{-(12543),-(123)\}$ $\{-(15342),(153)\}$ $\{-(13524),-(15342)\}$ $\{-(125),-(13524)\}$ $\{-(14253),(13254)\}$ $\{-(15243),-(14253)\}$ $\{(12435),-(13524)\}$ $\{-(12435),(253)\}$

With the above in mind, our goal is to display all those tensegrities that satisfy the conditions above that have a positive semi-definite stress matrix with a 4-dimensional kernel. There is choice of the ratio of the lengths of cables from the two transitivity classes, for example. For many of our pictures, we have arbitrarily decided to make the two cable force coefficient equal to each other.

Our method in creating the catalogue of pictures of the final tensegrities is as follows.

(i) List all the equivalence classes of tensegrity Cayley graphs as described in Section 11.6.6.
(ii) For each class, calculate the determinant polynomials for each of the irreducible representations of $\mathcal{G}$.
(iii) Calculate the winning representation, say j as described in Section 11.6.3.
(iv) If the winning representation is 3-dimensional, and the stresses that correspond to the representation are $\omega_1, \omega_2, \omega_{-1}$, compute the 3-by-3 matrix $\mathbf{\Omega}_j = \mathbf{\Omega}_j(\omega_1, \omega_2, \omega_{-1})$.
(v) By construction $\mathbf{\Omega}_j$ is singular, and so has a non-zero (column) vector p_1 such that $\mathbf{\Omega}_j p_1 = 0$. Then the configuration for the desired tensegrity is given by

$$(\rho_j(g_1)p_1, \rho_j(g_2)p_1, \ldots, \rho_j(g_n)p_1)$$

for all the elements $g_1, \ldots, g_n$ of $\mathcal{G}$. The rule for determining which pairs of elements determine cables or struts is given by the rule in Section 11.6.4.

11.6.8 Results for the Dihedral Groups

We can give a complete description of the tensegrities for the conditions in Subsection 11.6.7.

We know from Subsections 11.6.5 and 11.6.6 that the only C-equivalence class for $\mathcal{D}_n, n \geq 3$, is $\{r, s\}$. However, we must make a choice of a group element s_1 corresponding to a strut. It turns out that if $s_1 = r^j$ for some j, then the resulting critical configuration will be either 2-dimensional or it will be 4-dimensional. So we will concentrate on the case when $s_1 = r^j s$, for $j = 2, \ldots, n - 1$. We calculate local stress matrices for each representation, ψ_j, for $j = 1, \ldots$.

$$\mathbf{\Omega}_j = \omega_1 \left(\rho_j (1 - r^k) + \rho_j (1 - r^{-k}) \right) + \omega_2 \rho_j (1 - s) + \omega_{-1}(1 - r^k s)$$

$$= \omega_1 \begin{bmatrix} 2\left(1 - \cos\left(\frac{2\pi j}{n}\right)\right) & 0 \\ 0 & 2\left(1 - \cos\left(\frac{2\pi j}{n}\right)\right) \end{bmatrix} + \omega_2 \begin{bmatrix} 0 & 0 \\ 0 & 2 \end{bmatrix} \omega_{-1} \begin{bmatrix} 1 - \cos\left(\frac{2\pi jk}{n}\right) & -\sin\left(\frac{2\pi jk}{n}\right) \\ -\sin\left(\frac{2\pi jk}{n}\right) & 1 + \cos\left(\frac{2\pi jk}{n}\right) \end{bmatrix}$$

$$= \begin{bmatrix} 2\left(1 - \cos\left(\frac{2\pi j}{n}\right)\right)\omega_1 + \left(1 - \cos\left(\frac{2\pi jk}{n}\right)\right)\omega_{-1} & -\sin\left(\frac{2\pi jk}{n}\right)\omega_{-1} \\ -\sin\left(\frac{2\pi jk}{n}\right)\omega_{-1} & 2\left(1 - \cos\left(\frac{2\pi j}{n}\right)\right)\omega_1 + 2\omega_2 + \left(1 + \cos\left(\frac{2\pi jk}{n}\right)\right)\omega_{-1} \end{bmatrix}$$

The local stress matrix for the representation ρ_2 is $[\omega_2 + \omega_{-1}]$, and so its determinant polynomial is $\omega_2 + \omega_{-1}$. This 1-dimensional representation must contribute to the kernel of $\mathbf{\Omega}$, if we want to have a 3-dimensional super stable tensegrity. So we assume that $\omega_2 + \omega_{-1} = 0$ and thus $\omega_2 = -\omega_{-1}$. Then

$$\Omega_j = \begin{bmatrix} 2\left(1 - \cos\left(\frac{2\pi j}{n}\right)\right)\omega_1 + \left(1 - \cos\left(\frac{2\pi jk}{n}\right)\right)\omega_{-1} & -\sin\left(\frac{2\pi jk}{n}\right)\omega_{-1} \\ -\sin\left(\frac{2\pi jk}{n}\right)\omega_{-1} & 2\left(1 - \cos\left(\frac{2\pi j}{n}\right)\right)\omega_1 - \left(1 - \cos\left(\frac{2\pi jk}{n}\right)\right)\omega_{-1} \end{bmatrix}.$$

So the corresponding determinant polynomial for the representation ψ_j is

$$\Delta_j = 4\left(1 - \cos\left(\frac{2\pi j}{n}\right)\right)^2 \omega_1^2 - \left(1 - \cos\left(\frac{2\pi jk}{n}\right)\right)^2 \omega_{-1}^2 - \sin\left(\frac{2\pi jk}{n}\right)^2 \omega_{-1}^2$$

$$= 4\left(1 - \cos\left(\frac{2\pi j}{n}\right)\right)^2 \omega_1^2 - 2\left(1 - \cos\left(\frac{2\pi jk}{n}\right)\right)\omega_{-1}^2.$$

The critical ratio of force coefficients that will imply that $\Delta_j = 0$ is when

$$\left(\frac{\omega_1}{\omega_{-1}}\right)^2 = 2\frac{\left(1 - \cos\left(\frac{2\pi j}{n}\right)\right)^2}{\left(1 - \cos\left(\frac{2\pi jk}{n}\right)\right)} = 4\frac{\sin\left(\frac{\pi j}{n}\right)^4}{\sin\left(\frac{\pi jk}{n}\right)^2}.$$

Equivalently,

$$\left|\frac{\omega_1}{\omega_{-1}}\right| = 2\frac{\sin\left(\frac{\pi j}{n}\right)^2}{\left|\sin\left(\frac{\pi jk}{n}\right)\right|}. \tag{11.5}$$

The following Lemma from Connelly and Terrell (1995) finishes this classification for dihedral groups.

Lemma 11.6.2. *For fixed n and $k = 1 \ldots n - 1$, the minimum value of $\dfrac{\sin\left(\frac{\pi j}{n}\right)^2}{\left|\sin\left(\frac{\pi jk}{n}\right)\right|}$ for $j = 1 \ldots n - 1$ occurs only when $j = 1$ or $j = n - 1$.*

Proof. The statement of the Lemma is equivalent to the following inequality for $j = 2, \ldots, n - 2$.

$$\left|\frac{\sin\left(\frac{\pi jk}{n}\right)}{\sin\left(\frac{\pi k}{n}\right)}\right| < \frac{\sin\left(\frac{\pi j}{n}\right)^2}{\sin\left(\frac{\pi}{n}\right)^2}, \tag{11.6}$$

and since $\dfrac{\sin\left(\frac{\pi jk}{n}\right)}{\sin\left(\frac{\pi k}{n}\right)} > 1$, the inequality (11.6) follows from the inequality

$$\left|\frac{\sin\left(\frac{\pi jk}{n}\right)}{\sin\left(\frac{\pi k}{n}\right)}\right| \leq \left|\frac{\sin\left(\frac{\pi j}{n}\right)}{\sin\left(\frac{\pi}{n}\right)}\right|, \tag{11.7}$$

for $j = 2, \ldots, n - 2$. By replacing j by $n - j$ and k by $n - k$ when necessary, it is enough to show Equation (11.7) when $2 \leq j \leq n/2$, and $1 \leq k \leq n/2$. Using the fact that $0 \leq t \leq \pi/2$ implies that $2t/\pi \leq \sin t \leq t$, we see that

$$\left|\frac{\sin jk\frac{\pi}{n}}{\sin k\frac{\pi}{n}}\right| \leq \frac{1}{\frac{2k}{\pi n}\pi} = \frac{n}{2k} \quad \text{and} \quad \frac{2j}{\pi} = \frac{\frac{2}{\pi}j\frac{\pi}{n}}{\frac{\pi}{n}} \leq \frac{\sin j\frac{\pi}{n}}{\sin \frac{\pi}{n}}.$$

Thus Equation (11.7) holds when $n/2k \leq 2j/\pi$, in other words when $\pi n/4 \leq jk$.

For the remaining cases $jk < \pi n/4 < n$, write $\sin t = t \prod_{m=1}^{\infty}(1 - \frac{t^2}{m^2\pi^2})$, which converges absolutely for all real t. Thus

$$\frac{\sin j\frac{\pi}{n}}{\sin \frac{\pi}{n}} = \frac{\frac{j}{n}\pi \prod_{m=1}^{\infty}\left(1 - \frac{j^2}{n^2m^2}\right)}{\frac{1}{n}\pi \prod_{m=1}^{\infty}\left(1 - \frac{1}{n^2m^2}\right)} = j \prod_{m=1, j\nmid m}^{\infty}\left(1 - \frac{j^2}{n^2m^2}\right)$$

and

$$\frac{\sin jk\frac{\pi}{n}}{\sin \frac{k\pi}{n}} = \frac{\frac{jk}{n}\pi \prod_{m=1}^{\infty}\left(1 - \frac{j^2k^2}{n^2m^2}\right)}{\frac{k}{n}\pi \prod_{m=1}^{\infty}\left(1 - \frac{k^2}{n^2m^2}\right)} = j \prod_{m=1, j\nmid m}^{\infty}\left(1 - \frac{k^2j^2}{n^2m^2}\right).$$

The quotients are indexed over the positive integers m which are not divisible by j, i.e. $j \nmid m$. When $jk < n$, it follows that for each m,

$$0 < \left(1 - \frac{k^2j^2}{n^2m^2}\right) < \left(1 - \frac{k^2}{n^2m^2}\right).$$

This implies Equation (11.7). □

This representation corresponds to the tensegrity where the c_1 cables form the edges of two convex polygons in parallel planes. An example is shown in Figure 1.2. The parameter k refers to the number of steps between the end of the lateral cable and strut that are adjacent to the same vertex. This analysis shows that for all values of $k = 1, \ldots, n - 1$, the corresponding stress matrix Ω has four zero eigenvalues, with all the rest positive. It is easy to show that there are no affine motions preserving all the stressed directions. So these tensegrity structures are prestress stable, and super stable.

11.7 Non-Transitive Examples

The essential observation in the calculations in this chapter is that the stress matrix is a linear combination of permutation matrices on the nodes of the tensegrity. This is still true for the non-transitive and non-free case. We do one example of that sort in this section. Here we use an example due to Grünbaum and Shephard (1975). This tensegrity has symmetry group, in Schoenflies notation, S_4, which is isomorphic to C_4, the cyclic group of order 4, but is generated by an improper rotation S_4 in $\mathbb{R}^3$ — however, to avoid the confusion with the permutation group, we will refer to the group as $\mathbb{Z}_4$. The action of $\mathbb{Z}_4$ is not transitive on the nodes. There are two transitivity classes. Figure 11.10 shows the tensegrity in a top view. Until you have a model in your hand, it is hard to believe that the crossings are as indicated. Note that all the struts are disjoint.

The permutation of the vertices in this representation is such that the generator g corresponds to the permutation $g \to (1234)(5678)$ in disjoint cycle notation. This is given by

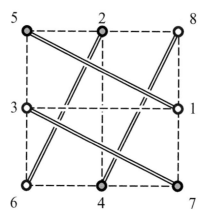

Figure 11.10 A tensegrity with $\mathbb{Z}_4$ symmetry ($\mathcal{S}_\triangle$ in Schoenflies notation) in $\mathbb{R}^3$. The nodes lie in two planes parallel with the paper, with the shaded nodes below. Nodes $1, 2, 3, 4$ form one transitivity class of nodes; nodes $5, 6, 7, 8$ form the other.

the improper rotation by $90°$ on the vertices with those labels. More generally, consider the two-by-two matrix

$$\mathbf{\Omega}(g) = \begin{bmatrix} \omega_1 + \omega_2 + \omega_4 + \omega_3 - g^2\omega_3 & -g^{-1}\omega_1 - \omega_2 - g\omega_4 \\ -g\omega_1 - \omega_2 - g^{-1}\omega_4 & \omega_1 + \omega_2 + \omega_4 \end{bmatrix}, \qquad (11.8)$$

where

$$\omega_1 = \omega_{17} = \omega_{28} = \omega_{35} = \omega_{48}$$
$$\omega_2 = \omega_{18} = \omega_{25} = \omega_{36} = \omega_{47}$$
$$\omega_3 = \omega_{13} = \omega_{24}$$
$$\omega_4 = \omega_{15} = \omega_{26} = \omega_{37} = \omega_{48},$$

we consider each entry of $\mathbf{\Omega}(g)$ to be in a $\mathcal{G}$-algebra, and it is understood that an entry without a $\mathcal{G}$ coefficient implicitly is multiplied by the identity group operation.

Let ρ_R to be the regular representation of the cyclic group of four elements $\mathbb{Z}_4$ as a set of 4-by-4 permutation matrices. Then the crucial observation is that $\mathbf{\Omega}(\rho_R(g))$ can be identified as the standard stress matrix for the tensegrity graph of Figure 11.10, and that $\mathbf{\Omega}(\rho_j(g))$ are its components, where ρ_j, for $j = 1, \ldots, 4$, are the irreducible representations of ρ_R. Since $\mathbb{Z}_4$ is an abelian group, all its representations are linear, but into the field of complex numbers. The description of $\mathbf{\Omega}(g)$ is obtained by choosing a particular representative node for each transitivity class, in this case node 1 and node 8. Each force coefficient ω_j contributes a term of the form $\omega_j(1 - g^k)$ to $\mathbf{\Omega}(g)$, where the identity term and the g term appear in the appropriate column and row of $\mathbf{\Omega}(g)$ depending on which pair of transitivity classes is being connected and which element g^k is needed to connect them using nodes 1 and 8 as base nodes.

The irreducible representations ρ_j of $\mathbb{Z}_4$ are such that $g \to 1, i, -1, -i$, for $j = 1, 2, 3, 4$, respectively. This gives the following by substituting into (11.8):

$$\Omega(\rho_1) = \begin{bmatrix} \omega_1 + \omega_2 + \omega_4 & -\omega_1 - \omega_2 - \omega_4 \\ -\omega_1 - \omega_2 - \omega_4 & \omega_1 + \omega_2 + \omega_4 \end{bmatrix},$$

which has rank 1 and is positive semi-definite as long as $\omega_1 + \omega_2 + \omega_4 > 0$. For ρ_2 we get:

$$\Omega(\rho_2) = \begin{bmatrix} \omega_1 + \omega_2 + 2\omega_3 + \omega_4 & i\omega_1 - \omega_2 - i\omega_4 \\ -i\omega_1 - \omega_2 + i\omega_4 & \omega_1 + \omega_2 + \omega_4 \end{bmatrix},$$

which has determinant $(\omega_1 + \omega_2 + \omega_4)^2 + 2(\omega_1 + \omega_2 + \omega_4)\omega_3 - (\omega_1 - \omega_4)^2 - \omega_2^2$. Note that $\Omega(\rho_4)$ will have the same determinant, since it is the conjugate transpose of $\Omega(\rho_2)$. For ρ_3 we get:

$$\Omega(\rho_3) = \begin{bmatrix} \omega_1 + \omega_2 + \omega_4 & \omega_1 - \omega_2 + \omega_4 \\ \omega_1 - \omega_2 + \omega_4 & \omega_1 + \omega_2 + \omega_4 \end{bmatrix},$$

which has determinant $(\omega_1 + \omega_2 + \omega_4)^2 - (\omega_1 - \omega_2 + \omega_4)^2$, which is 0 only when $\omega_1 + \omega_2 + \omega_4 = \pm(\omega_1 - \omega_2 + \omega_4)$. So assuming, in addition, that $\omega_2 \neq 0$, then $\omega_4 + \omega_1 = 0$.

We are looking for a tensegrity that is 3-dimensional and has a positive semi-definite stress matrix, which means that it must have at least four zero eigenvalues. It always happens that $\Omega(\rho_1)$ has at least one zero eigenvalue. If it has another, then $\omega_1 + \omega_2 + \omega_4 = 0$. If that happens then $\Omega(\rho_2)$ will have a negative value or have $\omega_2 = 0$, neither of which is possible. So $\omega_1 + \omega_2 + \omega_4 > 0$. Thus there must be at least one other zero eigenvalue from $\Omega(\rho_3)$ which implies that $\omega_4 + \omega_1 = 0$. Finally, to pick up two more zero eigenvalues, we must have $\omega_2^2 + 2\omega_2\omega_3 = (2\omega_1)^2 + \omega_2^2$. In other words, $\omega_3 = 2\omega_1^2/\omega_2$. So ω_1 and ω_2 can be chosen arbitrarily as positive numbers, while $\omega_4 = -\omega_2$, and $\omega_3 = 2\omega_1^2/\omega_2$ determine the tensegrity. In other words, there is a one-parameter family of 3-dimensional super stable tensegrities.

It is also interesting to determine the configuration that corresponds to these tensegrities. Nodes $1, 2, 3, 4$ form a square when projected into the xy plane, say. This is the action of the ρ_2 and ρ_4 representations. The ρ_1 representation reflects the nodes about the xy plane. Since $\omega_4 = -\omega_1$, the four nodes with the same colour are in the same plane parallel to the xy plane. For all choices of the parameters, the $\{1, 8\}$ cable remains perpendicular to the $\{1, 3\}$ cable, while the $\{1, 8\}$ cable extends while fixing the $1, 2, 3, 4$ nodes.

11.8 Comments

There are several interesting things to observe from the pictures in the catalogue. When the cable graph is planar, that is it has a topological embedding in the plane (or equivalently in the 2-dimensional surface of a sphere), then it appears that the cables in most of the winning representations are edges of the convex polytope determined by its vertices. For many of these examples the winning representation is determined by the choice of the cable generators.

On the other hand, for the fifth cable class for the group $\mathcal{A}_5$, both of the two irreducible 3-dimensional representations appear, seemingly at random. But for that cable class the two generators are of order 5 and are in distinct conjugacy classes. Indeed, there is an automorphism of $\mathcal{A}_5$ that interchanges the two degree 5 generators. So which representation is the winner depends on the choice of the strut group element s_1. One can see that the order 5 elements are in distinct conjugacy classes because the corresponding cables form a convex pentagon in one case, and a self-intersecting pentagram in the other case.

The action of the automorphisms on the group $\mathcal{A}_5$ is important. When both of the generators for $\mathcal{A}_5$ are of the same order, which turns out to be both of order 3 or both of order 5, then there is an automorphism that interchanges the two generators. There are two pairs of non-C-equivalent generators of order 5 and one pair of order 3. So when we look for representatives for each of the C-equivalence classes of pairs of generators for the group $\mathcal{A}_5 \times \mathcal{S}_2$, there are three possibilities for each of the eight C-equivalence classes of pairs of generators for $\mathcal{A}_5$, plus or minus each generator, where at least one generator for $\mathcal{A}_5 \times \mathcal{S}_2$ has a minus. But for the three cases of pairs of generators of order 3 and 5 for $\mathcal{A}_5$, when one generator receives a minus, it is C-equivalent to the case when the other generator receives a minus. This accounts for three cases that are C-equivalent. So there are $3 \cdot 8 - 3 = 21$ C-equivalence classes in all for $\mathcal{A}_5 \times \mathcal{S}_2$.

When the cables form a pentagram, which is necessarily self-intersecting, it is possible to create other tensegrities that are also stable with a positive semi-definite, maximal rank stress matrix. The idea is to replace the pentagram with a pentagon, which will evidently show up in the other C-equivalence class, or replace the pentagram with a star figure as indicated below. One is simply adding positive semi-definite quadratic forms together to get other positive semi-definite quadratic forms. See Figure 5.7. We apply that idea in Figure 11.11 and Figure 11.12. Note that it is not always possible to take a cable polygon and replace it with a star figure as in the case of a dihedral tensegrity in Zhang et al. (2010).

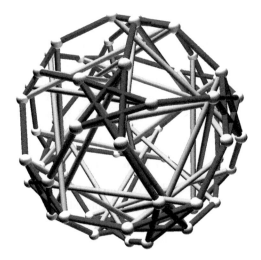

Figure 11.11 A super stable highly symmetric tensegrity in three-space. The red and blue members are cables, and the yellow members are struts.

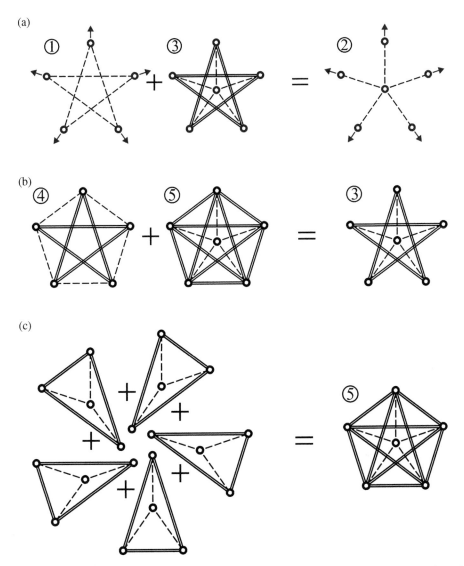

Figure 11.12 If a super stable tensegrity incorporates a pentagram of cables (such as the red cables of the tensegrity in Figure 11.11), then this figure shows that the pentagram can be replaced with a star of five cables connected to a central node without affecting the super stability. In (a), the pentagram of cables is shown as ①, and the forces that the rest of the tensegrity used to apply are shown as arrows. The tensegrity ③ is super stable. It is chosen with a symmetric stress, and then added this to ① canceling the overlapping cables. The tensegrity ② is the result, where the five cables are connected to a central node, carrying the same forces. In (c), tensegrity ⑤ is shown to be super stable as it can be formed by the superposition of five super stable (0,2)-tensegrities (see Figure 5.8). In (b), tensegrity ③ is shown to be the sum of the super stable symmetrized Cauchy polygon ④ as in subsection 5.14.2, and the tensegrity ⑤, thus showing that tensegrity ③ is super stable.

Figure 11.13 A super stable rotationally symmetric tensegrity with a star on the top and a regular hexagon on the bototm.

Figure 11.11 shows an example of a super stable highly symmetric tensegrity, where one of the cable orbits, in red, is a pentagram instead of a pentagon, as verified by a computer. Figure 11.12 shows how to add the stresses of previously known super stable tensegrities. If the stresses are scaled appropriately when cables and struts overlap, those stresses will cancel in the sum. The end result is that the red cables are replaced by star figures.

Another situation is the super stability for star figures replacing at some of the convex polygons for tensegrities with dihedral symmetry as in Connelly and Terrell (1995). In that case, for example, when the pentagonal polygon is replaced by the self-intersecting pentagram, the resulting tensegrity is definitely not super stable. So the argument above will not work to imply that, when the star replaces the pentagon, the resulting tensegrity is super stable. Nevertheless, in Zhang et al. (2010); Zhang and Ohsaki (2007, 2012), that kind of star replacement and many others are shown, and their super stability and prestress stability are determined.

For example for the tensegrity with 6-fold rotational symmetry in Figure 11.13, the top is a star and the bottom is a regular hexagon, and it is super stable. If the top is replaced by a regular hexagon, it is also super stable as in Connelly and Terrell (1995). If the bottom is replaced by a star to get two stars, the resulting tensegrity is not even rigid, since the whole tensegrity breaks up into two pieces that rotate relative to each other about the line through the two star vertices. This tensegrity is what is called *divisible* by Zhang and Ohsaki.

Appendix A

Useful Theorems and Proofs

A.1 Basic Rigidity

We provide another proof of Theorem 3.8.1, Theorem A.1.2 here, due to Walter Whiteley. But first we provide a useful lemma that seems to be part of the folklore for many years in many contexts.

Lemma A.1.1. *Suppose that* $(G, \mathbf{p})$ *and* $(G, \mathbf{q})$ *are two tensegrities in* $\mathbb{E}^d$, *each configuration with affine span all of* $\mathbb{E}^d$. *Then* $\mathbf{p} - \mathbf{q}$ *is an infinitesimal displacement (i.e. an infinitesimal flex) of* $(G, (\mathbf{p} + \mathbf{q})/2)$ *if and only if* $\mathbf{q}$ *satisfies the tensegrity constraints with respect to* $\mathbf{p}$ *of Section 4.2. Furthermore* $\mathbf{p} - \mathbf{q}$ *is a non-trivial infinitesimal flex of* $(G, (\mathbf{p} + \mathbf{q})/2)$ *if and only if* $\mathbf{p}$ *and* $\mathbf{q}$ *are congruent.*

Proof. We calculate that for each $1 \le i < j \le n$,

$$(\mathbf{p}_i - \mathbf{p}_j)^2 - (\mathbf{q}_i - \mathbf{q}_j)^2 = [(\mathbf{p}_i - \mathbf{p}_j) + (\mathbf{q}_i - \mathbf{q}_j)] \cdot [(\mathbf{p}_i - \mathbf{p}_j) - (\mathbf{q}_i - \mathbf{q}_j)]$$

$$= [(\mathbf{p}_i + \mathbf{q}_i) - (\mathbf{p}_j + \mathbf{q}_j)] \cdot [(\mathbf{p}_i - \mathbf{q}_i) - (\mathbf{p}_j - \mathbf{q}_j)]$$

$$= 2 \left[\frac{(\mathbf{p}_i + \mathbf{q}_i)}{2} - \frac{(\mathbf{p}_j + \mathbf{q}_j)}{2} \right] \cdot [(\mathbf{p}_i - \mathbf{q}_i) - (\mathbf{p}_j - \mathbf{q}_j)].$$

Thus corresponding lengths from $(G, \mathbf{p})$ to $(G, \mathbf{q})$ increase, stay the same, or decrease according to whether the inner product for the infinitesimal rigidity of $(G, (\mathbf{p} + \mathbf{q})/2)$ also increase, stay the same, or decrease. An infinitesimal flex of a framework whose nodes span $\mathbb{E}^d$ is trivial if and only if it is an infinitesimal flex for the complete graph on the same set of nodes. This shows the last statement of the lemma. $\qquad \square$

Theorem A.1.2 (Basic Rigidity). *If a (bar or tensegrity) framework is infinitesimally rigid in* $\mathbb{E}^d$, *then it is rigid in* $\mathbb{E}^d$.

Proof. Let $(G, \mathbf{p})$ be the bar or tensegrity framework in $\mathbb{E}^d$ that is infinitesimally rigid, and suppose that $\mathbf{q}(i)$, $i = 1, 2, \ldots$ is a sequence of configurations each not congruent to $\mathbf{p}$ such that $\mathbf{q}(i) \to \mathbf{p}$ as convergent vectors, while each $\mathbf{q}(i)$ satisfies the tensegrity constraints with respect to $\mathbf{p}$ as in Section 4.2. Then each $(G, (\mathbf{p} + \mathbf{q}(i))/2)$ has a non-trivial infinitesimal flex $\mathbf{p} - \mathbf{q}(i)$, which can be replaced by a unit length infinitesimal flex that is in the orthogonal complement of the trivial infinitesimal flexes. Then the limit of those infinitesimal flexes will be a non-zero unit length non-trivial flex for $(G, \mathbf{p})$, as desired. $\qquad \square$

A.2 Proof for the Cusp Mechanism

(Section 8.8.3)

Proof. This is the proof in Connelly and Servatius (1994) that the mechanism there is third-order rigid by the definition in Section 8.8.3, and yet it is a finite mechanism. Referring to Figure 8.25, the points $\mathbf{p}_0, \mathbf{p}_4, \mathbf{q}_4$ are pinned, and we assume that all n-th order flexes of these three vertices are zero for $n \geq 1$. Since the two triangles connecting $\mathbf{p}_1, \mathbf{p}_2, \mathbf{p}_3$ and the two triangles connecting $\mathbf{q}_1, \mathbf{q}_2, \mathbf{q}_3$ are infinitesimally rigid, and that $\mathbf{p}_3 = (\mathbf{p}_1 + \mathbf{p}_2)/2$, and $\mathbf{q}_3 = (\mathbf{q}_1 + \mathbf{q}_2)/2$, the same linear condition holds for all the higher-order flexes. Since $\mathbf{p}_0$ and $\mathbf{p}_4$ are pinned, the following are the most general third-order flex of the framework for $\mathbf{p}_1$ and $\mathbf{p}_2$:

$$\mathbf{p}_1' = (0, a_1), \mathbf{p}_1'' = (a_1^2, b_1), \mathbf{p}_1''' = (-3a_1 b_1, c_1)$$

$$\mathbf{p}_2' = (0, a_2), \mathbf{p}_2'' = (a_2^2, b_2), \mathbf{p}_2''' = (-3a_2 b_2, c_2).$$

The bar $\{1, 2\}$, with $\mathbf{p}_2 - \mathbf{p}_1 = (1, 1)$, gives three equations:

$$(1, 1) \cdot (0, a_2 - a_1) = 0,$$

$$(1, 1) \cdot (2a^2, b_2 - b_1) = 0,$$

$$(1, 1) \cdot (3a(b_2 + b_1), c_2 - c_1) = 0,$$

where $a_1 = a_2 = a$ from the first equation. Thus

$$b_2 - b_1 = -2a^2, c_2 - c_1 = -3a(b_2 - b_1).$$

Since $(\mathbf{p}_3', \mathbf{p}_3'', \mathbf{p}_3''')$ is the average of $(\mathbf{p}_1', \mathbf{p}_1'', \mathbf{p}_1''')$ and $(\mathbf{p}_2', \mathbf{p}_2'', \mathbf{p}_2''')$, we get that

$$\mathbf{p}_3' = (0, a), \quad \mathbf{p}_3'' = (0, (b_1 + b_2)/2), \quad \mathbf{p}_3''' = (-6a^3, (c_1 + c_2)/2.$$

A similar analysis for the third-order flex for $(\mathbf{q}, \mathbf{q}'', \mathbf{q}''')$ gives

$$\mathbf{q}_3' = (0, \bar{a}), \quad \mathbf{q}_3'' = (0, (\bar{b}_1 + \bar{b}_2)/2), \quad \mathbf{q}_3''' = (6\bar{a}^3, (\bar{c}_1 + \bar{c}_2)/2.$$

The $\mathbf{p}_3, \mathbf{q}_3$ bar gives three equations, where $\mathbf{q}_3 - \mathbf{p}_3 = (2, 0)$. The first-order equation is satisfied, and the second-order equation is

$$(2, 0) \cdot (0, (\bar{b}_2 + \bar{b}_1 - b_2 - b_1)/2) + (0, a - \bar{a}) \cdot (0, a - \bar{a}) = 0,$$

implying that $a = \bar{a}$, and so $\mathbf{p}_3' - \mathbf{q}_3' = 0$. The third-order equation is

$$(2, 0) \cdot (12a^3, (\bar{c}_2 + \bar{c}_1 - c_1 - c_2)/2) = 0.$$

So $a = 0$, and the first-order flex is zero as desired.   □

References

Adriaenssens, S., Block, P., Veenendaal, D., and Williams, C., editors (2014). *Shell Structures for Architecture: Form Finding and Optimization*. Routledge, London and New York. 37

Alexandrov, A. D. (2005). *Convex polyhedra*. Springer Monographs in Mathematics. Springer-Verlag, Berlin. Translated from the 1950 Russian edition by N. S. Dairbekov, S. S. Kutateladze, and A. B. Sossinsky, With comments and bibliography by V. A. Zalgaller and appendices by L. A. Shor and Yu. A. Volkov. 51, 54

Alfakih, A. Y. (2007). On dimensional rigidity of bar-and-joint frameworks. *Discrete Appl. Math.*, 155(10):1244–1253. 132

Alfakih, A. Y. (2017). Graph connectivity and universal rigidity of bar frameworks. *Discrete Appl. Math.*, 217(part 3):707–710. 150

Alfakih, A. Y. and Ye, Y. (2013). On affine motions and bar frameworks in general position. *Linear Algebra Appl.*, 438(1):31–36. 102

Altmann, S. L. and Herzig, P. (1994). *Point-Group Theory Tables*. Clarendon Press, Oxford. 215, 229

Asimow, L. and Roth, B. (1978). The rigidity of graphs. *Trans. Amer. Math. Soc.*, 245: 279–289. 145

Asimow, L. and Roth, B. (1979). The rigidity of graphs. II. *J. Math. Anal. Appl.*, 68(1): 171–190. 145

Atkins, P. W., Child, M. S., and Phillips, C. S. G. (1970). *Tables for Group Theory*. Oxford University Press, Oxford. 215, 229

Bang, S.-J., Williams, J. C., Isaacs, I. M., Connelly, R., Hubbard, J. H., Whiteley, W., Rudin, W., Canfield, E. R., Golomb, S. W., Zha, H., Krafft, O., and Schaefer, M. (1993). Problems and solutions: Problems: 10306-10313. *Amer. Math. Monthly*, 100(5): 498–499. 108

Barvinok, A. I. (1995). Problems of distance geometry and convex properties of quadratic maps. *Discrete Comput. Geom.*, 13(2):189–202. 109

Belk, M. (2007). Realizability of graphs in three dimensions. *Discrete Comput. Geom.*, 37(2):139–162. 134

Belk, M. and Connelly, R. (2007). Realizability of graphs. *Discrete Comput. Geom.*, 37(2):125–137. 134

Berg, A. R. and Jordán, T. (2003). A proof of Connelly's conjecture on 3-connected circuits of the rigidity matroid. *J. Combin. Theory Ser. B*, 88(1):77–97. 152

Bezdek, K. and Connelly, R. (2006). Stress matrices and m matrices. *Oberwolfach Reports*, 3(1):678–680. 109

Biedl, T., Demaine, E., Demaine, M., Lazard, S., Lubiw, A., O'Rourke, J., Robbins, S., Streinu, I., Toussaint, G., and Whitesides, S. (2002). A note on reconfiguring tree linkages: trees can lock. *Discrete Appl. Math.*, 117(1-3):293–297. 170

Bishop, D. M. (1973). *Group Theory and Chemistry.* Clarendon Press, Oxford. 6, 187, 203, 229, 230

Bochnak, J., Coste, M., and Roy, M.-F. (1998). *Real algebraic geometry*, volume 36 of *Ergebnisse der Mathematik und ihrer Grenzgebiete (3) [Results in Mathematics and Related Areas (3)].* Springer-Verlag, Berlin. Translated from the 1987 French original, Revised by the authors. 11, 64, 176

Bolker, E. D. and Crapo, H. (1979). Bracing rectangular frameworks. I. *SIAM J. Appl. Math.*, 36(3):473–490. 60

Bolker, E. D. and Roth, B. (1980). When is a bipartite graph a rigid framework? *Pacific J. Math.*, 90(1):27–44. 59, 152, 154

Bottema, O. (1960). Die bahnkurven eines merkwürdigen zwölfstabgetriebes. *Osterr. Ingen Archiv*, 14:218–222. 159

Bricard, R. (1897). Mémoire sur la theórie de l'octaédre articuleé. *J. Math. Pures Appl.*, 5:113–148. 157, 159, 162

Calladine, C. R. (1978). Buckminster Fuller's "Tensegrity" structures and Clerk Maxwell's rules for the construction of stiff frames. *Int. J. Solids Struct.*, 14:161–172. 36, 116, 208

Calladine, C. R. (1982). Modal stiffnesses of a pretensioned cable net. *Int. J. Solids Struct.*, 18(10):829–846. 116

Cantarella, J., Demaine, E., Iben, H., and O'Brian, J. (2004). An energy-driven approach to linkage unfolding. *Proceedings of the Twentieth Annual Symposium on Computational Geometry*, pages 134–143. SGC'04. 175, 176

Carwardine, G. (1935). Improvements in equipoising mechanism. UK Pat. 433 617 (filed 10 February and 7 March 1934, granted 12 August 1935). 80

Castigliano, C. A. P. (1879). *Théorie de l'équilibre des systèmes élastiques et ses applications.* A. F. Negra, Turin. 78

Cauchy, A. L. (1813). Sur les polygones et polyèdres. *second mémoire, Journal de l'Ecole Polytechnique*, 19:87–98. 49, 51

Chern, S. S. (1967). *Studies in global geometry and analysis.* Number 4. The Mathematical Association of America, Washington, DC. 106

Coates, R., Coutie, M., and Kong, F. (1988). *Structural Analysis.* Chapman and Hall, London, third edition. 37

Connelly, R. (1974). An attack on rigidity I, II. Preprint, available at http://www .math.cornell.edu/~connelly/Attack.I.eps and http://www.math .cornell.edu/~connelly/Attack.II.eps . Russian translation available in the book: A. N. Kolmoogorov, S. P. Novikov (eds.) Issledovaniya po metricheskoj teorii poverkhnostej, Moscow: Mir, 1980, pp. 164–209. 161

Connelly, R. (1978). The rigidity of suspensions. *J. Differential Geom.*, 13(3):399–408. 162

Connelly, R. (1979a). A flexible sphere. *Math. Intelligencer*, 1(3):130–131. 166

Connelly, R. (1979b). The rigidity of polyhedral surfaces. *Math. Mag.*, 52(5):275–283. 166

Connelly, R. (1980). The rigidity of certain cabled frameworks and the second-order rigidity of arbitrarily triangulated convex surfaces. *Adv. in Math.*, 37(3):272–299. 131

Connelly, R. (1982). Rigidity and energy. *Invent. Math.*, 66(1):11–33. 106

Connelly, R. (1991). On generic global rigidity. In *Applied geometry and discrete mathematics*, volume 4 of *DIMACS Series in Discrete Mathematics and Theoretical Computer Science*, pages 147–155. American Math Society, Providence, RI. 152

Connelly, R. (1993). Rigidity. In *Handbook of convex geometry, Vol. A, B*, p. 223–271. North-Holland, Amsterdam. 131

Connelly, R. (2005). Generic global rigidity. *Discrete Comput. Geom.*, 33(4):549–563. 5, 147, 148

Connelly, R., Demaine, E. D., Demaine, M. L., Fekete, S. P., Langerman, S., Mitchell, J. S. B., Ribo, A., and Rote, G. (2010). Locked and unlocked chains of planar shapes. *Discrete Comput. Geom.*, 44(2):439–462. 181

Connelly, R., Demaine, E. D., and Rote, G. (2003). Straightening polygonal arcs and convexifying polygonal cycles. *Discrete Comput. Geom.*, 30(2):205–239. U.S.-Hungarian Workshops on Discrete Geometry and Convexity (Budapest, 1999/Auburn, AL, 2000). 171, 172, 173, 176

Connelly, R., Fowler, P. W., Guest, S. D., Schulze, B., and Whiteley, W. J. (2009). When is a symmetric pin-jointed framework isostatic? *Int. J. Solids Struct.*, 46:762–773. 225

Connelly, R. and Gortler, S. J. (2015). Iterative universal rigidity. *Discrete Comput. Geom.*, 53(4):847–877. 95, 132

Connelly, R. and Gortler, S. J. (2017). Prestress stability of triangulated convex polytopes and universal second-order rigidity. *SIAM J. Discrete Math.*, 31(4):2735–2753. 131

Connelly, R., Gortler, S. J., and Theran, L. (2018). Affine rigidity and conics at infinity. *Int. Math. Res. Not. IMRN*, (13):4084–4102. 101

Connelly, R., Gortler, S. J., and Theran, L. (2020). Generically globally rigid graphs have generic universally rigid frameworks. *Combinatorica*, 40(1):1–37. 150

Connelly, R. and Henderson, D. W. (1980). A convex 3-complex not simplicially isomorphic to a strictly convex complex. *Math. Proc. Cambridge Philos. Soc.*, 88(2): 299–306. 184

Connelly, R., Hendrickson, B., and Terrell, M. (1994). Kaleidoscopes and mechanisms. In *Intuitive geometry (Szeged, 1991)*, volume 63 of *Colloquia Mathematica Societatis János Bolyai*, pages 67–75. North-Holland, Amsterdam. 182

Connelly, R., Rybnikov, K., and Volkov, S. (2001). Percolation of the loss of tension in an infinite triangular lattice. *J. Statist. Phys.*, 105(1–2):143–171. 131

Connelly, R., Sabitov, I., and Walz, A. (1997). The bellows conjecture. *Beiträge Algebra Geom.*, 38(1):1–10. 170

Connelly, R. and Schlenker, J.-M. (2010). On the infinitesimal rigidity of weakly convex polyhedra. *European J. Combin.*, 31(4):1080–1090. 54

Connelly, R. and Servatius, H. (1994). Higher-order rigidity—what is the proper definition? *Discrete Comput. Geom.*, 11(2):193–200. 180, 270

Connelly, R. and Terrell, M. (1995). Tenségrités symétriques globalement rigides. *Structural Topology*, (21):59–78. Dual French–English text. 262, 268

Connelly, R. and Whiteley, W. (1996). Second-order rigidity and prestress stability for tensegrity frameworks. *SIAM J. Discrete Math.*, 9(3):453–491. 128, 129

Coxeter, H. S. M. and Greitzer, S. L. (1967). *Geometry Revisited*. The Mathematical Association of America, Washington, DC. 177

Crapo, H. (1979). Structural rigidity. *Structural Topology*, (1):26–45, 73. 145

Crapo, H. and Whiteley, W. (1982). Statics of frameworks and motions of panel structures, a projective geometric introduction. *Structural Topology*, (6):43–82. With a French translation. 14, 40

Crapo, H. and Whiteley, W. (1993). Autocontraintes planes et polyèdres projetés. I. Le motif de base. *Structural Topology*, (20):55–78. Dual French-English text. 40

Crapo, H. and Whiteley, W. (1994a). Spaces of stresses, projections and parallel drawings for spherical polyhedra. *Beiträge Algebra Geom.*, 35(2):259–281. 40, 183

Crapo, H. and Whiteley, W. (1994b). Spaces of stresses, projections and parallel drawings for spherical polyhedra. *Beiträge Algebra Geom.*, 35(2):259–281. 57

Cremona, L. (1890). *Graphical Statics*. Oxford University Press, London. English Translation. 38

Danzer, L. (1963). Endliche punktmengen auf der 2-sphäre mit möglichst grossem minimalabstand. *Habilitationsschrift, Universität Göttingen*. 76

Davis, P. J. (1983). *The Thread: A Mathematical Yarn*. Birkhauser, Basel. 177

Dehn, M. (1916). Über die Starrheit konvexer Polyeder. *Math. Ann.*, 77(4):466–473. 51

Demaine, E. (1993). Web page. pages 1–100. Animation at http://theory.lcs.mit.edu/~edemaine/linkage/. 172

Demaine, E. D. and O'Rourke, J. (2007). *Geometric Folding Algorithms*. Cambridge University Press, Cambridge. Linkages, origami, polyhedra. 181

Donev, A., Connelly, R., Stillinger, F. H., and Torquato, S. (2007). Underconstrained jammed packings of nonspherical hard particles: ellipses and ellipsoids. *Phys. Rev. E (3)*, 75(5):051304, 32. 76

Donev, A., Torquato, S., Stillinger, F. H., and Connelly, R. (2004). A linear programming algorithm to test for jamming in hard-sphere packings. *J. Comput. Phys.*, 197(1): 139–166. 3, 76

Dudeney, H. E. (1902). Puzzles and prizes. *Weekly Dispatch*. The puzzle appeared in the April 6 issue of this column, and the solution appeared on May 4. 181

Edmondson, A. C. (1987). *A Fuller explanation*. Design Science Collection. Birkhäuser Boston Inc., Boston, MA. The synergistic geometry of R. Buckminster Fuller, A Pro Scientia Viva Title. 3

Fogelsanger, A. (1988). Web page. pages 1–60. A scanned version of his thesis www.armadillodanceproject.com/AF/Cornell/rigidity.htm. 147

Fowler, P. W. and Guest, S. D. (2000). A symmetry extension of Maxwell's rule for rigidity of frames. *International Journal of Solids and Structures*, 37(12):1793–1804. 187, 223

Frank, S. and Jiang, J. (2011). New classes of counterexamples to Hendrickson's global rigidity conjecture. *Discrete Comput. Geom.*, 45(3):574–591. 152

French, M. J. and Widden, M. B. (2000). The spring-and-lever balancing mechanism, George Carwardine and the Anglepoise lamp. *Proc. Inst. Mech. Eng. C – J. Mech. Eng. Sci.*, 214(3):501–508. 80

Gaifullin, A. A. (2014). Generalization of Sabitov's theorem to polyhedra of arbitrary dimensions. *Discrete Comput. Geom.*, 52(2):195–220. 170

Gale, D. (1960). *The Theory of Linear Economic Models*. McGraw-Hill Book Co., Inc., New York. 69

Gluck, H. (1975). Almost all simply connected closed surfaces are rigid. In *Geometric topology (Proc. Conf., Park City, Utah, 1974)*, pages 225–239. Lecture Notes in Math., Vol. 438. Springer, Berlin. 39, 51, 155

Gomez-Jauregui, V. (2010). *Tensegrity Structures and their Application to Architecture [Perfect Paperback]*. PUbliCan; 1st edition. 3

Gordon, J. E. (1978). *Structures; Or Why Things Don't Fall Down*. Da Capo Press. 1

Gortler, S. J., Healy, A. D., and Thurston, D. P. (2010). Characterizing generic global rigidity. *Amer. J. Math.*, 132(4):897–939. 5, 148

Gortler, S. J. and Thurston, D. P. (2014). Characterizing the universal rigidity of generic frameworks. *Discrete Comput. Geom.*, 51(4):1017–1036. 149

Graver, J., Servatius, B., and Servatius, H. (1993). *Combinatorial rigidity*, volume 2 of *Graduate Studies in Mathematics*. American Mathematical Society, Providence, RI. 15

Graver, J. E. (2001). *Counting on frameworks*, volume 25 of *The Dolciani Mathematical Expositions*. Mathematical Association of America, Washington, DC. 135

Grünbaum, B. and Shephard, G. C. (1975). Lectures on lost mathematics. pages 1–79. Lecture notes at https://digital.lib.washington.edu/researchworks/handle/1773/15700. 61, 263

Guest, S. and Fowler, P. (2007). Symmetry conditions and finite mechanisms. *Journal of Mechanics of Materials and Structures*, 2(2):293–301. 231, 232, 233

Guest, S. D. (1999). Mechanisms of the icosahedral compound of ten tetrahedra. *Period. Math. Hungar.*, 39(1-3):213–223. Discrete geometry and rigidity (Budapest, 1999). 182

Guest, S. D. (2006). The stiffness of prestressed frameworks: a unifying approach. *International Journal of Solids and Structures*, 43(3–4):842–854. 115

Guest, S. D. (2011). The stiffness of tensegrity structures. *IMA Journal of Applied Mathematics*, 76(1):57–66. 116

Haas, R., Orden, D., Rote, G., Santos, F., Servatius, B., Servatius, H., Souvaine, D., Streinu, I., and Whiteley, W. (2005). Planar minimally rigid graphs and pseudo-triangulations. *Comput. Geom.*, 31(1-2):31–61. 175

Hatcher, A. (2002). *Algebraic Topology*. Cambridge University Press, Cambridge. 50

Hendrickson, B. (1995). The molecule problem: exploiting structure in global optimization. *SIAM J. Optim.*, 5(4):835–857. 151

Henneberg, L. (1911). *Die graphische Statik der starren Systeme*. Leipzig. 26

Herder, J. L. (2001). *Energy-Free Systems. Theory, Conception and Design of Statically Balanced Spring Mechanisms*. PhD thesis, Delft University of Technology. 80

Heyman, J. (1998). *Structural Analysis: a Historical Approach*. Cambridge University Press. 78

Hilbert, D. and Cohn-Vossen, S. (1981). *Naglyadnaya geometriya*. "Nauka", Moscow, third edition. Translated from the German by S. A. Kamenetskiǐ. 100

Izmestiev, I. and Schlenker, J.-M. (2010). Infinitesimal rigidity of polyhedra with vertices in convex position. *Pacific J. Math.*, 248(1):171–190. 54

Jackson, B. and Jordán, T. (2005). Connected rigidity matroids and unique realizations of graphs. *J. Combin. Theory Ser. B*, 94(1):1–29. 5, 152

Jackson, B. and Jordán, T. (2010). Globally rigid circuits of the direction-length rigidity matroid. *J. Combin. Theory Ser. B*, 100(1):1–22. 152

Jackson, B., Jordán, T., and Szabadka, Z. (2006). Globally linked pairs of vertices in equivalent realizations of graphs. *Discrete Comput. Geom.*, 35(3):493–512. 152

Jackson, B., Servatius, B., and Servatius, H. (2007). The 2-dimensional rigidity of certain families of graphs. *J. Graph Theory*, 54(2):154–166. 152

Jacobs, D. J. and Hendrickson, B. (1997). An algorithm for two-dimensional rigidity percolation: the pebble game. *J. Comput. Phys.*, 137(2):346–365. 5, 143, 145

Jacobs, D. J. and Thorpe, M. F. (1996). Generic rigidity percolation in two dimensions. *Physical Review E*, 53(4):3682–3693. 5, 153

James, G. and Liebeck, M. (2001). *Representations and Characters of Groups*. Cambridge University Press. 6, 87, 187, 197, 200, 203, 206

Jordán, T., Király, C., and Tanigawa, S.-i. (2014). Generic global rigidity of body-hinge frameworks. Technical Report TR-2014-06, Egerváry Research Group, Budapest. www.cs.elte.hu/egres/www/tr-14-06.html. 152

Jordán, T. and Szabadka, Z. (2009). Operations preserving the global rigidity of graphs and frameworks in the plane. *Comput. Geom.*, 42(6-7):511–521. 151

Juan, S. H. and Tur, J. M. M. (2008). Tensegrity frameworks: Static analysis review. *Mechanism and Machine Theory*, 43(7):859–881. 3

Kangwai, R. D. and Guest, S. D. (1999). Detection of finite mechanisms in symmetric structures. *International Journal of Solids and Structures*, 36(36):5507–5527. 231, 232

Kangwai, R. D. and Guest, S. D. (2000). Symmetry-adapted equilibrium matrices. *International Journal of Solids and Structures*, 37(11):1525–1548. 187, 231

Kangwai, R. D., Guest, S. D., and Pellegrino, S. (1999). An introduction to the analysis of symmetric structures. *Computers & Structures*, 71(6):671–688. 187

Katoh, N. and Tanigawa, S.-i. (2011). A proof of the molecular conjecture. *Discrete Comput. Geom.*, 45(4):647–700. 153

Kirby, R. (1995). *Problems in Low-Dimensional Topology*. Website. Problem 5.18 in http://math.berkeley.edu/~kirby/. 170

Krantz, S. G. and Parks, H. R. (2002). *The implicit function theorem*. Birkhäuser Boston Inc., Boston, MA. History, theory, and applications. 48

Kuznetsov, E. N. (1991). *Underconstrained Structural Systems*. Springer-Verlag, New York. 127

Lakatos, I. (1976). *Proofs and Refutations*. Cambridge University Press, Cambridge. The logic of mathematical discovery, Edited by John Worrall and Elie Zahar. 50

Laman, G. (1970). On graphs and rigidity of plane skeletal structures. *J. Engrg. Math.*, 4:331–340. 140

Levy, M. and Salvadori, M. (2002). *Why Buildings Fall Down–How Structures Fail*. W. W. Norton & Company, Castle House, 75/76 Wells Street, London WIT 3QT. 2

Livesley, R. K. (1964). *Matrix Methods of Structural Analysis (Commonwealth Library) [Paperback]*. Pergamon Press. 6

Lovász, L. (2001). Steinitz representations of polyhedra and the Colin de Verdière number. *J. Combin. Theory Ser. B*, 82(2):223–236. 108

Lovász, L., Saks, M., and Schrijver, A. (1989). Orthogonal representations and connectivity of graphs. *Linear Algebra Appl.*, 114/115:439–454. 150

Lovász, L., Saks, M., and Schrijver, A. (2000). A correction: "Orthogonal representations and connectivity of graphs" [Linear Algebra Appl. **114/115** (1989), 439–454; MR0986889 (90k:05095)]. *Linear Algebra Appl.*, 313(1-3):101–105. 150

Lovász, L. and Yemini, Y. (1982). On generic rigidity in the plane. *SIAM J. Algebraic Discrete Methods*, 3(1):91–98. 145

Maxwell, J. C. (1864a). On reciprocal figures and diagrams of forces. *Philosophical Magazine Series 4*, (27):250–261. 37

Maxwell, J. C. (1864b). On the calculation of the equilibrium and stiffness of frames. *Philosophical Magazine*, 27:294–299. 12, 36, 209

Maxwell, J. C. (1869–1872). On reciprocal figures, frames and diagrams of forces. *Trans Royal Soc. Edinburgh*, (26):1–40. 37, 38

Menshikov, M. V., Rybnikov, K. A., and Volkov, S. E. (2002). The loss of tension in an infinite membrane with holes distributed according to a Poisson law. *Adv. in Appl. Probab.*, 34(2):292–312. 131

Motro, R. (2003). *Tensegrity: Structural Systems for the Future*. A Butterworth-Heinemann Title. 3

Owen, J. C. and Power, S. C. (2010). Frameworks symmetry and rigidity. *International Journal of Computational Geometry & Applications*, 20(6):723–750. 231

Pandia Raj, R. and Guest, S. D. (2006). Using symmetry for tensegrity form-finding. *Journal of the International Association for Shell and Spatial Structures*, 47(3): 245–252. xii

Parkes, E. W. (1965). *Braced Frameworks: An Introduction to the Theory of Structures (Commonwealth and International Library. Structures and Solid Body Mechanics Division) [Hardcover]*. Pergamon. 6

Pellegrino, S. (1993). Structural computations with the singular value decomposition of the equilibrium matrix. *International Journal of Solids and Structures*, 30(21): 3025–3035. 209

Pellegrino, S. and Calladine, C. R. (1986). Matrix analysis of statically and kinematically indeterminate frameworks. *International Journal of Solids and Structures*, 22(4): 409–428. 116

Penne, R. and Crapo, H. (2007). A general graphical procedure for finding motion centers of planar mechanisms. *Adv. in Appl. Math.*, 38(4):419–444. 57

Petroski, H. (2008). *Design Paradigms–Case Histories of Error and Judgment in Engineering*. W. W. Norton & Company, Cambridge University Press. 2

Pollaczek-Geiringer, H. (1927). über die gliederung ebener fachwerke. *ZAMM – Journal of Applied Mathematics and Mechanics / Zeitschrift für Angewandte Mathematik und Mechanik*, 7(1):58–72. 140

Pollaczek-Geiringer, H. (1932). Zur gliederung räumlicher fachwerke. *ZAMM – Journal of Applied Mathematics and Mechanics / Zeitschrift für Angewandte Mathematik und Mechanik*, 12(6):369–376. 140

Pugh, A. (1976). *Introduction to Tensegrity [Paperback]*. University of California Press. 3

Ramana, M. V. (1997). An exact duality theory for semidefinite programming and its complexity implications. *Math. Programming*, 77(2, Ser. B):129–162. Semidefinite programming. 121

Ramar, P. R. and Guest, S. D. (2011). Minimizing the self-weight deflection of tensegrity structures. In *Proceedings of IABSE-IASS 2011: Taller, Stronger, Lighter*. xii

Rankine, W. J. M. (1858). *A Manual of Applied Mechanics*. Richard Griffin and Company, Publishers to the University of Glasgow, London and Glasgow. 37

Roth, B. (1980). Questions on the rigidity of structures. *Structural Topology*, 4:67–71. Special issue on research objectives and open problems, Part 1. 129

Roth, B. and Whiteley, W. (1981). Tensegrity frameworks. *Trans. Amer. Math. Soc.*, 265(2):419–446. 71

Sabitov, I. K. (1998). The volume as a metric invariant of polyhedra. *Discrete Comput. Geom.*, 20(4):405–425. 170

Schek, H. J. (1974). The force density method for form finding and computation of general networks. *Computer Methods in Applied Mechanics and Engineering*, 33:115–134. 28, 88

Schenk, M., Guest, S. D., and Herder, J. L. (2007). Zero stiffness tensegrity structures. *International Journal of Solids and Structures*, 44:6569–6583. 80

Schoenberg, I. J. and Zaremba, S. C. (1967). On Cauchy's lemma concerning convex polygons. *Canad. J. Math.*, 19:1062–1071. 106

Schulze, B. (2010a). Block-diagonalized rigidity matrices of symmetric frameworks and applications. *Contributions to Algebra and Geometry*, 51(2):427–466. 231

Schulze, B. (2010b). Symmetric laman theorems for the groups c_2 and c_s. *Electronic Journal of Combinatorics*, 17(1). 229

Schulze, B. (2010c). Symmetric versions of Laman's theorem. *Discrete & Computational Geometry*, 44(4):946–972. 229

Schulze, B. (2010d). Symmetry as a sufficient condition for a finite flex. *SIAM Journal on Discrete Mathematics*, 24(4):1291–1312. 231

Schulze, B. and Tanigawa, S.-i. (2015). Infinitesimal rigidity of symmetric bar-joint frameworks. *SIAM Journal on Discrete Mathematics*, 29(3):1259–1–286. 231

Servatius, B., Shai, O., and Whiteley, W. (2010). Geometric properties of Assur graphs. *European J. Combin.*, 31(4):1105–1120. 60

Skelton, R. E. and de Oliveira, M. C. (2010). *Tensegrity Systems [Paperback]*. Springer. 3

Snelson, K. (1948). www.kennethsnelson.net/. 3

Snelson, K. (2009). *Forces made Visible*. Hard Press Editions, Lenox, MA. 110

Spivak, M. (1965). *Calculus on Manifolds. A Modern Approach to Classical Theorems of Advanced Calculus*. W. A. Benjamin, Inc., New York-Amsterdam. 48, 137

Spivak, M. (1979). *A Comprehensive Introduction to Differential Geometry. Vol. V*. Publish or Perish Inc., Wilmington, Del., second edition. 14

Stachel, H. (1992). Zwei bemerkenswerte bewegliche Strukturen. *J. Geom.*, 43(1-2):14–21. 182

Stoker, J. J. (1968). Geometrical problems concerning polyhedra in the large. *Comm. Pure Appl. Math.*, 21:119–168. 54

Strang, G. (2009). *Introduction to Linear Algebra*. Wellesley-Cambridge Press, 4th edition edition. 6, 203, 206, 208

Streinu, I. (2005). Pseudo-triangulations, rigidity and motion planning. *Discrete Comput. Geom.*, 34(4):587–635. 173

Sturmfels, B. and Whiteley, W. (1991). On the synthetic factorization of projectively invariant polynomials. *J. Symbolic Comput.*, 11(5-6):439–453. Invariant-theoretic algorithms in geometry (Minneapolis, MN, 1987). 57

Sullivan, D. (1971). Combinatorial invariants of analytic spaces. In *Proceedings of Liverpool Singularities—Symposium, I (1969/70)*, pages 165–168, Berlin. Springer. 176

Tanigawa, S.-i. (2015). Sufficient conditions for the global rigidity of graphs. *J. Combin. Theory Ser. B*, 113:123–140. 152

Tarnai, T. (1980). Simultaneous static and kinematic indeterminacy of space trusses with cyclic symmetry. *International Journal of Solids and Structures*, 16:347–359. 232

Tarnai, T. and Gáspár, Z. (1983). Improved packing of equal circles on a sphere and rigidity of its graph. *Math. Proc. Cambridge Philos. Soc.*, 93(2):191–218. 76

Tay, T.-S. (1984). Rigidity of multigraphs. I. Linking rigid bodies in *n*-space. *J. Combin. Theory Ser. B*, 36(1):95–112. 153

Tay, T.-S. and Whiteley, W. (1984). Recent advances in the generic rigidity of structures. *Structural Topology*, (9):31–38. Dual French-English text. 153

Tay, T.-S. and Whiteley, W. (1985). Generating isostatic frameworks. *Structural Topology*, (11):21–69. Dual French–English text. 153

Tibert, A. G. and Pellegrino, S. (2003). Review of form-finding methods for tensegrity structures. *International Journal of Space Structures*, 18:209–223. 88

Tutte, W. T. (1963). How to draw a graph. *Proc. London Math. Soc. (3)*, 13:743–767. 85

Vandenberghe, L. and Boyd, S. (1996). Semidefinite programming. *SIAM Rev.*, 38(1): 49–95. 121

White, N. L. and Whiteley, W. (1983). The algebraic geometry of stresses in frameworks. *SIAM J. Algebraic Discrete Methods*, 4(4):481–511. 135

Whiteley, W. (1982). Motions and stresses of projected polyhedra. *Structural Topology*, (7):13–38. With a French translation. 39

Whiteley, W. (1984). Infinitesimal motions of a bipartite framework. *Pacific J. Math.*, 110(1):233–255. 58

Whiteley, W. (1988a). Infinitesimally rigid polyhedra. II. Modified spherical frameworks. *Trans. Amer. Math. Soc.*, 306(1):115–139. 182

Whiteley, W. (1988b). Problems on the realizability and rigidity of polyhedra. In *Shaping space (Northampton, Mass., 1984)*, Design Sci. Collect., pages 256–257, 260. Birkhäuser Boston, Boston, MA. 146, 147

Whiteley, W. (1989). Rigidity and polarity. II. Weaving lines and tensegrity frameworks. *Geom. Dedicata*, 30(3):255–279. 102

Wunderlich, W. (1976). On deformable nine-bar linkages with six triple joints. Indagationes Mathematicae (Proceedings), 79(3):257–262. 159

Zhang, J. Y., Guest, S. D., Connelly, R., and Ohsaki, M. (2010). Dihedral 'star' tensegrity structures. *International Journal of Solids and Structures*, 47(1):1–9. 266, 268

Zhang, J. Y. and Ohsaki, M. (2007). Stability conditions for tensegrity structures. *Internat. J. Solids Structures*, 44(11-12):3875–3886. 268

Zhang, J. Y. and Ohsaki, M. (2012). Self-equilibrium and stability of regular truncated tetrahedral tensegrity structures. *J. Mech. Phys. Solids*, 60(10):1757–1770. 268

Index